P9-ARK-044

BEYOND ENGINEERING

THE SLOAN TECHNOLOGY SERIES

Dark Sun: The Making of the Hydrogen Bomb by Richard Rhodes

Dream Reaper: The Story of an Old-Fashioned Inventor in the High-Stakes World of Modern Agriculture by Craig Canine

Turbulent Skies: The History of Commercial Aviation by Thomas A. Heppenheimer

Tube: The Invention of Television by David E. Fisher and Marshall Jon Fisher

The Invention That Changed the World: How a Small Group of Radar Pioneers Won the Second World War and Launched a Technological Revolution by Robert Buderi

Computer: A History of the Information Machine by Martin Campbell-Kelly and William Aspray

Naked to the Bone: Medical Imaging in the Twentieth Century by Bettyann Kevles

A Commotion in the Blood: A Century of Using the Immune System to Battle Cancer and Other Diseases by Stephen S. Hall

Beyond Engineering: How Society Shapes Technology by Robert Pool

The One Best Way: Frederick Winslow Taylor and the Enigma of Efficiency by Robert Kanigel

Crystal Fire: The Birth of the Information Age by Michael Riordan and Lillian Hoddesen

Robert Pool

BEYOND ENGINEERING

How

Society

Shapes

Technology

New York Oxford
OXFORD UNIVERSITY PRESS
1997

Oxford University Press

Oxford New York
Athens Auckland Bangkok Bogotá Bombay
Buenos Aires Calcutta Cape Town Dar es Salaam Delhi
Florence Hong Kong Istanbul Karachi
Kuala Lumpur Madras Madrid Melbourne
Mexico City Nairobi Paris Singapore
Taipei Tokyo Toronto

and associated companies in
Berlin Ibadan

Published by Oxford University Press, Inc.
198 Madison Avenue, New York, New York 10016

Oxford is a registered trademark of Oxford University Press

Library of Congress Cataloging-in-Publication Data

Pool, Robert, 1955-
Beyond engineering : how society shapes technology / by Robert Pool.
p cm.—(Sloan technology series)
Includes bibliographical references and index.
ISBN 0-19-510772-1
1. Technology. 2. Technological innovations. 3. Nuclear energy.
I. Title. II. Series.
T45.P66 1997
600—DC20 96-34364

3 5 7 9 8 6 4

Printed in the United States of America
on acid-free paper

To my mother

Contents

Preface to the

SLOAN TECHNOLOGY SERIES

Technology is the application of science, engineering, and industrial organization to create a human-built world. It has led, in developed nations, to a standard of living inconceivable a hundred years ago. The process, however, is not free of stress; by its very nature, technology brings change in society and undermines convention. It affects virtually every aspect of human endeavor: private and public institutions, economic systems, communication networks, political structures, international affiliations, the organization of societies, and the condition of human lives. The effects are not one-way; just as technology changes society, so too do societal structures, attitudes, and mores affect technology. But perhaps because technology is so rapidly and completely assimilated, the profound interplay of technology and other social endeavors in modern history has not been sufficiently recognized.

The Sloan Foundation has had a longstanding interest in deepening public understanding about modern technology, its origins, and its impact on our lives. The Sloan Technology Series, of which the present volume is a part, seeks to present to the general reader the stories of the development of critical twentieth-century technologies. The aim of the series is to convey both the technical and human dimensions of the subject: the invention and effort entailed in devising the technologies, and the comforts and stresses they have introduced into contemporary life. As the century draws to an end, it is hoped that the Series will disclose a past that might provide perspective on the present and inform the future.

The Foundation has been guided in its development of the Sloan Technology Series by a distinguished advisory committee. We express deep gratitude to John Armstrong, Simon Michael Bessie, Samuel Y. Gibbon, Thomas P. Hughes, Victor McElheny, Robert K. Merton, Elting E. Morison (deceased), and Richard Rhodes. The Foundation has been represented on the committee by Ralph E. Gomory, Arthur L. Singer Jr., Hirsh G. Cohen, and Doron Weber.

Alfred P. Sloan Foundation

Acknowledgments

At the top of a long list of people who deserve thanks for their help on this book is Arthur L. Singer Jr. of the Sloan Foundation, without whom none of it would have been possible.

Four people were particularly important in shaping this book. Michael Golay, head of the nuclear engineering department at MIT, first described to me how factors beyond engineering shaped the development of nuclear power. Thomas Hughes, the Mellon Professor Emeritus of History and Sociology of Science at the University of Pennsylvania, taught me (through his writings) to see technology as a historian does, looking for the defining events that send it down one path or another. Don Kash, the Hazel Professor of Public Policy at George Mason University, impressed upon me the importance of complexity in influencing technological development. And Thelma Z. Lavine, the Clarence J. Robinson Professor of Philosophy at George Mason University, instructed me in the importance and the limitations of the social sciences in knowing our world.

A number of people were kind enough to read through the entire manuscript and comment upon it, suggesting improvements and pointing out mistakes and inaccuracies. Michael Golay deserves special thanks for a most thorough reading and many thoughtful suggestions, but others were also very helpful: Don Kash; Thomas Hughes; Todd La Porte of the political science department at the University of California at Berkeley; Richard Rhodes, the author of *The Making of the Atomic Bomb* and *Dark Sun: The Making of the Hydrogen Bomb*; W. Brian Arthur, an economist now at the Santa Fe Institute; Paul Schulman, a political scientist at Mills College in Oakland, California; Robin Cowan, an economist at the University of Western Ontario; and James Pfiffner, a political scientist at George Mason University.

During the four years I was researching and writing this book, many scientists, engineers, and people in the business world spoke with me about their work. Any list will inevitably leave some off, but nonetheless I must offer as many names as I can call to mind, because their generosity in sharing their time and knowledge made this book possible. Thanks to: Charles Forsberg, Alvin Weinberg, Henry Jones, Howard Kerr, William Fulkerson, Claude Pugh, John Jones Jr., Frank Homan, and Elizabeth Peelle, all at Oak

Ridge National Laboratory; Lawrence Lidsky and Henry Kendall at MIT; Ted Marston, John Taylor, Chauncey Starr, and Gary Vine at the Electric Power Research Institute; Gernot Gessinger, Regis Matzie, Charles Bagnal, and Paul Kramarchyk at ABB-Combustion Engineering's nuclear power division; Kåre Hannerz, Lars Nilsson, and Dusan Babala at ABB in Västerås, Sweden; Bruno Baumgartl, Dieter Schneider, and Michel Prévost of Nuclear Power International in Paris; Charles Till, Yoon I. Chang, and Bob Avery at Argonne National Laboratory; Jerry Goldberg, Dave Sager, Tom Plunkett, and Ray Golden at Florida Power & Light; Edward Davis at the American Nuclear Energy Council; Phillip Bayne and Carl Goldstein at the U.S. Council for Energy Awareness; Kathleen Reil of Holistic Impacts in Toronto; Harold Lewis at the University of California at Santa Barbara; Stanley Rothman of Smith College; Dave Fiorelli at TU Electric, a subsidiary of Texas Utilities; Jan Beyea of the National Audubon Society; Tony Wallace and Peter Murray, consultants with Westinghouse Electric Corporation's nuclear group; M. Granger Morgan and Baruch Fischhoff at Carnegie Mellon University; Eugene Rosa at Washington State University; Bob Berglund at General Electric's nuclear division and Bertram Wolfe, retired vice president of nuclear power at GE; Will Kaefer, Mike Angus, and Jim Molden at Pacific Gas & Electric; Robert Uhrig at the University of Tennessee; and Hank Jenkins-Smith at the University of New Mexico. Special thanks to Linda Royal and Andrea Williamson, librarians at the Virginia Electric Power Company and the Nuclear Energy Institute, respectively, who opened their files to me.

Finally, I thank my wife, Amy Bunger Pool, for reading various versions of the book and for sharpening my thinking in a number of areas, particularly those concerning courts and technology.

Introduction

UNDERSTANDING TECHNOLOGY

This is a very different book from the one I began writing four years ago. That happens sometimes, usually when an author doesn't really understand a subject or when he discovers something else more interesting along the way, and in my case it was both. Allow me to explain.

In 1991 the Alfred P. Sloan Foundation provided grants to some two dozen writers to create a series of books on technology. Because technology has shaped the modern world so profoundly, Sloan wanted to give the general, nontechnical reader some place to go in order to learn about the invention of television or the history of X-rays or the development of birth control pills. This would be it. Sloan asked that each book in the series focus on one particular technology and that all of the books be accessible to readers with no background in science or engineering, but otherwise the foundation left it up to the writers to decide what to write about and how. Various authors agreed to produce books on vaccines, modern agriculture, radar, fiber optics, the transistor, the computer, software, biotechnology, commercial aviation, the railroads, and other modern technologies. I took on nuclear power.

At the time, I planned to produce a straightforward treatment of the commercial nuclear industry—its history, its problems, and its potential for the future. I knew that nuclear power was controversial, and I believed that both sides in the debate over its use were shading the truth somehow. My job would be to delve into the technical details, figure out what was really going on, and report back to the readers. To keep the book as lively and readable as possible, I would sprinkle anecdotes and colorful characters throughout, but the book's heart would be a clear, accurate account of the engineering practices and scientific facts that underlie nuclear technology. Given this information, readers could then form their own opinions on the nuclear conundrum.

There was nothing original about this approach. Nearly everyone who writes seriously about technology for the general public—journalists, engineers, or technological critics—focuses on the technical details. It is here that the truth lies; it is from here that a real understanding of a technology will emerge. People may not always agree on the implications of those details—the same data are used as evidence of nuclear power's safety by supporters and as proof of its peril by opponents—but everyone believes in their primacy. I did too, at first.

In the beginning, it seemed that the hardest part about a book on nuclear power would be giving some structure to this diffuse and complicated subject, lest readers get bogged down in the minutiae. To that end, I proposed that the book be told from a particular point of view—the development of a coming second generation of nuclear power. Some next-generation nuclear plants were already under construction elsewhere in the world, although there were no plans for building any in the United States, and it seemed an ideal way to tie everything together. By writing from the perspective of the developers of this next generation, I could recall the history of the first generation and detail its problems—the technical shortcomings, the regulatory failings, the inadequate management, and the public opposition. I could describe how the nuclear industry is trying to mend these problems. And I could offer some conjectures about the future. According to the original proposal I sent to the Sloan Foundation, the main goal of the book would be "to trace the intellectual development of a new technology, describing not only how scientists and engineers solve problems and meet goals but also the more interesting question of how they decide which goals they should set in the first place." In short, I was setting out to understand the technology through the eyes of the technical specialists who were responsible for developing it.

After this book was under way, two other books appeared with similar ambitions. The first was *Nuclear Choices*, by Richard Wolfson, which offers a readable layman's description of the technical details of nuclear power and explains many of the disagreements surrounding its use. Then Richard Rhodes released *Nuclear Renewal*, which describes work on the next generation of reactors. It is a slim book, but it manages to summarize nicely the problems that hit the U.S. nuclear industry and to give a sense of what is needed if the industry is to be revived. Between them, the two books covered much of the territory I had been aiming at.

Even before these two books appeared, however, I had been rethinking my own. The rethinking process started with a suggestion from the Sloan Foundation's Arthur Singer that I beef up the parts of the book devoted to the history of nuclear power. He thought it was important that readers get a better idea of why the technology ended up as it has, and I agreed. And so began a journey that took me on a far different route than I had planned and led me to see nuclear power—and all of technology—in a new and unfamiliar way.

As I dug into the story of nuclear power, seeking the forces that had shaped it, I discovered that the key lay not so much with scientists and engineers as with a host of nontechnical influences. Why, for instance, was nuclear power such a Jekyll-and-Hyde technology? In France, nuclear power generates nearly 80 percent of the country's electricity and is generally accepted as safe, economical, and reliable. In the United States, on the other hand, nuclear power has experienced one major accident and a number of near-misses, cost the utilities that invested in it billions of dollars, and precipitated a large antinuclear movement. Although nuclear reactors generate 20 percent of this country's electricity, no new nuclear plants have been ordered in the United States since 1978, and none seem likely to be ordered any time soon. And Italy, next door to France, has sworn off nuclear power altogether, closing its existing plants and vowing to build no more, at least until the technology is changed significantly. The reasons for the different outcomes cannot be found in the engineering, since all three countries use essentially the same reactor technology. (Indeed, France began its nuclear program in the early 1970s by borrowing reactor designs from the United States.) Instead, I learned, one must look past the technology to the broader "sociotechnical system"—the social, political, economic, and institutional environments in which the technology develops and operates. The United States, France, and Italy provided very different settings for their nuclear technologies, and it shows.

No one is more aware of this than the engineers who developed nuclear power. It frustrates them terribly. Here, they say, is a perfectly good way to solve many of our energy problems, yet some societies screw it up. In response, a number of engineers have written books and magazine articles and addressed town meetings in attempts to educate the public. If people could only understand the facts as engineers do, these engineers believe, everything would be okay. The technology would work as planned.

Yet the hard, objective facts that engineers talk about can be quite difficult to find. Every time I investigated what I thought was a technical question—Why was light-water the dominant reactor choice? What breakthrough led to the broad commercialization of nuclear power in the 1960s? Is there a feasible solution to the problem of storing nuclear wastes?—I found the answers taking me beyond the realm of engineering. The line between the technical and the nontechnical that at first had seemed so clear slowly dissolved, and I came to see the development of nuclear power as a collaboration—albeit an unwitting and often uncomfortable one—between engineers and the larger society. And in that partnership, society's role proved to be surprisingly deep and complex. I had expected it to be little more than a hand on the throttle—speeding the development when a technology was seen as desirable, or slowing it down or stopping it when the cons overtook the pros. But throughout the history of nuclear power, society kept a strong hand on the tiller as well.

The nice, simple story I first envisioned had disappeared. In its place was a complex, often convoluted tale of how the technology had been shaped by a host of nontechnical factors in addition to the expected technical ones. To explain why nuclear power evolved as it did, I would have to paint a broad picture, one that showed the technology in the context of the society in which it developed.

I could have stopped there. Perhaps I should have. But the more I read, the more I realized that I had stumbled onto just one example of a larger pattern. Although nuclear power is unique in many ways—its birth in the Manhattan Project, the early government control, its potential for catastrophic accident, its production of highly radioactive nuclear waste—it is quite typical in the way that its development was influenced by factors beyond engineering. Any modern technology, I found, is the product of a complex interplay between its designers and the larger society in which it develops.

Consider the automobile. In the early part of this century, gas-powered cars shared the roads with those powered by boilers and steam engines, such as the Stanley Steamer. Eventually, internal combustion captured the market and the old steamers disappeared. Why? The usual assumption is that the two contenders went head to head and the best technology won. Not at all.

Although the internal combustion engine did have some advantages in performance and convenience, steam-powered cars had their own pluses: they had no transmission or shifting of gears, they were simpler to build, and they were smoother and quieter to operate. Experts then and now have called it a draw—the "better" technology was mostly a matter of opinion. Instead, the steamers were killed off by several factors that had little or nothing to do with their engineering merits. For one, the Stanley brothers, builders of the best steam-powered cars of the time, had little interest in mass production. They were content to sell a few cars at high prices to aficionados who could appreciate their superiority. Meanwhile, Henry Ford and other Detroit automakers were flooding the country with inexpensive gas-powered cars. Even so, the steamers might well have survived as high-end specialty cars were it not for a series of unlucky breaks. At one point, for example, an outbreak of hoof-and-mouth disease caused public horse troughs to be drained, removing a major source of water for refilling the cars' boilers. It took the Stanley brothers three years to develop a closed-cycle steam engine that didn't need constant refilling, but by then World War I had begun, bringing strict government limits on the number of cars that businesses could build for the consumer market. The Stanley company never recovered, and it folded a few years later. The remnants of the steam automobile industry died during the Depression, when the market for high-priced cars all but disappeared.

Nonengineering factors play a role in the development of all technologies, even the simplest. In *The Pencil*, Henry Petroski tells how pencil designers in the late 1800s, in order to get around a growing shortage of red cedar, devised a pencil with a paper wrapping in place of the normal wood. It "worked well technically and showed great promise," Petroski writes, "but the product failed for unanticipated psychological reasons." The public, accustomed to sharpening pencils with a knife, wanted something that could be whittled. The paper pencil never caught on.

Today, particularly for such sophisticated creations as computers, genetic engineering, or nuclear power, nontechnical factors have come to exert an influence that is unprecedented in the history of technology. Invention is no longer, as Ralph Waldo Emerson's aphorism had it, simply a matter of "Build a better mousetrap and the world will beat a path to your door." The world is already at your door, and it has a few things to say about that mousetrap.

The reasons for this are several, some grounded in the changing nature of technology itself and others arising from transformations in society. A hundred years ago, people in western nations generally saw technological development as a good thing. It brought prosperity and health; it represented "progress." But the past century has seen a dramatic change in western society, with a resulting shift in people's attitudes toward technology. As countries have become more prosperous and secure, their citizens have become less concerned about increasing their material well-being and more concerned with such aesthetic considerations as maintaining a clean environment. This makes them less likely to accept new technologies uncritically. At the same time, citizens of western democracies have become more politically savvy and more active in challenging the system with lawsuits, special interest groups, campaigns to change public opinion, and other weapons. The result is that the public now exerts a much greater influence on the development of technologies—particularly those seen as risky or otherwise undesirable—than was true one hundred, or even fifty, years ago.

Meanwhile, the developers of technology have also been changing. A century ago, most innovation was done by individuals or small groups. Today, technological development tends to take place inside large, hierarchical organizations. This is particularly true for complex, large-scale technologies, since they demand large investments and extensive, coordinated development efforts. But large organizations inject into the development process a host of considerations that have little or nothing to do with engineering. Any institution has its own goals and concerns, its own set of capabilities and weaknesses, and its own biases about the best ways to do things. Inevitably, the scientists and engineers inside an institution are influenced —often quite unconsciously—by its culture.

A closely related factor is the institutionalization of science and engineering. With their professional societies, conferences, journals, and other means of communion, scientists and engineers have formed themselves into relatively close-knit—though large—groups that have uniform standards of practice and hold similar ideas. Today, opinions and decisions about a technology tend to reflect a group's thinking more than any given individual's.

The existence of large organizations and the institutionalization of the professions allow a technology to build up a tremendous amount of momentum in a relatively short time. Once the choice has been made to go a certain way, even if the reasons are not particularly good ones, the institutional machinery gears up and shoves everybody in the same direction. It can be tough to resist.

But the most important changes have come in the nature of technology itself. In the twentieth century, the power of our machines and devices has grown dramatically—along with their unanticipated consequences. When DDT was introduced, it seemed an unallayed good: a cheap, effective way to kill insect pests and improve crop yields. It took years to understand that the pesticide made its way up the food chain to weaken the shells of birds' eggs and wreak other unintended havoc. Similarly, chlorofluorocarbons, or CFCs, were widely used for decades—as refrigerants, as blowing agents in making foams, and as cleaners for computer chips—before anyone realized they were damaging the ozone layer.

Even normally benign technologies can take on different complexions when multiplied to meet the needs of a world with five billion people. Burning natural gas is an economical, safe, and clean way to heat homes and generate electricity. Its only major waste material is carbon dioxide, the same gas that humans exhale with each breath. But carbon dioxide, in the quantities now being produced worldwide by burning fossil fuels (coal and oil as well as natural gas), is exaggerating the greenhouse effect in earth's atmosphere and threatening major changes in the global climate.

Modern technology is like a Great Dane in a small apartment. It may be friendly, but you still want to make sure there's nothing breakable within reach. So to protect the china and crystal, government bodies, special interest groups, businesses, and even individuals are demanding an increasing say in how technologies are developed and applied.

Besides its power, modern technology has a second feature—more subtle, but equally important—that makes it qualitatively different from earlier technologies: its complexity. The plow, the cotton gin, even the light bulb—these are simple devices. No matter how much they are changed and improved, it is still easy to understand their functions and capabilities. But for better or worse, technology has reached the point where no individual can understand completely how, say, a petrochemical plant works, and no team of experts can anticipate every possible outcome once a technology is put to work. Such complexity fundamentally changes our relationship with technology.

Consider the accident that destroyed the space shuttle *Challenger.* Although the cause was eventually established as the failure of O-rings at low temperatures, which allowed the escape of hot gases and led to an explosion of a fuel tank, the real culprit was the complexity of the system. Space shuttle engineers had been concerned about how the O-rings would behave in below-freezing weather, and some even recommended the launch be postponed to a warmer day, but no one could predict with any certainty what might happen. There were too many variables, too many ways in which the components of the system could interact. Management decided to proceed with the launch despite the engineers' disquiet, and it was only months later that experts pieced together the chain of events that led to the explosion.

Complexity creates uncertainty, limiting what can be known or reasonably surmised about a technology ahead of time. Although the shuttle engineers had vague fears, they simply did not—could not—know enough about the system to foresee the looming disaster. And in such cases, when there is not a clear technical answer, people fall back on subjective, often unconscious reasoning—biases and gut feelings, organizational goals, political considerations, the profit motive. In the case of the *Challenger,* NASA was feeling pressure to keep its shuttles going into space on a regular basis, and no one in the organization wanted to postpone a launch unless it was absolutely necessary. In this case it was, but no one knew.

For all these reasons, modern technology is not simply the rational product of scientists and engineers that it is often advertised to be. Look closely at any technology today, from aircraft to the Internet, and you'll find that it truly makes sense only when seen as part of the society in which it grew up.

The insight is not a particularly new one. Thoughtful engineers have discussed it for some time. As early as the 1960s, Alvin Weinberg, the longtime director of Oak Ridge National Laboratory, was writing on the relationship between technology and society, particularly in regard to nuclear power. There have been others. But until recently no one had studied the influence of society on technology in any consistent, comprehensive way. Philosophically inclined engineers like Weinberg did not have the time, the temperament, or the training to make careful studies. They reported what they saw and mused about the larger implications, but nothing more. And social scientists, when they noticed technology at all, viewed it primarily in terms of how it shapes society. Sociologists, economists, and others have long seen technology as the driving force behind much of history—a theory usually referred to as "technological determinism"—and they have happily investigated such things as how the invention of the printing press triggered the Reformation, how the development of the compass ushered in the Age of Exploration and the discovery of the New World, and how the cotton gin created the conditions that led to the Civil War. But few of these scientists turned the question around and asked how society shapes technology.

In just the past decade or two, however, that has begun to change. Indeed, it has now become almost fashionable for economists, political scientists, and sociologists to bring their analytical tools to bear on various technologies, from nuclear power and commercial aviation to medical instruments, computers, even bicycles. Part of the reason for this is, I suspect, the increasing importance of technology to our world, and another part is the realization by social scientists that science and technology are just as amenable to social analysis as politics or religion. Whatever the reason, the result has been to put technology in a whole new light. Scholars now talk about the push and pull between technology and society, rather than just the push of technology on society. Engineers have been brought down from the mountain to take their place as one—still very important—cog in the system by which technology is delivered to the world.

Unfortunately, very little of this has filtered down. It remains mostly specialists talking to other specialists in books and journals that few outside their fields ever see. This book aims to change that. From its original design as a study of how engineers are creating a new generation of nuclear power, it has metamorphosed into a more general—and more ambitious—look at how nontechnical forces shape modern technologies. In it I collect and synthesize work from a wide variety of disciplines: history, economics, political science, sociology, risk analysis, management science, psychology. At various points, the book touches on personal computers, genetic engineering, jet aircraft, space flight, automobiles, chemical plants, even steam engines and typewriters. Such a book obviously cannot be comprehensive. Instead, my goal is to introduce a different way to think about technology and to show how many things make much more sense when technology is viewed in this way.

Throughout all its changes, the book has remained at its core a story about one particular technology: nuclear power. There are several reasons for this. One is that I had agreed to write about nuclear power for the Sloan Technology Series. Although that seems very long ago now, I still feel committed to deliver such a book. But even if that were not the case, there are several good arguments for giving nuclear power a major role here. Because technology is such a broad and varied field, a book like this runs the risk of seeming too disconnected. Each element may seem clear and convincing in its own right, but when they're piled one on top of the other, the total may seem less than the sum of the parts. To avoid this, it is important to have one technology to which the reader returns again and again, seeing its story unfold from beginning to end and gradually building up a coherent picture of how its various pieces fit together. For that purpose, there is no better choice than nuclear power. Not because nuclear power is somehow typical or representative of other technologies—it is not. But rather because nuclear power, perhaps more than any other technology, reflects the times and the societies in which it evolved. Like Moby Dick, whose scars told of a lifetime of battles with whalers, nuclear power carries with it a record of encounters with the larger society, a record that is not at all subtle for those who know how to decipher it. Furthermore, thanks to nuclear power's importance and its controversial history, there is a rich literature to draw on. Social scientists of every stripe have tried to understand how the promise of nuclear power fell so flat, each of them applying a different lens and supplying different insights. Their work is scattered throughout dozens of different scholarly journals and hundreds of books, and, as far as I know, no one has tried to weave together those many threads into a coherent whole. Here I do.

Lurking just beneath the surface of this book is one of the most intriguing and frustrating questions of our time, although it seldom gets much attention outside universities and other scholarly places: How do we know what we know? Or, to put it differently, What is the nature of human knowledge? This may sound like the sort of abstract question that only a philosopher could love, but its answer has practical—and important—ramifications for how we deal with science and technology.

There are two schools of thought on the nature of human knowledge, and they have little common ground. One has its roots in the physical sciences and goes by such names as positivism or objectivism or rationalism. Positivism accepts as knowledge only those things that have been verified by the scientific method of hypothesis formation and testing. It is, of course, impossible to verify anything absolutely—no matter how many times the sun comes up right on schedule, there is no guarantee it won't be a couple of hours late tomorrow—but positivists are generally content with verifying something beyond a reasonable doubt. Karl Popper, the influential philosopher of science, put a slightly different spin on it: Scientific statements, he said, are those that can be put to the test and potentially proven wrong, or falsified. It's not possible to prove a hypothesis is true, but if the hypothesis is tested extensively and never proven false, then one accepts it as provisionally true. If it later turns out to be false in certain situations, it can be modified or replaced. By this method, one hopes to get better and better approximations to the world's underlying physical reality. Absolute knowledge is not attainable. This provisional knowledge is the best we can do.

The strength of positivism—its insistence on verification—is also its weakness, for there is much that people think of as "knowledge" that cannot be verified in the same way that theories in physics or biology can. "The United States is a capitalist country" is a statement most would accept as true, but how does one prove it? Or how about "Santa Claus wears a red suit and rides in a sleigh pulled by eight reindeer"? This is clearly knowledge of a sort—everyone in our culture older than two or three "knows" it—and though it may not be in the same intellectual league as, say, the general theory of relativity, it's much more important to most people than anything Einstein came up with. Yet the positivist approach has no place for such folderol.

For many years social scientists, impressed with the success of the physical sciences, modeled their methods along the same positivist lines. They made observations, formed hypotheses, and tested their theories, attempting to make their research as objective as possible. Many social scientists still do, but in the past few decades an influential new school of thought has appeared, one that offers a different take on human knowledge. This approach, often referred to as "social construction" or "interpretation," is designed explicitly to deal with social reality—the web of relationships, institutions, and shared beliefs and meanings that exist in a group of people—instead of physical reality. It sees knowledge not as something gleaned from an underlying physical reality but as the collective product of a society. Social constructionists speak not of objective facts but only of interpretations of the world, and they set out to explain how those interpretations arise. They make no attempt to judge the truth or falsity of socially constructed knowledge. Indeed, they deny that it even makes sense to ask whether such knowledge is true or false.

Thus positivism and social construction offer diametrically opposed views of knowledge. Positivists see knowledge as arising from nature, social constructionists see it as a product of the human mind. Positivists speak of proof, social constructionists of interpretation. Positivists assume knowledge to be objective, social constructionists believe it to be subjective. In general, positivists have been willing to defer to the social constructionists in the case of social knowledge. After all, information about Santa Claus or the importance of capitalism is not really what a positivist has in mind when he speaks about "knowledge." But the social constructionists have been unwilling to return the favor, and so have triggered a sharp, if so far bloodless, battle.

All human knowledge is social knowledge, the social constructionists say, even science. After all, scientific knowledge is created by groups of people—the scientific community and its various subsets—and so it inevitably has a collective character. There is no such thing as a scientific truth believed by one person and disbelieved by the rest of the scientific community; an idea becomes a truth only when a vast majority of scientists accept it without question. But if this is so, the argument goes, then science is best understood as socially constructed rather than derived in some objective way from nature.

The earliest and best known example of this approach to science is Thomas Kuhn's *The Structure of Scientific Revolutions*. In it, Kuhn depicts most science as taking place inside a "paradigm"—a set of beliefs and expectations that guide the research, defining which questions are important and designating the proper ways to go about answering them. Scientific revolutions—such as the shift from the Ptolemaic to the Copernican view of the universe—occur when a paradigm breaks down and the scientific community collectively settles on a new paradigm in which to work. Kuhn argued that because such a paradigm shift is a change of the rules by which science is done, there can be no objective reasons for making the shift or for deciding on one paradigm over another. The choice is a subjective one. Ironically, much of the positivist scientific community has accepted the general idea of paradigms without understanding the deeper implications of Kuhn's work. It's not unusual to hear a scientist speak of "working within a paradigm" even though that scientist would be appalled by Kuhn's claim that scientific paradigms have no objective basis.

Today, many social scientists agree with Kuhn that the positivist claims for science were a myth. For example, in 1987 the sociologists Trevor Pinch and Wiebe Bijker wrote:

> [T]here is widespread agreement that scientific knowledge can be, and indeed has been, shown to be thoroughly socially constructed. . . . The treatment of scientific knowledge as a social construction implies that there is nothing epistemologically special about the nature of scientific knowledge: It is merely one in a whole series of knowledge cultures (including, for instance, the knowledge systems pertaining to "primitive" tribes). Of course, the successes and failures of certain knowledge cultures still need to be explained, but this is to be seen as a sociological task, not an epistemological one.

Such comments anger many scientists. Physicists especially dispute the conclusions of the social constructionists. Yes, they admit, of course the creation of scientific knowledge is a joint effort, but nonetheless it *is* epistemologically special. Quantum mechanics, for instance, provides predictions that are consistently accurate to a dozen or more decimal places. This is no accident, they insist, but instead reflects the fact that science is uncovering and explaining some objective reality.

On the whole, the physicists get the better of this particular argument. Social construction theory is useful in explaining social knowledge and belief, but it does little to explain why physicists accept general relativity or quantum mechanics as accurate portrayals of physical reality. Social constructionists like Pinch and Bijker ignore a key difference between science and other knowledge cultures: science's insistence that its statements be falsifiable. As long as science restricts itself in this way and continues testing its theories and discarding those that disagree with experiments, it is indeed epistemologically special. This and nothing else explains why science has been so much more successful than other knowledge cultures. Science may not be as objective as the positivists would like to believe, but the positivist approach comes closer than any other to capturing the essence of science.

This all may seem to be a tempest in an academic teapot, a question of interest to no one besides philosophers and the handful of scientists with an interest in epistemology, but it underlies many of the debates on scientific issues that affect the public. Consider the controversy in the mid-1980s over the use of recombinant bovine growth hormone in dairy cattle. Although the scientific community pronounced it safe, opponents of its use suggested that the researchers were influenced more by the sources of funding for their research and by their own inherent biases than by evidence. This argument depended implicitly upon the assumption that the scientific opinions were not objective but rather were socially constructed. In general, if science is accepted as objective, people will give a great deal of weight to the conclusions of the scientific community and to the professional opinions of individual scientists. But if science is seen as a social construct, vulnerable to biases and fashion, people will question or even dismiss its conclusions.

Which brings us back to our original subject. At its heart, this is a book about technological knowledge: What is it? and, How is it formed? Traditionally, engineers have seen their work in positivist terms. Like scientists, they take it for granted that their work is objective, and they believe that to understand a technology, all one needs are the technical details. They see a strict dichotomy between the pure logic of their machines and the subjectivity and the irrationality of the world in which they must operate. On the other hand, a growing school of social scientists sees technology as socially constructed. Its objectivity, they say, is a myth created and propagated by engineers who believe their own press. As with science, this is no mere academic debate. Our attitudes toward technology hinge, in large part, on what we believe about the nature of the knowledge underlying it.

To understand technological knowledge, this book argues, it is necessary to marry the positivist and the social constructionist perspectives. Technology combines the physical world with the social, the objective with the subjective, the machine with the man. If one imagines a spectrum with scientific knowledge on one end and social knowledge on the other, technological knowledge lies somewhere in the middle. It is falsifiable to a certain extent, but not nearly to the same degree as science. Conversely, much of technological knowledge is socially constructed, but there are limits—no matter what a group of people thinks or does, an airplane design that can't fly won't fly. In short, the physical world restricts technology. Some things work better than others, some don't work at all, and this leads to a certain amount of objectivity in technological knowledge. But unlike scientists, engineers are working with a world of their own creation, and the act of creation cannot be understood in positivist terms.

Ultimately, any understanding of technological knowledge must recognize the composite nature of that knowledge. Our technological creations carry with them the traces of both the engineer and the larger society. In the following pages, we will watch those agents as they interact to produce the machines of our lives.

My hope is that anyone reading this book will never see technology in quite the same way again. For those who assume, as I did, that all the important things about a technology can be found in engineering textbooks and manuals, I hope this book makes them more aware of the limits of rationality. For those who are already aware of the limits of rationality—the social scientists, the philosophers, the poets—I hope they come to appreciate the unique ability of science and engineering to tap into the physical world. And for everyone else, I hope this book offers a way of thinking about technology that helps make better sense out of today's technological world.

one

HISTORY AND MOMENTUM

It was New Year's Eve 1879, and the small community of Menlo Park, New Jersey, was overrun. Day after day the invaders had appeared, their numbers mounting as the new decade approached. When the *New York Herald* dispatched a man into the New Jersey countryside to report on the scene, he described a spectacle somewhere between a county fair and an inauguration: "They come from near and far, the towns for miles around sending them in vehicles of all kinds—farmers, mechanics, laborers, boys, girls, men and women—and the trains depositing their loads of bankers, brokers, capitalists, sightseers, hungry agents looking for business." At first it had been hundreds, but by the last evening of the year, some 3,000 had gathered.

They were here to see the future. Thomas Alva Edison, inventor of the phonograph, master of the telephone and telegraph, was said to have a new marvel, and it was his most amazing yet. If the newspapers could be believed, the Wizard of Menlo Park was lighting up the night with a magic lamp that ran on electricity.

The news had first broken ten days earlier. A reporter from the *Herald* had spoken with Edison, who showed off his latest success: a light bulb that would glow for dozens of hours without burning out. On December 21, the *Herald* trumpeted the achievement, taking an entire page plus an extra column to describe the bulb ("Complete Details of the Perfected Carbon Lamp") as well as Edison's trial-and-error search for it ("Fifteen Months of Toil") and the electrical system that would power it ("Story of His Tireless Experiments with Lamps, Burners and Generators"). Other newspapers soon picked up the story, and Edison, never one to pass up good publicity, announced he would open his laboratory after Christmas. Members of the public could come see the marvel for themselves.

And what a marvel it was. Perhaps in this age, when city folk must travel miles into the country to *not* see an electric light, it's hard to appreciate the wonder of that night. But in 1879, if people wanted light after the sun went down, they still performed that age-old ritual: they lit a fire. Outdoors, it might be a wood fire; indoors, a candle or kerosene lamp. In the larger cities, gas companies had become big business by providing a cheaper, more convenient flame. Their gas lamps, fueled by underground pipes, lit up city streets, factories, and shops, and, increasingly, the homes of the well-to-do. But none of these options was a satisfactory substitute for daylight. Wood fires put out more heat than light, while the light from candles and kerosene lamps was weak and flickering. The gas lamps were the best, but they were dim by today's standards, and they gave off heat, smoke, and sometimes noxious fumes.

A new, high-tech alternative—arc lighting—was just making its way into cities, but only for specialized applications. The arc lamps worked by passing a powerful electric current between two carbon electrodes, which slowly incinerated the carbon and produced an intense blue-white light. These "electric candles" had first been used in lighthouses, and by the late 1870s they were replacing gaslights on some city streets. A few were even installed in buildings, but they were too bright, too harsh, and too expensive for anything smaller than a factory or large store.

Indeed, electricity itself was a new and unfamiliar technology. Although it coursed through arc lights, danced along telegraph wires, and tinkled inside the newfangled telephone (invented less than four years earlier), it had not yet moved into the home or even into many businesses. It remained remote and mysterious.

So Edison's visitors can be forgiven for thinking there was something magical about his exhibition, even though by modern standards—or even the standards of a few years later—it was nothing special. As the callers approached, they could see a glow from the Menlo Park complex. Coming nearer, they could count the individual globes: two over the entrance of a small building at the gate of the complex, and eight more set on poles along the road and on the grounds. After pausing to admire the lights up close, guests would cross the yard and step inside the big laboratory for the main display. Here thirty electric lamps illuminated the rooms with, in the words of the *Herald*, "a bright, beautiful light, like the mellow sunset of an Italian autumn." Perhaps even more novel than the lamplight itself was the fact that it could be turned on and off with a switch. Over and over a member of the crowd would darken a room and light it up again, delighted with the easy control. And, as Edison biographer Matthew Josephson writes, a few sightseers got more than they expected: "Several persons who ventured into the dynamo room, despite warnings not to do so, had their pocket watches magnetized. A well-dressed lady, who bent down to examine something on the floor near one of the generators, found all the hairpins leaving her head." Despite the mishaps, the visitors—and the newspapers—deemed Edison's show a success. "Nearly all who came acknowledged that they were satisfied that they had seen 'progress on the march.'"

And so they had. The next few years saw the beginning of a technological revolution that transformed society more than almost any other. Much of it was driven by Edison himself. He improved his primitive bulbs to the point that they would burn for hundreds of hours before fizzling out, and he devised a system to power them: efficient generators, inexpensive ways to carry electricity from generating plant to customers, safety circuits, electric meters, and more. In one year—1882—Edison applied for 141 patents, most of them associated with his electric-lighting system.

On September 4, 1882, the Edison Electric Illuminating Company opened its first central power station, on Pearl Street in Manhattan. The service area, in the middle of the financial district, was about a half-mile wide and half-mile long. On that first day it claimed eighty-five customers burning about four hundred lamps. After that the technology spread quickly, driven both by Edison and a host of competitors, many of whom borrowed quite shamelessly from his patented inventions. Today, thanks to decades of innovations, that first primitive system of dynamo and lamps has evolved into a vast electrical network which powers a collection of machines and devices whose diversity would have surprised even the Wizard of Menlo Park himself. Not just incandescent and fluorescent lighting, but motors of various types, radios, TVs, computers, air conditioners, refrigerators, washing machines, microwave ovens, mixers, blenders, bread makers, and cappuccino machines—all these and more can trace their ancestry back to a laboratory where the public oohed at electrical switches and aahed at the mellow glow of artificial sunlight.

So goes the story of Edison and the light bulb. And a neat story it is. Too neat, in fact. As it's usually told, the development of electrical technology is reduced to little more than a series of inventions and technical advances fleshed out with descriptions of memorable events (such as New Year's Eve at Menlo Park) and historic firsts (the opening of the Pearl Street station). Sometimes, for good measure, you might even get a personal anecdote about Edison—his ability to work for days with only occasional catnaps, or his comment that "Genius is one percent inspiration and ninety-nine percent perspiration."

And the light bulb is not alone. In general, writers tend to present the development of any technology as a progression of technical innovations: A led to B, which allowed C to happen, then D and E opened the door to

F, and so on. It is, on the surface, a sensible approach. Edison's first commercial lighting network could not have opened without the light bulb and many other devices having been developed first. Today's electrical power systems in turn rely on a host of more recent technical advances, such as transformers that turn the high-voltage current of transmission wires into the low-voltage current used in businesses and homes. Clearly, any technology depends critically on the ideas and innovations of inventors and engineers.

But there is more—much more—to the story of a technology than this. Look closely at the history of any invention and you'll find coincidence and serendipity and things that happen for no reason at all but which affect everything that comes after. In the personal computer industry, for instance, it's well known that the Microsoft Corporation's dominating influence on software got its start by pure chance. When IBM decided to create its personal computer, the PC, the group in charge of its development first approached a company called Digital Research about providing the PC's operating system. When Digital Research put IBM off, the group interviewed Bill Gates and his still-small company, Microsoft. Recognizing what a tremendous opportunity it was, Gates did everything he could to convince IBM that Microsoft would be the right choice, but the ultimate selection of Microsoft came down at least in part to a personal connection: IBM chairman John Opel had served on the board of the United Way with Gates's mother and thought it would be nice to do business with "Mary Gates's boy's company." The operating system that Microsoft developed, MS-DOS, became the industry standard—not so much on its own merits as because it was part of the IBM PC package—and Gates was on his way to becoming a multibillionaire.

Besides such random but formative events, the development of any technology has much else that doesn't fit into the neat logical progression of the textbook. There are the wrong turns and blind alleys that in retrospect seem hardly worth mentioning but that at the time played a large role in engineers' thinking. There are the quirks and biases that can influence individuals' opinions as much as any rational arguments. There are the external circumstances that constrain and shape the inventors and their inventions. And there are the details of how a society adopts a technology and adapts it to its own needs.

To truly understand why a technology developed as it did, we must look past the ideas and engineering choices and put them in some sort of context. Who were the inventors? What were their strengths and weaknesses, and how did they interact with their peers and rivals? What was going on in the world outside the laboratory? Did the government weigh in? The media? What sorts of organizations did the inventors work in?

How did business factors influence the development and marketing of the innovation? Answering such questions is the job of the historian of technology, a specialization that has come into its own in just the past decade or so. Such historians trace the factors, technical and otherwise, that shape a particular innovation.

One of these scholars, Thomas Hughes at the University of Pennsylvania, has examined the early history of electrical power, and through his lens the saga takes on a very different appearance. To see just how different, let us return to the tale of Edison and his bulb.

THE BULB, THE POWER BUSINESS, AND THE ELECTROCUTIONER

The roots of the electric revolution lay—as do most technological revolutions—in scientific research and discovery. A century of work on electricity (going back to Ben Franklin and his kite) had raised the possibility of replacing gas lamps with electric ones, and by the 1870s a number of inventors had tried to create a light bulb. It was relatively simple to make a bulb that emitted light—all it took was the right sort of filament in a glass globe from which most of the air had been removed—but making a practical bulb was another matter. The first bulbs generally gave off too little light, consumed too much electricity, and burned out quickly, usually within a matter of seconds or minutes. The problem lay with the filament, the tiny element that emitted light, which would break or disintegrate when heated by the electric current.

In 1878 Edison decided to join the chase. He went after it with his customary thoroughness, trying many different designs and materials and recording the results of each. It was a textbook example of good engineering, although Edison would have called it "inventing." He decided what the characteristics of a good bulb would be—it should be relatively long lasting, able to be built relatively inexpensively, able to operate on low current, and so on—and set out systematically to discover how to make one. It took him just over a year. By the end of October 1879, he had produced bulbs that burned for at least forty hours, and he was confident he could increase that to several hundred. The first workable filament was a piece of carbonized cotton thread—thread that had been heated until it was mostly carbon.

On one level, that bulb was purely a product of engineering; its success lay in scientific insight, state-of-the-art fabrication techniques, and hard-won empirical knowledge about the characteristics of different filament materials and designs. But the work did not take place in a vacuum, and on another level the bulb was a product both of Edison's ability to sell himself and of his establishment of an "invention factory."

Invention is expensive, particularly when it involves repeated trial-and-error testing, and Edison was not wealthy. Instead, he relied on his reputation as a successful, money-making inventor to attract businessmen to underwrite his ventures, which in turn made Edison very aware of the value of good public relations. In September 1878, shortly after beginning serious work on electric lighting and more than a year before he first found a practical filament, Edison spoke to a reporter from the *New York Sun* as if the invention were nearly done: "With the process I have just discovered, I can produce a thousand [lights]—aye, ten thousand—from one machine. Indeed, the number may be said to be infinite." He went on to describe how he would light up Lower Manhattan with a system of dynamos, underground wiring, and electric lamps in the place of gaslights. The story, picked up by other newspapers, not only whetted the public's appetite for such a system but also got the attention of the financiers whose support Edison needed. Within a month, he had incorporated the Edison Electric Light Company with money from a dozen backers. Without these sponsors, his creation of a practical bulb would have come along much more slowly or perhaps not at all.

But the most important factor in Edison's success—outside of his genius for invention—was the organization he had set up to assist him. By 1878, Edison had assembled at Menlo Park a staff of thirty scientists, metalworkers, glassblowers, draftsmen, and others working under his close direction and supervision. With such support, Edison boasted that he could turn out "a minor invention every ten days and a big thing every six months or so." The phonograph—devised in late 1877—was one such big thing, the light bulb another. Both were team efforts, although the idea behind each was Edison's.

This "invention factory," as it was called, changed the nature of innovation. Menlo Park was the first of its kind, a forerunner of the modern industrial laboratory which combined invention, development, and sometimes even production under one roof, and it allowed Edison to turn out increasingly more complex devices at a pace that would have been impossible without it.

At the time of the New Year's Eve exhibition, for instance, the electric light was little more than a laboratory demonstration. Edison had learned how to make filaments that reliably lasted a hundred hours or more, but the production of the bulbs remained a slow, labor-intensive process. Indeed, the lights on display at his laboratory and a few more in his house were most of the world's supply at that time—and fourteen of them were stolen by visitors during the week the lab was open. But within half a year Edison would have a simple system ready for commercial installation, and in less than three years he would light up the heart of Manhattan. It was a spurt of creation unsurpassed in the annals of engineering, as Edison and his invention factory churned out not just improvements in the bulb but also innovations in electrical generation, transmission, and distribution. When it was done, he had created an entire new technological system.

Early on, for example, he revolutionized the dynamo, the device that transforms mechanical power into electricity. Previously, the best dynamos had managed only 40 percent efficiency, and some experts thought 50 percent was the theoretical maximum. But by experimenting with various configurations, Edison built a dynamo that surpassed 80 percent, cutting the fuel cost for generating electricity in half. Later he developed a "feeder and main" system for distributing electricity, which sliced the amount of copper needed in transmission lines to one-eighth of previous estimates. (The copper wires were the single most expensive component of the distribution system.) Not satisfied, Edison devised a "three-wire" refinement that chopped another 62½ percent off the copper requirements. All of these refinements were crucial to making electric lights economically competitive with the gas lamps.

Edison also came up with ways of insulating the current-carrying wires so that they could be run under the streets of New York without endangering people above; he invented a meter to record how much electricity a customer used; and he crafted a fuse to prevent too much current from surging through a circuit.

Less than five months after that bulb-lit night at Menlo Park, Edison had sold his first commercial lighting system, to the passenger ship S.S. *Columbia.* After being equipped with generator and lights, the steamship left New York in May 1880 and sailed around Cape Horn to San Francisco. The lights functioned perfectly throughout. Edison followed up that success with similar systems in factories, office buildings, even private homes. The financier J.P. Morgan bought one of the first systems for his house, as did the widow of railroad magnate Cornelius Vanderbilt.

But Edison's real goal was—and had been from his first experiments in electric lighting—the creation of "central station" power plants that would generate electricity with large dynamos and distribute it to the surrounding area over wires. He correctly saw that these, not individual systems, would be the key to a revolutionary change in society. Initially these electrical networks would provide mainly light, in competition with the gas companies, but eventually they would power motors and other machinery, doing jobs that natural gas could not.

The gas companies, of course, did not appreciate the competition, and their wealth gave them plenty of clout with local government officials. Edison knew that and knew also that the success of an electric lighting system, with its wires running along city streets, depended on avoiding active opposition from the authorities. With this in mind, when planning his first lighting system for Manhattan, he invited some of New York City's aldermen to his home for dinner and a demonstration of the new technology. By the light of ornate electric chandeliers, waiters in white gloves and black coats served the aldermen a fancy catered meal and fine wines. By the time the evening was topped off with champagne and cigars, the aldermen were much better disposed to consider Edison's requests. The lobbying worked, and in September 1882 Edison opened the Pearl Street station, but his fight to get the new technology accepted was just beginning.

To the general public, for instance, Edison's new technology seemed a mysterious and potentially dangerous servant. To begin with, the power for his generators came from steam engines, machines that had a reputation even worse than nuclear reactors have now. The high-pressure boilers in some steam engines—particularly on fast riverboats—had a habit of exploding, and they killed dozens of people before better engineering and tighter regulation settled them down. The lower-pressure boilers that Edison used were probably safe, but the distinction was lost on some people. Mrs. Cornelius Vanderbilt, in particular, had contracted for a home lighting system without realizing that a boiler was part of the package. As Edison later wrote, Mrs. Vanderbilt "became hysterical" upon learning of the monster in her cellar and declared that she "could not live over a boiler. We had to take the whole installation out."

Later, as the electric-light industry began to lure away a growing number of the gas companies' customers, those companies fought back by playing on a different fear. As more and more wires were strung along city streets, in buildings, and in homes, the gas companies warned that accidental electrocutions were becoming a threat to the public. The Edison Electric Light Company responded in kind, with bulletins describing fatal gas explosions in gory detail. The battle was a draw.

Edison's fiercest fight was not over the acceptance of electrical systems but over the question of which type of electrical system would dominate. On the surface it would appear to have been a purely technical dispute, but the details prove otherwise. Indeed, at times, the technical concerns seemed to disappear completely in a fog of business, legal, political, and personal issues.

The roots of the struggle lay in Edison's choice of a technology that had a major technical shortcoming. For a variety of reasons, including safety considerations, Edison had designed his light bulbs to work at the relatively low voltage of 110 volts. This in turn forced him to distribute his electricity at either 110 or 220 volts. (With his three-wire system, a single 220-volt circuit could be "split" into two 110-volt circuits, each of which could power 110-volt bulbs or appliances.) But that created a problem. Sending electricity through wires at 110 or 220 volts is terribly inefficient, and the power loss is so great that it's impractical to transmit the electricity more than a mile or two. The only way around this is to reduce the resistance by using very thick copper wires, an option that would make the cost of the distribution system prohibitively expensive.

Having chosen to work at 110 volts, Edison was thus forced to put his generating plants very close to his customers. This wasn't a problem in a place like Manhattan, which offered enough potential customers inside a couple of square miles to buy all the electricity even a big plant could produce. But in less densely populated areas the generating plants could not benefit from the economies of scale that come with very large installations. Furthermore, Edison couldn't take advantage of the water power available in many parts of the country. A hydroelectric plant could generate electricity much more cheaply than his standard system of coal-fired boilers, steam engines, and dynamos, but its hydro power could not be transported.

In 1883 a French inventor, Lucien Gaulard, and his English business partner, John Gibbs, offered a way out. They built transformers—devices to increase and decrease voltage—so that the current could be boosted to high voltage for efficient long-distance transmission and then stepped back down to low voltage for use by customers. Unfortunately, the system worked only for a type of electrical current different from what Edison was using—alternating current (AC) instead of his direct current (DC).

This innovation made AC an attractive alternative to DC, especially outside large cities and in areas with water power, but AC had its own big failing. Although it worked well with light bulbs, no one knew how to make an AC motor, and motors of various sorts were fast becoming major users of central-station electrical power. In factories, individual electric motors were replacing systems in which a single large steam engine turned dozens or hundreds of devices attached to it by belts. Meanwhile, electric trolleys were springing up in American cities. As long as AC couldn't turn a motor, its use was sharply limited, no matter how well it transmitted power over long distances. Finally, in 1888, the Serbian-born electrical engineer Nikola Tesla patented an AC motor, and within a few years practical AC motors were on the market. Suddenly AC could do everything DC could, and more.

If only technical issues were at stake, that would have been the end of it. AC was clearly the logical choice. But, as the historian Thomas Hughes points out, technical solutions alone are rarely enough to settle technological problems. Technological problems demand *technological solutions*. It was not enough simply to develop a new and superior electrical system using alternating current. A host of social, political, economic, business, and personal issues had to be settled along with the technical ones. And so the "battle of the currents" would drag on for several decades.

Edison, who had never hesitated to embrace the new and unproven when it was one of his own ideas, had no use for AC. Part of the reason may have been overfondness for his own DC system, on which he had worked so hard and long. Part of it, too, may have been the complexity of designing and working with AC components. While the DC system was developed by a group of inventors who were more old-fashioned tinkerers than scientists, work on AC demanded complex calculations. The AC design work had been tackled by a new generation of mathematically trained electrical engineers, people like Tesla and the German-born genius Charles Steinmetz.

Still another reason Edison distrusted AC was his fear that high-voltage transmission at 1,000 volts and more was too dangerous. In the early 1880s, when some cities had installed high-voltage DC lines to power arc lights, there had been a series of fatal accidents. Several linemen were electrocuted on their poles, in clear sight of the public. Edison, by contrast, had always been careful to make his systems as safe as possible, even running the lines from his Pearl Street station underground despite the extra cost. He didn't believe that the high-voltage AC systems could be made both safe and economical.

But apart from these understandable objections to getting involved with AC, there was a more personal, even bitter aspect to Edison's opposition. When Edison first announced in 1878 that he intended to develop a system of electrical lighting to replace the gaslights, the establishment was dubious. For decades, a string of scientists and inventors had tried to make an incandescent lamp and had failed. Why should Edison be any different? One electrical engineer, W.H. Preece, even argued that Edison's scheme was physically impossible. Supposing he did manage to make workable bulbs, Preece said, scientific theory proved that only a few of them at a time could ever be powered on a single, dynamo-driven circuit. Edison was chasing an *ignis fatuus*—a will-o'-the-wisp or, literally, "foolish fire."

Once Edison succeeded, however, many of the former doubters leaped into what they now recognized would be a lucrative business. Edison had patented his lamp and other devices, of course, but no matter. Some of the new competitors were other inventors who had themselves been attempting to make an incandescent bulb without success. After seeing his bulb, they modified their own efforts or even copied his work directly, adding one or two minor changes, then applied for patents. In other cases, Edison's competitors would find a patent vaguely similar to something he had built, then buy the rights to that patent and claim prior invention. The court battle over the patent for the carbon-filament lamp alone lasted until 1891 and eventually cost Edison's company $2 million.

Of all his competitors in the new electric-lighting industry, Edison may have liked George Westinghouse least. Westinghouse had originally made his name and his fortune as the inventor and manufacturer of the air brake, which had made high-speed rail travel safe. But after Edison's invention, Westinghouse moved quickly into the electrical equipment business, making and selling many items that were almost direct copies of Edison's inventions. Westinghouse covered himself by buying patent rights from inventors who were stealing from Edison. One Westinghouse advertisement even touted this method of doing business for the savings it offered to customers:

> We regard it as fortunate that we have deferred entering the electrical field until the present moment. Having thus profited by the public experience of others, we enter ourselves for competition, hampered by a minimum of expense for experimental outlay. . . . In short, our organization is free, in large measure, of the load with which [other] electrical enterprises seem to be encumbered. The fruit of this . . . we propose to share with the customer.

But Westinghouse was not simply a copycat. He recognized early the advantage that AC offered over DC, and he moved to develop electrical systems using the new alternating current. Westinghouse became the foremost American proponent of AC, giving Edison one more reason to stick stubbornly with DC.

Throughout the 1880s, a number of companies competed to install electrical lighting systems in the United States, and by the end of the decade three dominant firms had emerged. There was the Edison Electric Light Company, the oldest and largest—and the holder of the all-important carbon-filament patent. Arrayed against it were Westinghouse and Thomson-Houston, which between them had bought most of the non-Edison patents, some of which they used as a basis for infringing upon the Edison patents. Beginning in 1885, Edison's lawyers filed a series of lawsuits against the infringers, but the complexity of the cases—and a variety of imaginative legal maneuvers—prolonged the court battles for years. Meanwhile, all three companies were rushing to install as many systems as they could. Orders were coming in so quickly that they all had trouble keeping up.

While the fight was going on in the courts and the marketplace, Edison opened a third front: public opinion. He had often used the media to his own advantage, and now he set out to convince the public that AC was dangerous. In his laboratory in West Orange, New Jersey, just outside Newark, he set up a gruesome performance for reporters and other invited guests: the electrocution, by 1,000-volt alternating current, of stray dogs and cats. Day after day the demonstrations went on. One Edison biographer dryly noted, "The feline and canine pets of the West Orange neighborhood were purchased from eager schoolboys at twenty-five cents each and executed in such numbers that the local animal population stood in danger of being decimated."

The following year, in February 1888, the Edison group published a garish pamphlet, bound in red, called "A Warning." It described the dangers of high voltages, listing a variety of fatal accidents in factories, theaters, and along high-voltage power lines used for arc lighting, and even providing names of the victims. In the same pamphlet, Edison attacked Westinghouse and Thomson-Houston for their infringements, calling them "patent pirates."

Simultaneously, Edison was campaigning for laws to limit electrical circuits to 800 volts or less. At one point, for instance, he traveled to Richmond to urge the Virginia legislature to protect its constituents from the AC menace. The legislature, uncomfortable with this attempted Yankee influence, declined.

The low point of Edison's campaign against Westinghouse was a complex and bizarre plot involving Harold Brown, a former Edison laboratory assistant. Apparently under the direction of Edison and his assistants, Brown carried out a number of "experiments" on cats and dogs to prove the effectiveness of high-voltage alternating current as an "instantaneous, painless and humane" form of execution. He then launched a campaign to convince the state of New York to replace hanging with an "electric chair" as punishment for capital crimes. In the fall of 1888, the legislature assented, and it hired Brown as a consultant to provide the chair. Not surprisingly, he determined that the most appropriate source of the killing current would be Westinghouse AC dynamos, and he bought three of them (without telling Westinghouse what they were for) to do the job. In August 1890, convicted murderer William Kemmler became the first person to be executed by electrocution. Someone—it may have been Edison himself—suggested calling the process "to be Westinghoused." The term never quite caught on, but many people did come to associate Westinghouse's electrical system with death.

But slowly, despite Edison's best efforts, AC overtook DC in popularity. High-voltage AC motors, much more powerful than those that ran on low-voltage DC, began to take the place of steam engines for running large machines in factories. And high-voltage AC transmission lines allowed generating plants to stretch their service areas wider and wider. In 1895, the first large-scale hydroelectric plant in the United States went into service at Niagara Falls—with a 5,000-horsepower AC dynamo made by Westinghouse.

Still, there was no easy path to an all-AC electrical system. In the early 1890s, the United States was served by a hodgepodge of electrical systems, some DC, some AC. Many large cities, such as Chicago, had several utilities operating inside their boundaries with incompatible systems. A factory or large building might have its own DC power system, while the area around it was served by an AC central station. And the streets in that area might be lit by a separate arc lamp system not compatible with either.

To confuse matters even further, Edison finally won his patent fight in July 1891, and his competitors were ordered to stop making carbon-filament bulbs by 1892. The patent would expire in 1894, but Westinghouse and Thomson-Houston were facing two years in which they would have to scramble to find ways to fulfill their contracts for lighting systems.

The muddle was gradually resolved through a combination of technical and nontechnical fixes. The opposition to AC in Edison's company gradually waned after Edison himself lost power. In 1889, the various companies that Edison had formed to manufacture his electrical equipment were merged into the Edison General Electric Company, with control put into the hands of a syndicate of financiers who had injected new capital into the cash-poor business. (As often happens with rapidly expanding firms, Edison had plenty of orders but often had difficulty scraping together the money to build new facilities or, sometimes, even to meet the payroll.) Edison retained 10 percent of the new company's stock and had a seat on its board, along with three of his associates from the old days, but the rest of the seats belonged to the financiers. What little influence Edison exerted on the company disappeared entirely in 1892, when J.P. Morgan, who had gained control of the firm, merged it with rival Thomson-Houston to form General Electric. Edison's name was gone, and soon so was he. Edison sold his stock and resigned his position on the board, bitterly declaring that he would have no more to do with electric lighting.

The merger created a giant company that controlled about three-fourths of the electric business in the United States. It brought Edison's DC know-how and Thomson-Houston's AC expertise under one roof, at the same time combining their patent holdings. In the new company, AC would gradually come to predominate, as a new generation of engineers took over from the Edison old guard.

With Edison gone, General Electric was also able to reach a truce with Westinghouse. In 1896, the companies signed a patent-exchange agreement that would make electrical products more uniform and consistent.

The problem remained of what to do with the mess of diverse systems that had been installed throughout the 1880s and 1890s. A utility that had made major investments in one type of equipment couldn't afford to junk it all and start from scratch with a new, state-of-the-art AC system. Nor could it expect its customers to be willing to buy all new equipment. Eventually, Westinghouse engineers came up with a method for consolidating AC and DC equipment into one integrated system using synchronous generators, which could transform AC into DC, and phase converters, which brought the different AC systems into harmony. In these new systems, large AC generating plants provided power that was sent out at high voltage over a widespread transmission system. Old DC generating stations, which had provided power to nearby customers, were converted into substations, where the high-voltage AC current was stepped down to low voltages and transformed to DC, so that the electricity provided to the customers was indistinguishable from what they had received under the old regime. Gradually, the customers using DC were switched over to AC in a relatively painless transition.

European countries faced similar problems. England in particular had a difficult time. London was served by a jumble of small companies using DC and AC in a variety of currents, voltages, and frequencies. But because these power companies were each serving some small political subdivision jealous of its independence, it proved impossible to consolidate the systems. It was not until 1927, when Parliament created the Central Electricity Board to build a national grid connecting all the stations, that the country finally began to put its electrical house in order.

This expanded version of the tale of Edison and the light bulb has a simple moral: There is more to the history of a technology than the technical details. From the vantage point of the present, it would be easy to characterize the AC-versus-DC episode as a simple engineering matter: AC was better suited for electric power systems, so it became the technology of choice. And indeed, both sides in the debate tried to make it out to be a purely rational dispute: AC is more practical; DC is safer. But, as we've seen, resolving the dispute involved much more than reaching an engineering consensus on the best design.

This is the rule rather than the exception for any technology. Thomas Hughes emphasizes this point by drawing a sharp distinction between a technology and its technical features. To him, a technology is "a complex system of interrelated factors," both technical and nontechnical. "Technical refers primarily to tools, machines, structures, and other devices. Other factors embedded in technology, besides the technical, are the economic, political, scientific, sociological, psychological, and ideological." The technology of electric power, for example, would include not just power plants, transmission and distribution systems, and all the tools and devices powered by electricity, but also the companies that produce and maintain all this, the people who design it, the government agencies that regulate the industry, and the banks that finance it.

The distinction between a technology and its technical components is particularly important in the modern world. As Hughes points out, modern technology has a tremendous amount of "momentum"—the tendency to keep going in the direction it's already traveling—and this momentum arises mainly from the nontechnical factors in a technology. Companies that have invested heavily in a particular technical approach are reluctant to drop it in a favor of a new one, even if the new one might be better. Engineers who have spent years honing their design skills are loath to start over and develop a whole new set of skills. Managers prefer a known commodity to a gamble with unknown risks and rewards. Banks are cautious about lending money for a new, unproven product. Regulators who have become comfortable with the strengths and weaknesses of a particular technology resist new approaches. And customers are cautious about buying anything too different from what they know works.

Because of this momentum, historical circumstances can be quite important in the development of a technology, particularly in its early days. Like a snowball rolling down a snow-covered hill and growing in size as it goes, a new technology can be pushed in one direction or another by relatively minor factors—a personality conflict here, a lack of funds there—but once it picks up size and speed it's much harder to divert from its course. Nowhere is this more evident than in the history of nuclear power.

BUILDING THE BOMB

It was late 1938 when the German chemists Otto Hahn and Fritz Strassmann made the discovery that would lead to both atomic power and the atomic bomb. At the time Europe was in turmoil. Hitler's Germany had reclaimed the Rhineland, annexed Austria, and—with the bullied acquiescence of France and England—swallowed the German-speaking sections of Czechoslovakia. Many scientists and intellectuals were leaving Germany and Austria while they still could, especially Jews. Among them was Lise Meitner, an old friend and constant collaborator of Hahn's. Meitner had left only a few months before Hahn and Strassmann's breakthrough and thus missed taking part in one of the century's most important scientific findings. Had she stayed—and had she survived Hitler's pogroms—Meitner would almost certainly have shared the 1944 Nobel Prize for Chemistry with her old colleague.

As it was, Hahn and Strassmann were working alone when they found they could split atoms of uranium by shooting neutrons into their centers. It was completely unexpected, and the two chemists did not quite believe it at first. Scientific theory then had no place for atoms that could be broken in two.

Scientists did know that atoms consist of a dense nucleus of protons and neutrons surrounded by a diffuse cloud of electrons. The protons are tiny particles that carry a positive electrical charge, the electrons have a negative charge, and the neutrons possess no charge. An atom of hydrogen, for instance, has a single proton circled by one electron. An oxygen-16 atom has eight protons and eight neutrons in its nucleus (the "16" referring to the total number of protons and neutrons) with eight electrons in orbit. A uranium-235 atom contains 92 protons and 143 neutrons in its nucleus, enveloped by 92 electrons. Research in the 1930s had shown that elements could be transmuted, or changed one into another, by adding protons and neutrons. For example, an alpha particle (two protons and two neutrons) could be shot into the nucleus of an aluminum atom (thirteen protons and fourteen neutrons) to produce an atom of phosphorus (fifteen protons and sixteen neutrons). Researchers also knew that some atoms were unstable—their nuclei would spontaneously spit out a neutron or some other particle.

Despite this, nuclei were considered relatively solid citizens, changing incrementally if they changed at all.

To the contrary, Hahn and Strassmann discovered that the uranium nucleus is just barely held together. It's constantly on the edge of a breakdown. Throwing a neutron at it causes it to "fission" into two smaller pieces, roughly equal in size, that fly apart with a great deal of energy.

The news spread quickly through the nuclear physics community. It was an astonishing discovery, and at any other time physicists would have happily spent the next several years exploring its scientific implications. But already many of them recognized that the finding's practical consequences might overshadow its scientific importance. If the energy liberated by the splitting of the uranium atom could be directed and controlled, it might be possible to build a bomb hundreds of times more powerful than any before.

This possibility particularly worried Leo Szilard, a Hungarian physicist who had immigrated to the United States. Szilard feared that the Nazis might grasp the significance of uranium fission and create a superweapon. After all, it had been German researchers who made the original discovery, and although many of the Continent's best scientists had fled, more than enough remained to follow up on the finding. If a uranium bomb were indeed possible, Szilard argued that everything must be done to keep Germany from developing it and, barring that, to make sure that the Germans were not the first to perfect it.

To help sound the warning, Szilard approached Albert Einstein. The recent findings were news to Einstein, but he quickly grasped the situation and agreed to help. Szilard had a friend of a friend who knew President Roosevelt personally and who had agreed to carry a letter from Einstein to the president. Roosevelt received the letter on October 11, 1939, six weeks after the Nazis had invaded Poland.

Within a week and a half, the Advisory Committee on Uranium held its first meeting to discuss uranium's potential military uses. In Germany, secret discussions had begun a month earlier on the same topic. Less than ten months after Hahn and Strassmann had first realized the uranium atom could be split, their research into fission had been turned toward the war effort.

It was inevitable. The discovery of fission offered a way to tap the energy of the atomic nucleus, and such "nuclear energies" are tremendously more powerful than the "chemical energies" that had previously been mankind's most potent source of power. A simple calculation showed that, in theory at least, a pound of uranium should provide a million times as much energy as a pound of coal or a pound of nitroglycerine. In its report to President Roosevelt on November 1, the Advisory Committee on Uranium concluded that, because of this, uranium might prove valuable as a source of power for submarines and might also provide "bombs with a destructiveness vastly greater than anything else now known."

With so much at stake, scientists in the United States began a crash program to study the physics of fissioning uranium atoms. In little more than a year, by early 1941, they had learned enough to know that it was theoretically possible to build an atomic bomb many thousands of times more powerful than the biggest bombs ever made. The researchers continued their work, and after Japan's attack on Pearl Harbor brought the United States into the war, the effort to develop and build an atomic bomb went into high gear. The Manhattan Project, as it was called, would become the largest technology program the world had seen.

It is hard to say where Hahn and Strassmann's discovery might have led if it had come at a different time—the early 1920s, say, when the great powers of Europe were still exhausted by the recent bloodletting and the United States wanted nothing more to do with war. Perhaps it might have remained a matter of scientific research only, at least for a time. But, by historical chance, the two scientists opened the door to nuclear fission as the world was about to embark upon the most destructive conflict in its history, and fission's fate was fixed. The earliest development of nuclear power would be as a weapon.

Coincidentally, the discovery of fission also came at a time when large technology projects were just coming into their own. The Manhattan Project would dwarf anything that had come before it, but its predecessors had provided critical experience in coordinating the efforts of large numbers of scientists, engineers, and workers. By the 1930s, large factories were common, a number of companies had major research and development labs, and the federal government was deeply into the business of building hydroelectric dams. Without the organizational and management skills developed in such enterprises, it would have been nearly impossible to create the atomic bomb.

When the story of the Manhattan Project is told, the scientists who designed and built the bomb usually get most of the attention. And that's understandable. The boys of Los Alamos made up the most impressive group of geniuses ever assembled to attack a single problem, and they punctuated their effort with an unforgettable exclamation point: the mushroom cloud. But for our purposes, the rest of the Manhattan Project was where the action was.

It would not be enough just to find a workable design for the bomb. The bomb would have to be built, and manufacturing it would demand the creation of an entire technological infrastructure. It was this atomic infrastructure—of knowledge, of tools and devices, and of organization—that would shape the development of nuclear power after the war.

Releasing the energy of uranium—either for a bomb or a nuclear power reactor—depends upon the creation of a chain reaction. When a uranium atom splits, its nucleus breaks into fragments, which include a number of neutrons. The neutrons can strike other uranium atoms, causing them to split and release yet more neutrons, and so on. The difference between a bomb and a nuclear reactor lies mainly in how quickly the chain reaction proceeds. In a bomb it goes so fast that much of the energy locked in the uranium is released in a split second, while in a reactor the chain reaction is kept at a carefully controlled level.

For either a bomb or a reactor, there is one overwhelmingly important fact, which was first grasped by the physicist Niels Bohr in February 1939: For the purposes of creating a chain reaction, not all uranium is created equal. Natural uranium appears mainly in two forms, or isotopes. Approximately 99.3 percent of natural uranium is uranium-238 (U-238), which contains 92 protons and 146 neutrons in its nucleus, while 0.7 percent is uranium-235 (U-235), with 92 protons and only 143 neutrons. The two isotopes respond to neutrons in very different ways. The U-238 nucleus is more stable and harder to split, so that some neutrons are too slow to do the job. The U-235, on the other hand, fissions each time it absorbs a neutron, no matter how slowly the neutron is moving. The bottom line is that it is much easier to maintain a chain reaction in U-235 than in U-238.

For the bomb makers in particular, this was a vital distinction. The nucleus of uranium-235 is so efficient at capturing neutrons and splitting up to create other neutrons that only a few pounds of it are enough to keep a chain reaction going. Such a "critical mass," if brought together in the right way, will explode with the force of millions of pounds of chemical explosives. Uranium-238, on the other hand, is essentially useless for making weapons.

The bomb makers had one other option, discovered in early 1941 by Glenn Seaborg, a young chemist at the University of California at Berkeley. Plutonium-239 (Pu-239), an artificial element created from uranium-238, could sustain a chain reaction in much the same way as uranium-235. Either U-235 or Pu-239 could make a superweapon.

Here, then, was the second challenge facing the scientists and engineers of the Manhattan Project, after the task of designing the bomb: accumulating enough U-235 or Pu-239 to actually make the bomb. At the time the project began, no one had ever accumulated more than a tiny fraction of an ounce of either uranium-235 or plutonium-239, and developing and building the bombs would take hundreds of pounds.

In naturally occurring uranium, only about one of every 140 atoms is U-235, with the rest almost exclusively U-238. Assembling pure, or almost pure, U-235 would demand finding some way to separate it out from the uranium-238—a difficult task because the isotopes have very similar properties. Chemically they are nearly identical, making it practically impossible to separate them by chemical means. They do respond differently to neutrons, but that's of no help in separating them, either. The only useful difference is that U-238 is slightly heavier than U-235, since nuclei of the first isotope contain three more neutrons. In 1941, chemists had several ways to separate isotopes by weight—spinning them in a centrifuge, for instance, so that the heavier isotope accumulated toward the outside—but the methods were designed to generate small quantities for research. No one had ever tried separation on an industrial scale.

Plutonium offered a different set of problems. Because it is an artificial element, every bit had to be created from scratch, by bombarding uranium-238 with neutrons. Sometimes when a neutron strikes a U-238 nucleus, the nucleus does not split but instead swallows the neutron and turns into the isotope U-239. This isotope is unstable, however, and quickly decays into plutonium-239. To produce large, pure quantities of Pu-239 would demand two steps: the transmutation of a portion of U-238 into Pu-239 in a nuclear chain reaction, and the isolation of the plutonium from the uranium. Theoretically, separating plutonium from uranium would be much easier than separating two isotopes of uranium, since plutonium and uranium are different elements, but the practical difficulties were daunting. For one thing, plutonium is a poison, fatal in extremely small quantities. Even worse are the isotopes that are produced along with plutonium when uranium is bombarded with neutrons; they include a variety of highly radioactive byproducts which no one in 1941 had dealt with before. To create large amounts of plutonium from uranium would produce radioactivity on a scale never before imagined.

When the Manhattan Project began, it was impossible to predict whether a uranium bomb or a plutonium bomb would be quicker to build. So the project's leaders decided to try both. One program would aim at separating out U-235 from natural uranium, while a second would target plutonium production. The two wartime programs would lay the foundation for the nation's peacetime nuclear power efforts.

To make enough plutonium for a bomb would demand the construction of large nuclear reactors—machines in which a controlled chain reaction in uranium would take place. At the time, of course, no one had built a reactor of any size, large or small, and a sustained chain reaction was still just a theoretical possibility, albeit one that physicists had a great deal of confidence in. (The minute amounts of Pu-239 that existed at the time had been produced not in reactors but by firing beams of neutrons into a uranium target.) So, beginning in early 1942, a group of physicists at the University of Chicago working under Enrico Fermi set out to build a nuclear reactor to test their theories.

That first reactor was a flattened sphere, 25 feet across at the equator and 20 feet high, constructed out of 4-by-4-by-12-inch graphite bricks. Since no one outside the small atomic bomb project knew of its existence, there was no need for a fancy name, and Fermi referred to it as simply "the pile." The name stuck, and for many years afterwards reactors of all shapes and designs were called "atomic piles." Uranium was distributed throughout the pile by drilling three-inch holes in the graphite bricks and dropping spheres of uranium into the holes. Most of the uranium was in the form of uranium oxide, but late in the construction process some pure uranium metal became available. The finished pile contained 385 tons of graphite, 40 tons of uranium oxide, and 6 tons of uranium metal.

The graphite was more than a structural material. It served as a "moderator," a substance that slows down neutrons emitted by fissioning uranium atoms and makes it easier for the neutrons to be absorbed by other uranium atoms. So, despite what the name seems to imply, a moderator actually speeds up a chain reaction. The Chicago pile was designed so that when a uranium atom split, the resulting neutrons would fly though the surrounding graphite, which would slow them down enough to trigger more splittings when they passed through other pockets of uranium.

Because uranium is radioactive, at any given time a few of the atoms in the pile would be splitting and releasing neutrons, some of which would cause other atoms to split. As long as the pile was "sub-critical," too many neutrons would escape the pile without striking other uranium atoms, and the chain reaction would not grow. But once the pile reached a certain size, there would be enough uranium for the spontaneous fission of a few atoms to snowball until large numbers of uranium atoms were splitting in tandem. Fermi had figured out precisely how big the uranium-graphite pile would have to be in order to produce such a self-sustaining chain reaction.

The world's first controlled nuclear chain reaction took place on December 2, 1942. There were none of the crowds or newspaper reports that accompanied Edison's first demonstration of an electrical lighting system. Only the physicists who had designed and built the pile were on hand, and it would not be until after the war was over that their tale could be told.

The reactor sat in a squash court beneath the stands of the university's old football stadium. Years before, the school had given up on intercollegiate football, so the stadium was available; the squash court offered a large space that was both out of sight and easily guarded. On December 1, researchers had set the last layer of graphite bricks and uranium in place and had run tests that indicated the reactor was nearly critical—it was almost to the point where the fissioning of the uranium atoms would be self-sustaining, and once the reaction got started it would continue under its own power. Indeed, the only thing that kept the reactor from going critical at that point was the presence of control rods—long rods of cadmium inserted in the reactor. Since cadmium absorbs neutrons, the control rods dampened the rate of fission by preventing some of the neutrons from splitting uranium atoms. One of the control rods was set up as an automatic safety. If radiation from the pile became too high, it would trip a solenoid and the rod would drop into the pile. As a backup, Fermi had positioned another control rod above the reactor, this one held by a rope. A physicist with an axe stood next to the rope, ready to cut it if necessary. He was the "safety control rod axe man," a name that still survives in acronym form. Today, an emergency shutdown of a reactor is referred to as a "scram."

Throughout the morning of December 2, the researchers ran tests. Fermi ordered all but one of the control rods pulled out and observed the performance of the pile. Although nothing visible was happening, the activity inside the reactor could be monitored by detectors that counted the number of neutrons escaping from the pile. The more neutrons, the faster the reaction in the pile was proceeding. As the last control rod was pulled out six or twelve inches at a time, the activity would increase and then level off. Each time Fermi would check the readings against his calculations before ordering the rod to be withdrawn yet farther. At 11:30 A.M., as the pile was nearly critical, Fermi ordered the control rods reinserted and told everyone to break for lunch.

Upon their return, the control rods were set to where they had been before lunch, and more measurements were taken. Then Fermi ordered the one remaining control rod pulled out twelve more inches. At that point, the neutron count on the recorder began to slowly increase and not level off. "The pile has gone critical," Fermi announced. For several minutes he let the reaction continue, slowly gaining strength, and then he ordered the control rods dropped back into place, stopping the reaction. The energy produced by the huge pile was negligible—about half a watt, less than a hundredth the output of a standard light bulb—but the event marked the beginning of the atomic age. For the first time, mankind had sustained and controlled a nuclear chain reaction.

Within a few months, construction had begun on three scaled-up versions of the Chicago pile, which would produce plutonium in large enough quantities to make a bomb. They were located on a 780-square-mile site in a desolate area of south-central Washington state, which would be called the Hanford Engineering Works. Each of the plutonium-producing reactors consisted of a rectangular block of graphite, 28-feet high by 28-feet wide by 36-feet deep, with 2,000 tubes running through it. Each tube was filled with small cylinders of uranium, about the size and shape of a roll of quarters. These slugs of uranium would sit in the reactor for several months, during which time about 0.025 percent of the uranium would be transmuted into plutonium. After removal from the reactor, the slugs would be sent to one of two processing plants built on the Hanford site, where the plutonium would be extracted.

Although the three Hanford reactors were designed and built to make plutonium, they had a number of features that later power-producing reactors would have. In particular, because the reactors each generated 250,000 kilowatts of heat energy, they needed to be cooled to keep the uranium fuel from melting and setting fire to the graphite moderator. The coolant of choice was water: 75,000 gallons a minute pumped from the Columbia River, through the 2,000 tubes and around the uranium slugs. By the end of 1944, all three reactors were creating plutonium.

Meanwhile, a parallel effort was under way to obtain uranium-235. In practice it wasn't actually necessary to get pure U-235; uranium that was 80 percent U-235 and 20 percent U-238 would be sufficient to make a bomb. Such uranium was said to be "enriched" in comparison to natural uranium's 0.7 percent U-235.

The Manhattan Project's enrichment complex was set up in a hilly, tree-covered stretch of Tennessee along the Clinch River, twenty miles from Knoxville. The ninety-two-square-mile reservation was named the Clinton Engineering Works, but later it would be known by the name of the city built nearby to house its tens of thousands of workers: Oak Ridge. Along with Los Alamos, New Mexico, where the atomic bombs were designed and built, Oak Ridge, Tennessee, would be one of America's new atomic cities, part of the atomic complex that would remain after the bomb was built and the war was won.

Because no one knew the best way to enrich uranium, Leslie Groves, the army general in charge of the Manhattan Project, decided to pursue different methods at the same time in the hope that at least one would work. The most promising approach seemed to be gaseous diffusion, in which the uranium is handled in the form of gaseous uranium hexafluoride, or hex. The gas is pushed against a porous barrier, a thin metal sheet pierced with microscopic holes. Since uranium-235 is lighter than uranium-238, the gas molecules containing U-235 move a little faster than those with U-238 and thus hit the barrier slightly more often. Proportionately more of the molecules with U-235 pass to the other side and the gas there is slightly enriched. The effect isn't large, however, and the process must be repeated thousands of times in order to get sufficient enrichment.

The Oak Ridge gaseous diffusion plant, known as K-25, was a massive complex of thousands of diffusion tanks joined by miles of pipes, and the four-story building that housed the plant covered more than forty acres. But it was the little things that caused the problems. Hex is a very corrosive gas, so finding a suitable barrier was hard. The pores in the barrier had to be just the right size: too large and the barrier would not separate the isotopes, but too small and the barrier material would not stand up well to the corrosive gas being pushed through it. Eventually, in late 1943, researchers figured out how to make a suitable barrier material based on a fine mesh of nickel, one of the few materials that resisted the corrosiveness of the hex.

The engineers also needed to keep the uranium hexafluoride from escaping the system of pipes, pumps, and tanks. To seal the pumps and other spots where the gas might leak out, they ultimately settled on a completely new material, one that would become a household name after the war: Teflon.

If gaseous diffusion didn't work, perhaps electromagnetic separation would. Originally developed by Ernest O. Lawrence at the University of California at Berkeley, the electromagnetic method uses a beam of electrically charged uranium atoms, or ions. A magnetic field steers the ions in a circle, the size of which depends on the strength of the magnetic field. (The circular orbit was about four feet across in some of Lawrence's early work; later machines would be much larger.) Since the size of a particle's orbit depends on its mass, and since the mass of U-235 is somewhat less than that of U-238, the U-235 ions would move in slightly tighter circles. So by placing a plate in the path that the U-235 atoms take, it was possible to accumulate enriched uranium. The samples weren't pure uranium-235, however, because the orbits were inexact and some of the U-238 inevitably was mixed in.

Lawrence called his device a "calutron"—"calu" for California University and "tron" referring to the cyclotron, a closely related device used by physicists to accelerate particle beams around a circle. At Oak Ridge, several miles from the gaseous diffusion plant, an area was set aside for the calutron complex. It was called Y-12, and like K-25, it was huge. Richard Rhodes describes it this way in *The Making of the Atomic Bomb*: "Eventually the Y-12 complex counted 268 permanent buildings large and small—the calutron structures of steel and brick and tile, chemistry laboratories, a distilled water plant, sewage treatment plants, pump houses, a shop, a service station, warehouses, cafeteria, gatehouses, change houses and locker rooms, a paymaster's office, a foundry, a generator building, eight electric substations, nineteen water-cooling towers—for an output measured in the best of times in grams per day."

Toward the end of the war, a third uranium-enrichment complex was added at Oak Ridge. Its thermal diffusion process involved running uranium hexafluoride in liquid form between two concentric pipes. The inner pipe was kept very hot by running steam through it, while the outer pipe was cooled. Because of their mass difference, uranium-235 concentrates closer to the hot inner surface, while uranium-238 prefers the cold outer surface. By tapping into the uranium liquid close to the inner pipe, one gets an enriched sample. As with gaseous diffusion, the enrichment is not large with a single run, but with many repetitions it is possible to get uranium that has significant amounts of U-235.

Ultimately, all three processes were combined to produce the uranium-

235 that went into Little Boy, the bomb that was dropped on Hiroshima. Uranium was enriched partly by either gaseous diffusion, thermal diffusion, or calutrons, and was then put into a second set of calutrons to further enrich it to the 80 percent necessary for the bomb.

On the other side of the country, plutonium production proved similarly successful. The reactors and chemical-processing plants at Hanford provided the material for Fat Man, used on Nagasaki.

MOMENTUM

After the war, the United States was left with a tremendous industrial complex devoted to building atomic bombs. But the knowledge and the physical infrastructure developed for the bomb could easily be turned to generating nuclear electricity. And so this new technology of nuclear power—which, strictly speaking, did not yet exist—had a tremendous amount of momentum as it headed into the post-war years.

This momentum came in various forms. Part of it was intellectual. The Manhattan Project had assembled an unprecedented collection of scientific talent, including three Nobel Prize winners and seven others who would later receive science's highest honor, and their crash program had created a body of knowledge that would likely have taken a couple of decades to accumulate in peacetime.

Furthermore, the Manhattan Project had produced a cadre of young scientists interested in nuclear physics and eager to apply it. Although after the war many scientists returned to the research they had pursued earlier, plenty of others were hooked on the power of the atom. Chauncey Starr, who was in charge of electromagnetic separation at Oak Ridge, recalls, "I was very enthusiastic. I was gung ho. I had run this big project during the war. . . . I had a mental machine ready to roll. My colleagues felt the same way."

The physical machine was also ready to roll. The manufacturing complex created by the Manhattan Project was approximately the same size as that of the U.S. automobile industry at the time. It comprised plutonium-production reactors, separation plants, test reactors, and the facilities to serve them. Along with them came the hard-won practical knowledge needed to run them. Members of the Manhattan Project were the trailblazers in reactor design, isotope-separation techniques, uranium processing, and many other skills that would eventually be essential parts of civilian nuclear power. Much of the equipment and the knowledge was concentrated at the national laboratories at Oak Ridge and Argonne, a site outside Chicago where much of the reactor design was done, and these would serve as the twin kernels of the nuclear industry after the war. Given all this, Starr says, the attitude among many scientists was, "What do we do now?" And the obvious answer was: put the war machine to work developing nuclear power. "I don't think we thought very deeply about it. Just 'Let's try it.'"

What if there had been no Manhattan Project? Starr, who spent much of his adult life working for the nuclear power industry and is today one of the most thoughtful commentators on the technology, believes that there might be no nuclear plants even now. "If there hadn't been a war, [Hahn and Strassmann's work] would have been written up in the scientific journals" and treated as a subject of mostly academic interest, he speculates. Nuclear physicists would have spent years forming theories and doing experiments while competing with scientists from other fields to get money for their work. Eventually, someone would have tried to make a power reactor, but given the relatively small level of funding for nonmilitary research, it would have taken a long time to get anything practical. Furthermore, in the years following the war the United States was the only country with the resources to devote to a major nuclear power program, and it didn't really need the energy. It had plenty of coal and oil. According to Starr's crystal ball, the 1990s would have had low-power nuclear reactors operating to produce medical isotopes, but nothing else.

We'll never know, of course, but certainly the wartime project gave peacetime nuclear power a big push. What few people realize, however, is that the push was not simply some generic acceleration, speeding up what would have happened anyway. It pointed nuclear power down a particular path that it might not otherwise have taken.

There are many possible ways to build a power reactor, and even before the end of the war Manhattan Project scientists were coming up with ideas and debating which might be best. "We occupied ourselves with speculations and designs of future nuclear power plants," recalls Alvin Weinberg, a physicist who would later direct the Oak Ridge National Laboratory for many years. "Crazy ideas and not-so-crazy ideas bubbled up, as much as anything because the whole territory was unexplored—we were like children in a toy factory. During this period of about a year, from spring 1944 to spring 1945, were born many of the concepts that evolved into today's nuclear technology."

All nuclear power reactors work essentially the same way—by creating a chain reaction, the heat from which is used to generate electrical power. But in designing a reactor, three basic choices must be made: Which fuel? Which moderator? And which coolant? The fuel can be anything that will support a chain reaction: natural uranium, enriched uranium, plutonium. The choice of moderator to slow neutrons down is a technical decision depending in part on the fuel and in part on the desired characteristics of the reactor. Similarly, a designer has great leeway in the choice of a coolant. It must flow through the reactor core and carry heat away to a secondary system that will generate electricity, but many substances can do this. In a boiling-water reactor, the coolant is water, which turns to steam when it comes in contact with the reactor core, and the steam drives turbines and creates electricity. Other reactors use gas or liquid metal as a coolant, but there is no clearly superior choice—all have problems of one sort or another.

The existence of the nuclear weapons complex in the United States (and later in the Soviet Union) opened up an option for nuclear power technology that might otherwise have remained closed: the ability to build reactors that ran on enriched uranium. By the end of the war, the U.S. government had spent hundreds of millions of dollars developing enrichment techniques and building enrichment plants. After the war, as the United States and the Soviet Union raced to build more and bigger atom bombs, both countries developed large enrichment capabilities. By the mid-1950s, the U.S. system comprised three massive enrichment complexes in three locations: at Oak Ridge; at Paducah, Kentucky; and at Portsmouth, Ohio. Although the major reason for building the plants had been to provide enriched uranium for bombs, there would be enough uranium to go around. The civilian nuclear power program would get a share.

This increased the options for types of reactors dramatically. If nonenriched uranium is the fuel, there are only two practical choices for a moderator: graphite or heavy water, a form of water in which deuterium atoms take the place of hydrogen. Both have their disadvantages. Graphite is not a particularly good moderator, so the reactor core must be made very large to compensate, which is why the Chicago pile and those at Hanford were so large. The necessity for a large core puts limits on the other elements of the design. Heavy water is a much better moderator, but it must be extracted from seawater in an expensive, time-consuming process.

On the other hand, if enriched uranium is available as fuel, the selection of moderator is nearly unlimited. With a greater percentage of the easily fissionable U-235 atoms in a reactor's core, each neutron produced has a greater chance of splitting a U-235 atom, and the efficiency of the moderator becomes less crucial. If, for example, the uranium fuel contains more than about 1 percent U-235, then normal water—H_2O, or "light water"—can be used as a moderator. Light water does not work as a moderator for a natural uranium reactor because it absorbs just enough neutrons to keep the chain reaction from becoming self-sustaining.

Uranium can be enriched anywhere from 1 or 2 percent up to nearly 100 percent, giving designers an almost unlimited number of ways to match fuel, moderator, and coolant. Highly enriched fuel, for instance, allows a reactor to be relatively small, since less volume is needed to sustain a chain reaction. Slightly enriched uranium is sometimes used in graphite or heavy-water reactors to improve their performance. And enriched uranium allows more exotic materials, such as liquid metals, to be used as moderator and coolant. With highly enriched uranium, it is even possible to build a reactor with no moderator at all.

Despite this flexibility, however, the evidence from the nuclear programs of other countries suggests that without the push from atomic weapons, no country would have ever built uranium-enrichment facilities. Instead, reactors would have been built to burn unenriched uranium with its 0.7 percent U-235. Canada, with no interest in atomic weapons, based its nuclear power program on reactors fueled by natural uranium, moderated by heavy water, and cooled by light water. Britain, although it did choose to build an atomic bomb, still shied away from the expense and difficulty of enriching uranium. Instead it settled on reactors with graphite moderators and high-pressure gas coolant, which could produce both electrical power and plutonium for bombs. Across the Channel, France took a similar path.

So, in the early years, only the United States and the Soviet Union had the option of building reactors that used enriched uranium as fuel. Eventually, the U.S. nuclear industry settled on light-water reactors, which in turn made light-water the dominant reactor design around the world. (Canada, which has stuck with its heavy-water reactors, is the lone holdout.) But without the momentum provided by the Manhattan Project and without the enriched uranium supplied by the postwar bomb-building program, it's unlikely that the light-water reactor would have been a serious contender, much less the design of choice.

Nor did the legacy of the Manhattan Project stop there. A more subtle but equally important example of the momentum it provided was the government control of nuclear power throughout its early years. The war had forced nuclear research to be conducted in strict secrecy, with the government in charge of all decisions. When the war was over, some Manhattan Project scientists—including Robert Oppenheimer, who had headed up the design of the atomic bombs—argued that the new atomic knowledge should be made public. Other countries would learn how to make nuclear weapons anyway, they said, and salvation from the atomic threat lay not in secrecy and competition but in openness and cooperation. That argument made few converts in government circles, however, and the 1946 Atomic Energy Act formalized the wartime setup: the government would have a monopoly on all atomic energy matters. The act did take control from the military and give it to civilians—the Atomic Energy Commission, overseen by Congress—but atomic policy would still be set in Washington.

And so for nearly two decades, government would shape atomic energy for its own purposes—not just in the United States but also in Russia, France, England, and other countries—and those purposes generally focused on national security and prestige. Early efforts were aimed mostly at developing nuclear weapons. Later came other military programs, such as nuclear propulsion for submarines and aircraft. Finally, civilian nuclear power got some attention, but even then the goal was as much to provide for the common defense as to promote the general welfare. Governments believed that a state-of-the-art nuclear power program would provide both political and economic payoffs: prestige and bargaining power when dealing with non-nuclear countries, and billions of dollars in profits from selling nuclear products. As we'll see in the next chapter, the government interest in nuclear power would push the technology much faster than it would normally have moved, and in somewhat different directions.

In addition, the early government control of nuclear power created a culture of secrecy that permeated the industry, even after commercial business had taken over all but the military aspects. Too often, the first instinct of both government officials and corporate managers was to hide information, to mislead, or even to lie. The result was predictable: eventually a portion of the public assumed that nuclear officials were hiding something or lying, even when they weren't.

Finally, the Manhattan Project left nuclear power with a psychological legacy, complex and contradictory. The people closest to the subject knew just how impressive the making of the atomic bomb had been. Many German scientists, for instance, had concluded during the war that building such a bomb was impossible, or practically so, and they were incredulous to learn that the Americans had actually done it. After this it seemed that, with hard work and clever engineering, anything was possible, and the brilliant success of the nuclear submarine program in the decade after the war reinforced that feeling. Like NASA after the Apollo program, the nuclear industry of the 1960s would be proud, confident, even cocky. And like NASA, it would be headed for a fall.

With the general public, the bombs at Hiroshima and Nagasaki left a different impression. Atomic energy was clearly powerful, but it was just as clearly dangerous, and the experts' assurances that nuclear power reactors had little in common with bombs made less of an impression than did the mushroom clouds. The genie in the bottle is a blessing only as long as it obeys its master, and the public would not be as optimistic as the experts that it would always be kept under control.

SUBMARINES AND SHIPPINGPORT

In 1946, the Atomic Energy Act took control of the atom from the Army and put it in civilian hands. The five-man Atomic Energy Commission, or AEC, would direct all research into atomic energy, whether for military or civilian purposes, and it would produce and own all fissionable materials. In particular, one of the AEC's tasks would be to develop nuclear power for commercial use.

The leaders of the AEC and the nation's nuclear laboratories, such as Oak Ridge, recognized the opportunity facing them. By building on the knowledge and physical infrastructure bequeathed by the Manhattan Project, they could create a new technology essentially from scratch, and they were determined to do it in the most logical, efficient way possible. Over the next fifteen years the commission funded research into various types of reactors: light-water, heavy-water, liquid-metal-cooled, gas-cooled and graphite-moderated, breeder reactors, even reactors moderated by organic (i.e., carbon-based) liquids. The idea was to test different designs, see what worked and what didn't, and eventually settle on the best type of reactor for commercial nuclear power plants. It was a good plan—exactly the sort of carefully thought-out research program that gives the best chance of making the right technological decisions.

But the civilian endeavor would be blindsided and overwhelmed by a second project that would shape commercial nuclear power much more strongly than all the AEC's prototypes, studies, and tests combined. The sneak attack would come by sea.

The Navy's interest in nuclear reactors dated back almost to the discovery of fission. When the splitting of the atom was announced at a George Washington University conference in January 1939, one of the most excited was Ross Gunn, a physicist at the Naval Research Laboratory. Gunn realized that atomic power could turn submarines into truly effective underwater weapons.

At the time, submarines were awkward composites, not fully comfortable either on the surface or below it. Their main engines were the same diesels that powered surface ships, but these couldn't be used when the ship was submerged because they needed a constant supply of fresh oxygen and produced a noxious exhaust that had to be vented from the ship. So instead of its diesels, a submarine relied on less efficient battery-powered electric engines when undersea. The electric engines were not nearly as powerful as diesels, so the submerged sub was relatively slow, and movement at any speed quickly discharged the batteries. A sub could stay under only a few hours before returning to the surface to recharge. Thus a submarine's strategy was to stay on the surface most of the time to find its prey and get into position, and then to submerge when action approached. Not only was this dangerous for the submarine, but it also created a design dilemma: Should the sub's hull be shaped for efficient operation on the surface, where the ship spent most of its time, or for best movement below the surface, where it hunted?

Nuclear power might offer a way out. A nuclear reactor would neither need oxygen nor discharge anything that had to be vented into the atmosphere, so a nuclear-powered sub could spend most of its time under water.

As a bonus, because a pound of uranium held a million times as much energy as a pound of coal or oil, a nuclear-powered submarine could travel much longer without refueling.

In 1939 and 1940, this was still speculation, but Gunn convinced the Navy it was worth looking into. Although the Navy was kept out of the Manhattan Project—that was the Army's show—it did fund research into uranium isotope separation. The thermal-diffusion method for enriching uranium was initially developed with Navy money, although once the leaders of the Manhattan Project got wind of it, they took over and moved the project to Oak Ridge.

After the war, the Navy was even more interested in developing a nuclear-powered submarine. Conventional subs had played an important role in the war, attacking both supply ships and warships, but sonar and antisubmarine technology were threatening to kill their effectiveness. A submarine that could move at high speeds under water and stay submerged for long periods of time would more than even the odds.

The history of the Navy's nuclear submarine program is, to an extent almost unprecedented in any modern technology, the history of one man: Hyman Rickover. In 1946 the Navy sent a group of officers and civilian workers to Oak Ridge to learn about nuclear energy. Rickover, a captain at the time, was the senior officer of the group. A twenty-seven-year veteran of the Navy, he had an outstanding performance record, but his way was seldom the "Navy way." He spoke his mind freely and often pointed out flaws in how things were done. This did not endear him to many of his fellow officers. When the Oak Ridge team was being assembled, some of his detractors managed to prevent him from getting official command of the group, despite his senior rank. They were worried that he might gain control of the fledgling nuclear program. He would.

At Oak Ridge, Rickover quickly became convinced that nuclear reactors were the technology of the future for submarines and, furthermore, that he was the right person to develop that technology. By force of personality, hard work, and his senior rank, he became the de facto leader of the team, pushing the members to learn as much as they could. He himself spent a year studying at Oak Ridge and other government atomic laboratories, and by the middle of 1947 he was the Navy's leading expert on what it would take to propel submarines with nuclear power.

After another year and a half of maneuvering, Rickover was in a position to do more than study. He had been put in charge of the newly created Nuclear Power Branch of the research division of the Navy's Bureau of Ships and had also been named head of the Naval Reactors Branch of the AEC's Division of Reactor Development. From then on, the drive to develop nuclear reactors for submarines was full speed ahead.

A submarine makes requirements of a reactor that are quite different from those made by a land-based reactor. First and most important, the reactor must be compact enough to fit into a submarine's hull. This is possible, but it's not easy, and the size requirement automatically eliminates some reactor types from consideration. When graphite is used as a moderator, for instance, the neutrons must travel a relatively long distance before they are slowed down enough to do a good job splitting uranium atoms. In water, the neutrons must travel only about one-tenth as far. Thus, graphite-moderated reactors are inevitably rather large, while water-moderated reactors can be relatively compact. From the beginning, Rickover narrowed the search to three options that could be made small enough to fit in a submarine: a gas-cooled reactor, a liquid-metal reactor, and a light-water reactor. All three choices would use highly enriched uranium for fuel, which would minimize the size of the reactor core.

Convinced that a nuclear submarine was vital to American security, Rickover pushed the development of the technology almost as if the United States were at war. Normally, a new propulsion system would evolve in several stages. First, a test device would be designed and built in order to observe the performance of its individual components. After it had been reworked to correct any problems, a full-size prototype would be assembled. This prototype would then be tested and evaluated over a number of years. Only after the successful completion of these trials would the propulsion system actually be installed in a submarine.

Rickover compressed all this into a two-step, concurrent process. First, he would build a land-based prototype propulsion system in which all the components—reactor, steam generator, pipes, pumps, valves, and so on—would be laid out exactly as if they were on a sub. Then, while that was still under construction, he would begin assembling the reactor for the submarine, modifying pieces of it in response to any problems that showed up in the prototype. Meanwhile, the submarine itself would be taking shape in a naval shipyard. The keel for the *Nautilus*, the world's first nuclear submarine, was laid on June 14, 1952. It was not until a year later that the prototype reactor—a pressurized light-water reactor built by Westinghouse—was finished and testing begun on it.

Normally such an approach would end in disaster. The nuclear propulsion systems were extremely complicated, and much of each reactor was designed and built from scratch. No ordinary mortal could expect to get it right the first time—certainly not right enough to put it all into a submarine and launch the sub with a crew aboard. But Rickover and his organization pulled it off. He demanded complete testing of everything that went into the reactor and was ruthless about reworking things or even starting over, if necessary, when problems appeared. From all accounts, it was both a harrowing and an exhilarating experience to work for Rickover. He gave people the freedom to do a job the way they thought best, but in return he demanded complete accountability. If something went wrong, there was always someone who bore ultimate responsibility. Rickover's management style had no room for finger-pointing or claims of "I was only doing my job according to the book."

On January 17, 1955, the *Nautilus* went to sea. Richard Hewlett and Francis Duncan described the results in *Nuclear Navy*:

> During the first trial, while the *Nautilus* was confined to surface runs, the ship ran into seas heavy enough to make her roll violently. Many of the crew and technicians aboard became seasick as they struggled to measure the performance of the ship and its propulsion plant, but both operated perfectly. Submerged tests a few days later were more comfortable. Again the nuclear propulsion plant functioned faultlessly. To some officers the performance of the *Nautilus* was almost unbelievable. No longer did submarines need two propulsion systems—electric for submerged runs and diesel for surface operation.

While Westinghouse had been developing a light-water reactor for the Nautilus, a second submarine reactor was under development at General Electric. This one would use beryllium as a moderator and a liquid metal (first sodium, then later a sodium-potassium alloy) for its coolant. Again, Rickover collapsed the schedule, building a prototype and then the submarine reactor while the submarine itself was under construction. On July 21, 1955, little more than six months after the *Nautilus* took to sea, the *Seawolf* was launched. The U.S. Navy now had two nuclear submarines, powered by two completely different reactors.

On paper, the sodium-cooled reactor had one major advantage over the water-cooled reactor of the *Nautilus*: liquid sodium does a much better job of transferring heat than does water. To generate power with a nuclear reactor, one must take the heat produced by splitting atoms in the reactor core and do something with it—for example, turn water into steam and let that steam turn the blades of a turbine. And purely from this standpoint, liquid metal was obviously superior to water as a coolant: it absorbs heat more easily than water does, a given volume of metal holds more heat than the same amount of water, and liquid metal can take much higher temperatures than water without boiling.

But, as Rickover discovered, a sodium-cooled reactor offers plenty of headaches. The biggest worry was keeping the sodium away from water, since the two react explosively. This is especially a problem in the steam generator, where a pipe carrying hot sodium from the reactor core comes in contact with water and turns it into steam to drive the turbine. A leak here could be disastrous. To guard against this, the steam generators were built with double-walled tubes, and mercury was placed between the walls. If a leak occurred, the presence of mercury in the steam or the liquid sodium would offer a warning.

Despite the precautions, leaks were a problem in both the prototype and the reactor installed on the *Seawolf.* In one instance, a month after launching, the *Seawolf* was tied up to the pier for reactor tests when a coolant leak caused two cracks in steam pipes and shut down the steam system. Rickover didn't dither. Having decided that the liquid-metal reactor was "expensive to build, complex to operate, susceptible to prolonged shut down as a result of even minor malfunctions, and difficult and time-consuming to repair," he ordered the reactor on the *Seawolf* to be replaced with a light-water reactor like the one in the *Nautilus.* The Navy's experimentation with liquid-metal reactors was over. (The third reactor under consideration, the gas-cooled reactor, never made it past initial studies.)

Building on the initial success of the *Nautilus,* Rickover oversaw the development of an entire nuclear underwater fleet, which eventually grew to more than a hundred submarines. All had pressurized-water reactors like the *Nautilus,* and the operational and safety records of those reactors have been nearly spotless. Judged solely as a submarine-propulsion technology, nuclear power has been remarkably successful.

Submarine propulsion, however, has little in common with electricity generation, and the success of light-water reactors in submarines did not necessarily mean they would be the best choice for commercial nuclear power. (Indeed, Rickover didn't claim they were intrinsically the best choice for submarines. They were workable and they could be developed faster than the other options, and that was enough.) For years after the Navy had settled on pressurized-water reactors, the nuclear industry debated which reactor type or types made the most sense for the commercial sector. Engineers drew up numerous designs, performed countless calculations, and constructed a number of test reactors and prototypes. But in the end it was not what the engineers said but what Rickover did that tipped the scales.

Once the submarine reactor program was well under way, the Navy turned its attention to surface ships. Nuclear power would not be the revolutionary factor for a carrier or a cruiser that it was for a submarine, but a nuclear ship would be able to cruise much farther and faster than a diesel-powered vessel. By late 1951 the Navy had decided its next project would be a reactor for an aircraft carrier. Westinghouse was asked to study which reactor types were suitable for a carrier, and it reported that at least five different reactors—including a light-water reactor scaled up from the one planned for the *Nautilus*—would work. This left the decision of which type to pursue up to Rickover. His choice was no surprise. The water-cooled submarine reactor was coming along nicely; the other possibilities were little more than paper reactors. The light-water reactor it would be. With the approval of the AEC, Rickover asked Westinghouse to begin design work on a carrier reactor.

Westinghouse had barely started when the carrier program ran aground. The AEC was becoming more and more interested in civilian nuclear power, and it wished to build a land-based power reactor as a step in that direction. Perhaps, the Commission thought, the prototype that Rickover would build for the carrier reactor could be modified slightly to produce electricity. Doing both at once would save money. Rickover saw it differently: Doing two things at once would inevitably mean neither was done well, and he fought against combining the projects. The point became moot when Eisenhower entered office in January 1953. He had campaigned on a platform of cutting federal spending, and one of the first things to go was the proposed nuclear-powered aircraft carrier.

This left the AEC in a bind. Its only other experimental power project, a proposed graphite-moderated, sodium-cooled reactor, was another victim of budget cuts. With both it and the prototype carrier reactor gone, the agency's civilian nuclear power program was reduced to nothing. Eisenhower was sympathetic and agreed to a compromise: the carrier prototype would be reincarnated as a reactor for a civilian power station. Rickover and Westinghouse would continue the development of the reactor—stripped of its naval features—while the AEC would find a utility willing to operate it to generate electricity. The Commission would get its civilian power reactor, but it would be essentially a scaled-up version of the machine in the *Nautilus.*

The company that made the best offer for the nuclear project was the Duquesne Light Company of Pittsburgh. Duquesne would provide a site at Shippingport, Pennsylvania, twenty-five miles from Pittsburgh, it would build a plant to generate electricity from the steam provided by the reactor, and it would even pay $5 million of the reactor's cost. The AEC would own the reactor and sell its steam to Duquesne at a fixed price. The generating power of the plant would be 60 megawatts, or 60 million watts—about the same power as a medium-sized coal-burning plant of that era.

Rickover built the Shippingport reactor the same way he did the reactor for the *Nautilus.* This meant everything was done right, but it also meant cost was not an important consideration. The nuclear navy was not in the habit of accepting an inferior product to save money, and when Shippingport went into full-power operation on December 23, 1957—on schedule—its price tag had grown considerably from the already generous initial estimate. The cost of generating electricity at Shippingport would turn out to be approximately six cents per kilowatt-hour, or about ten times the cost of producing electricity from coal.

Otherwise, Shippingport performed well. It proved that a nuclear plant could operate continuously for long periods of time—which was essential if it was to be useful for utilities. The plant could quickly increase or decrease its output in response to sudden changes in demands. It was faster to shut down or start up than conventional power plants. And it operated safely. No workers were exposed to hazardous levels of radioactivity, and there were no dangerous releases into the environment. In short, Shippingport did what it was intended to do. It demonstrated the feasibility of nuclear power for generating electricity commercially. It was too expensive, but costs should come down. The important thing was: it worked.

Shippingport, however, did much more than it was supposed to do. In the AEC's eyes, the Shippingport project was little more than a quick and dirty way to bring a nuclear power plant on line. The Commissioners had a number of reasons for wanting to rush a plant into service—keeping the nuclear power enthusiasts in Congress happy, demonstrating a peaceful use for nuclear energy, establishing a position in the expected international market—but they also had long-term plans for nuclear power, and Shippingport represented just one among several competing types of reactors. Over the next decade, the AEC would sponsor a series of reactor-development programs in which it worked with utilities to build power plants with reactors of various sizes and types. In 1963, for instance, Consumers Public Power District of Columbus, Nebraska, completed a 75-megawatt plant near Hallam, Nebraska, which used a sodium-graphite reactor built by North American Aviation. In 1964, at Monroe, Michigan, Detroit Edison christened a 100-megawatt fast-breeder reactor. (It used "fast," or unmoderated, neutrons to produce plutonium as well as electricity.) Also in 1964, Rural Cooperative Power Association opened a 25-megawatt boiling-water plant at Elk River, Minnesota, and a 12.5-megawatt organic-moderated reactor started operations in Piqua, Ohio. Each reactor type had its strengths and weaknesses, which the AEC wanted to assess. Only in this way would it be possible to make an objective judgment about the best reactor for commercial use.

But thanks to Shippingport, it was never an equal contest. As we'll see in more detail later, Rickover's influence and the connection with the naval reactor program gave the light-water option a momentum that the others were never able to overcome.

Part of this had only an indirect connection with Shippingport. In the early 1950s, the AEC was spending several times as much on the development of military reactors as on civilian reactors. In 1952, for instance, the military-civilian ratio was \$39.5 million to \$1.6 million; by 1955 the gap was smaller but still significant—\$53 million versus \$21.4 million. The military funding allowed Westinghouse and General Electric, the two main contractors for naval reactors, to accumulate a tremendous amount of expertise in building light-water reactors and to gain a head start in the civilian market.

When they began building commercial reactors, it was natural they would stick with light-water machines. Even if the AEC had stayed out of commercial nuclear power and Shippingport had never been built, it would have been difficult for other reactor types to catch up with light-water after the boost it had been given by the military.

But the construction and operation of Shippingport handed the light-water technology an almost insurmountable lead. When Rickover was assigned the task of building the plant, he saw it as more than a construction job. It was an opportunity to set the civilian nuclear program on a path as carefully thought-out and engineered as the naval program. So as Shippingport was being designed and built, Rickover made sure to document and distribute as much information as possible about it. In 1954 and 1955, for instance, Westinghouse, Duquesne, and the Naval Reactors Branch held four seminars to describe the technical aspects of Shippingport. Hundreds of engineers from both the AEC and industry attended. During and after construction, thousands of technical reports were distributed on both construction details and operating experience. Once the plant began operation, Duquesne offered a series of training courses on reactor operation and safety which ran for six years.

The end result was to create a large and widespread body of theoretical and practical knowledge about light-water reactors—their design, construction, and operation. If any other reactor design were to compete, it would have to overcome this head start that the Navy had given to light-water. The contest was not yet over, and it would not be over for more than a decade, but in the end the technology of the submarine program and of Shippingport would prevail. Light-water reactors would become the nearly universal choice for commercial nuclear power.

two

THE POWER OF IDEAS

When Edison introduced his new-fangled electric-lighting system, he found a receptive audience. The public, the press, and even his competitors—with the possible exception of the gaslight industry—recognized that here was a technology of the future.

Alexander Graham Bell, on the other hand, had a tougher time. In 1876, just three years before Edison would create a practical light bulb, Bell's invention of the telephone fell flat. "A toy," his detractors huffed. What good was it? The telegraph already handled communications quite nicely, thank you, and sensible inventors should be trying to lower the cost and improve the quality of telegraphy. Indeed, that's just what one of Bell's rivals, Elisha Gray, did—to his everlasting regret. Gray had come up with a nearly identical telephone some months before Bell, but he had not patented it. Instead, he had turned his attention back to the telegraph, searching for a way to carry multiple signals over one line. When Gray eventually did make it to the patent office with his telephone application, he was two hours behind Bell. Those two hours would cost him a place in the history books and one of the most lucrative patents of all time.

Some months later, Bell offered his patent to the telegraph giant Western Union for a pittance—$100,000—but company officials turned him down. The telephone, they thought, had no future. It wasn't until the next year, when Bell had gotten financing to develop his creation on his own, that Western Union began to have second thoughts. Then the company approached Thomas Edison to come up with a similar machine that worked on a different principle so that it could sidestep the Bell patent and create its own telephone. Eventually, the competitors combined their patents to create the first truly adequate telephones, and the phone industry took off. By 1880 there were 48,000 phones in use, and a decade later nearly five times that.

More recently, when high-temperature superconductors were first created in 1986, the experts seemed to be competing among themselves to forecast the brightest future for the superconductor industry. Superpowerful motors, superfast computers, magnetically levitated trains, electric transmission without the normal loss in the wires—all this and more was thought to be in the cards. Many compared the potential of the new materials with that of the transistor when it had been discovered forty years earlier. Companies rushed in to be part of the revolution. Now, a decade later, we're still waiting. High-temperature superconductors have been used in a few commercial devices, such as sensitive detectors of magnetic fields, but because of a host of practical difficulties most of the truly revolutionary applications, such as the superefficient, long-range transmission of electricity, seem further away now than they did ten years ago. Many may never get here at all.

And by contrast, when the Haloid Company finished development of a new type of plain-paper copying technology in 1959, it didn't stir much interest. Needing extra capital to bring its product to market, the company approached several other firms, including giant IBM, but couldn't get a dime. The consulting firm Arthur D. Little, hired by IBM to evaluate the potential of the new technology, saw no future in the new invention and recommended against investing in it. Thus scorned, Haloid ended up raising capital on its own by selling extra stock, and it went on to ship its first copier in 1960. Eight years later Haloid—now renamed Xerox—had sales of more than $1 billion and was well on its way to revolutionizing the way offices were run.

Why are some new technologies accepted immediately and others resisted or even rejected? The answer lies in what might be called a "momentum of ideas"—the way that opinions, attitudes, and beliefs take on a life of their own, both in individuals and in groups. The importance of this momentum is often underestimated by people who see ideas as weightless things, able to be changed or redirected at the drop of a new discovery. But, in truth, the momentum of ideas shapes technology as surely as does the momentum of historical circumstances, the momentum of technological infrastructure, or the momentum of scientific knowledge.

The most obvious example of this is the frequent failure by people to recognize the value of an innovation. The history of technology is littered with examples of valuable inventions that received the same sort of treatment the telephone and the copier did. When the first Remington typewriters went on the market in 1874, they were ignored. It took more than a decade for the business world first to accept them, then to decide they were essential. In the late 1950s, when scientists at Bell Labs did early work on the laser, now at the heart of fiber-optic long-distance telephone systems, company patent lawyers saw no need to apply for a patent on it. Light had never played a role in communications, so why should Bell be interested? And the computer, in its early years, was seen as a large, expensive machine that would be run by a priesthood of carefully trained operators and used to solve complicated mathematical problems or to keep track of and manipulate large amounts of data. No one foresaw that its most revolutionary role would be as a personal tool for communications, bookkeeping, word processing, information gathering, games, and dozens of other applications.

None of this is surprising. People get stuck in old ways of thinking and find it hard to get out. Faced with a new invention, their first reaction is to see how it fits in the old system. When Marchese Guglielmo Marconi began to work on radio communications, he saw it as a supplement to the existing system of telephones and telegraphs which would be used where it was impossible to string wires. The idea of broadcasting from a central transmitter to many receivers would not come until later. Such inertia explains why Bell's telephone was much harder to appreciate at first than Edison's light bulb. Electric lamps would replace an existing technology—gas lighting—that was widely recognized as unsatisfactory, but the telephone would do a job that had never been done before and whose value was hard to foresee. It takes more imagination than most people possess to recognize an unmet human need or want and to realize that a particular invention will satisfy it.

Then why are some innovations greeted with great fanfare? Take the Internet, for example. John Perry Barlow, a founder of the Electronic Frontier Foundation, a group dedicated to defending freedom in electronic communications, declared in 1995 that "we are in the middle of the most transforming technological event since the capture of fire. I used to think that it was just the biggest thing since Gutenberg, but now I think you have to go back farther." Barlow may have been extreme, but he was by no means alone. Everyone, it seemed, was predicting a great future for the Internet. As this is written, the Internet is becoming an increasingly important tool with a rapidly growing number of uses, but whether its effects will rival those of the printing press remains to be seen. Either way, this is one technology that did not sneak into our lives unnoticed.

Surprisingly, optimistic predictions such as those surrounding the Internet are grounded in the same sort of thinking that produced the conservative, unimaginative reception to the telephone or the copier—that is, people are simply projecting the past onto the future. In the case of the Internet, the hype has its roots in ideas and beliefs that have been rolling along for years. Anyone with a modest familiarity with history knows that the invention of the printing press in the 1400s and the resulting spread of information created unparalleled change in western society, change that no one could have predicted in detail at the time. More recently, people have seen firsthand how the personal computer and electronic communications have reworked society by putting the control of information in individual hands. Reasoning from these premises, it's not a large leap to conclude that the Internet—by providing more information to more people and by creating new ways of manipulating bits of information and drawing connections between them—will open up a completely new era in human history, an era whose characteristics we can only guess.

The excitement following the discovery of high-temperature superconductors had a similar explanation. Technology has a history of putting scientific discoveries to work. Research into electricity and magnetism, for example, led to the birth of the electric power industry in the late 1800s. An understanding of electromagnetic waves allowed the invention of radio. And the discovery of nuclear fission brought about the atomic bomb and nuclear power. So now, whenever a scientific breakthrough occurs, people look for ways that it will translate into technology. And in the case of high-temperature superconductors, they didn't have to look far. Low-temperature superconductors—which must be chilled close to absolute zero before they lose their resistance to electricity—had already been put to work in such devices as powerful magnets, and scientists had long speculated on the additional applications that would open up if superconductors were ever discovered that operated at higher temperatures. So once researchers announced that such high-temperature materials had been found, the explosion of speculation and optimistic predictions followed naturally.

In short, expectations for a technology—whether positive or negative—are generally the product of past experience. They arise from rational calculation sprinkled liberally with biases and gut feelings. Sometimes the biases and gut feeling seem to take over completely, with the rational calculation applied retroactively to justify an already determined opinion. But whatever the mixture between the rational and the irrational, it all springs from a knowledge of what has gone before and beliefs about what all that means—and knowledge and beliefs have a weight to them. Once someone "knows" something, it's hard to unlearn it. Once a belief has been formed, it takes a lot of contrary evidence to turn it around.

This is true for individuals, and it is magnified when the individuals are thinking and acting in groups. When a collection of people hold an opinion, they reinforce it in one another. Sociologists tell us that much of what we think and believe is socially constructed—it arises within the larger context of a group's view of the world and can only be understood from that point of view—and socially constructed beliefs have particular momentum. Not only do they have the weight of numbers, but they are also seldom scrutinized or questioned because they seem completely natural, having been absorbed unconsciously by a person as part of the socialization process.

The attitudes of individuals and groups of people shape the development of technology in many ways. A hundred years ago Americans generally considered technology to be a "good thing." It had been responsible for much of the improvement in material well-being over the previous decades, and people welcomed new advances as signs of continuing "progress." This, along with the wide-open capitalism of the times and the relative freedom from government intervention, created an environment in which technologies grew like kudzu, spreading into every corner of society. Today, on the other hand, people are more cautious—even suspicious—about new technologies, thanks to the environmental degradation, losses of traditional jobs, and economic inequalities that have accompanied the technological development of the West. In some countries, this caution and suspiciousness has slowed down or even stopped the development of technologies such as genetic engineering and nuclear power.

On a smaller scale, individuals and small groups of people are constantly making decisions about this technology or that based on beliefs about its future prospects. A company decides to invest in developing applications of high-temperature superconductors. A group of entrepreneurs forms a corporation to provide services over the Internet. An agricultural firm drops its genetic engineering program after the flap over bovine growth hormone.

As long as tomorrow remains inaccessible to us, decisions about technology will have to be guesses based on what we can see today. But the momentum—or, often, inertia—of ideas means that the decisions are often based on yesterday's facts, not today's. This may well be unavoidable, but it should not go unrecognized.

TECHNOLOGICAL REVOLUTION

Although predictions about how an innovation will be used provide the most obvious example of how a momentum of ideas shapes technology, there is actually a more basic way. The process of invention itself is guided by beliefs and practices that are created by years of trial-and-error experience and that are just as resistant to radical new ideas as was Western Union in 1876 or IBM in 1959.

Consider, for example, how aerodynamic engineers reacted to the jet engine in the 1930s. As the historian Edward Constant relates in his book *The Origins of the Turbojet Revolution,* four men—three Germans and one Englishman—independently arrived at the ideas that led to the invention of the jet engine. Each, to one degree or another, found himself at odds with the aeronautical establishment.

Frank Whittle was the Englishman. In the late 1920s, he was looking for a way around the limitations that the conventional piston engine and propeller placed on aircraft speed. By streamlining airplane bodies, building more powerful engines, and refining propeller design, aeronautical engineers had steadily improved airplane performance—the air speed record was 357 miles per hour by 1929—but calculations showed that further improvements would be harder and harder to make. The use of a propeller placed strong limits on just how fast a plane could go, since the propeller became less efficient at speeds over 500 or 600 miles per hour. So Whittle began looking for a new propulsion system. He considered and rejected rocket propulsion, and eventually came up with the idea for the turbojet. It would combine a compressor, a turbine, and a combustion chamber. Air would be pulled into the engine and compressed to several times atmospheric pressure, then combined with fuel and ignited; the resulting hot air would turn a turbine that powered the compressor; and the air would then be directed out the back of the engine, providing the thrust to push an aircraft.

The concept was radically different from anything that aircraft designers were familiar with, and Whittle, a young, unproven engineer, found it difficult to convince anyone that his idea was worth pursuing. Although he was working in the Royal Air Force at the time he came up with the turbojet, Whittle couldn't interest the air ministry, so in January 1930 he patented the invention himself. After five years of further engineering training and of refining his design, he still had no takers, and when the air ministry declined to pay the £5 renewal fee to continue his patent, Whittle let the patent lapse. The turbojet was free to anyone who could recognize its potential and develop it. Soon afterward, however, an old RAF friend put Whittle together with some financiers willing to put up £50,000. His enthusiasm renewed, Whittle came up with enough "improvements" in the turbojet concept to apply for and receive a new patent, and he set up a company to turn his blueprints into a jet engine. After two years, as World War II was approaching, Whittle had a prototype that worked well enough to interest the air ministry. The ministry began funding development of Whittle's engine in 1937, the first test flight came in 1941, and by mid-1944 the first RAF jet fighters went into service.

In Germany, Hans von Ohain invented the turbojet independently of Whittle and at almost the same time. He, too, found little support from the traditional aeronautics industry, but he was luckier than Whittle in that he came across an iconoclastic sponsor who was interested in the radical new airplane designs: the aircraft manufacturer Ernst Heinkel. Heinkel had already supported the efforts of Wernher von Braun to create a rocket-powered aircraft, and he was immediately interested in von Ohain's idea for a turbojet. With Heinkel's support, von Ohain built the first plane to fly on turbojet power alone, a single-engine test craft that took its maiden flight on August 27, 1939, four days before Germany invaded Poland to trigger World War II. Soon after, his work was merged by the German government with that of two other men who had also developed turbojets, Herbert Wagner and Helmut Schelp, to form a single jet aircraft program. Germany built more than 1,000 jet fighters before the end of the war, and they were far superior to anything available to the Allies, but most of them were squandered on missions for which they were not suited, such as low-level bombing, and they had almost no effect on the war's outcome.

The invention of the turbojet, Constant says, was a technological revolution, but it was a revolution in a sense that is different from how most people use the term. Constant is not referring to the changes that the new turbojet technology wreaked on society: faster and cheaper travel bringing distant places in much closer contact, the extra deadliness of warplanes, and so on. Instead, he is speaking of the radical change in thinking that had to take place in the engineering community before the turbojet could be accepted. An industry that had grown up with propeller-driven planes had to come to grips with the idea of throwing away the propeller and piston engine and replacing them with a propulsion system that worked on completely different principles. The process was so dramatic and so wrenching that it can only be described as a revolution.

Most technological change, Constant notes, is the result of a slow, steady accumulation of improvements in existing devices and systems—what he calls "normal technology." The 1920s and 1930s were a time of normal technology for the aircraft industry. Everyone knew what sorts of things had to be done to improve airplanes—make the craft more aerodynamic; build more powerful engines; create stronger, lighter materials for the plane itself and stronger, heat-resistant materials for the engine; find construction methods that reduced drag; and so forth—and it was the engineer's job to figure out how to do these things. Over the years, the aeronautical engineering profession had developed a set of accepted practices to guide engineers in how they approached such tasks.

But the invention of the turbojet threw much of that out the window. Once the aeronautical community—or at least a good chunk of it—accepted the turbojet as the engine of the future and set out to develop it, engineers found themselves in uncharted waters. The turbojet offered a completely new set of design problems, problems that demanded skills different from those that engineers had accumulated while working on propeller-driven planes. Gradually, the engineering community learned what those problems were and settled into a new pattern that was just as fixed and comfortable as the one that had existed before the revolution.

This is the usual pattern that innovation follows, Constant argues: normal technology, technological revolution, then more normal technology. What engineers do most of the time is to play by well-established rules, making relatively minor improvements and refinements in an existing device. These modifications may add up to major changes—consider the evolution of aircraft between World War I and World War II or of the personal computer between 1980 and 1995—but the fundamental principles underlying the machine remain the same. Every now and then, however, someone comes up with a completely new approach, one that involves a major conceptual leap. And if it proves successful, this new idea serves as the basis for a whole new round of normal technology.

A technological revolution often meets resistance because it runs counter to the momentum that has built up during the time of normal technology. This momentum—or inertia—has its roots both in the engineers who work on the normal technology and the companies that employ them. Consider, for instance, an aircraft manufacturer in the 1930s that has spent a couple of decades building up expertise on propeller-driven planes. It knows them inside and out, knows their strengths and weaknesses, knows how much it costs to make them and how much profit it can expect from them. If someone approaches that company with an idea for a radically new type of plane, he's not likely to be met with much enthusiasm, and the more radical the idea, the less enthusiasm there will be. Most companies dislike uncertainties, and there's nothing more uncertain than the prospects of a totally new type of technology seeking to replace an existing one.

Now consider an individual engineer who has spent the 1920s and 1930s improving the performance of the piston engine. Having devoted much of his professional career to that quest, the engineer might find it difficult to believe that some new, untested invention is going to make the piston engine obsolete. Even if he is willing to admit the long-term potential of the new approach, he will have a hard time convincing his bosses and their bosses that it is worthwhile to pursue. And he probably won't try too hard. In the short term, at least, it usually seems smarter to work on the older, well-understood technology, where the returns on investment are much more predictable.

Engineers can be imaginative, risk-taking people, but most of the time—whether because of employers' demands or because of their own inclinations—they color inside the lines. They stick with the tried and true and confine their innovations to the little things that make up normal technology. And usually this conservatism is a good thing. If engineers take too many chances, the tail may drop off an airplane or an engine may stop working in midflight. But a habit of playing by the rules can also create a bias, often unconscious, in favor of the tried-and-true and against anything that is radically different. And this bias can give a technology a momentum that tends to keep it going in the direction it has been moving.

So it's not surprising that technological revolutions tend to be triggered by outsiders—by people who are not tied too closely to the status quo. It was

Alexander Graham Bell who recognized the value of the telephone, not Elisha Gray, whose close association with the telegraph industry skewed his perspective. And of the four originators of the turbojet revolution, none was an insider in the aeronautical engineering establishment. Whittle was only twenty-two years old and fresh out of engineering school when he first imagined the turbojet. Von Ohain was somewhat older, but even less a part of the club—he had studied mathematics and physics in college and was working on a Ph.D. in physics when he came up with the turbojet concept. Wagner had done a great deal of work in aircraft design, but his specialty was the structure of the plane itself, not the engine, and he had a broad background that included a familiarity with steam turbines. Schelp was a generalist with a knowledge of gas turbines.

A revolution succeeds, Constant says, when a significant fraction of an engineering community accepts the innovation as worthy of development. By that measure, the turbojet revolution "was a *fait accompli* by the end of 1939: the most advanced aeronautical practice in England and in Germany was by that date firmly committed to development of the turbojet." It was not necessary to build a jet-powered aircraft that was superior to conventional planes. The success of the revolution derived from an acceptance of the limitations of propeller-driven planes coupled with a belief that jets would offer a way around those limitations; once calculations and preliminary testing indicated that the turbojet was indeed feasible, people began to get on board.

Constant's view of technological revolutions has a number of parallels with the more familiar work of Thomas Kuhn on scientific revolutions. In *The Structure of Scientific Revolutions*, Kuhn argues that periodic upheavals change the underlying assumptions on which science is done, but that between times, scientists concentrate on "normal science"—improving and extending the existing, well-accepted theories.

One of best known examples of this pattern is the shift from the Ptolemaic to the Copernican system in astronomy. Ptolemy, a Greek astronomer of the second century A.D., set down a model for the motions of the sun, planets, and stars that had been developed gradually over the previous several centuries. In his earth-centered universe, all the heavenly bodies revolved around the earth on epicycles—circles on circles. The predictions made from his model matched astronomical observations of the time relatively well, given the limited accuracy of those observations. In the following centuries, as celestial measurements improved, astronomers fiddled with the Ptolemaic model to keep pace. The model became more and more complicated, however, and by the sixteenth century it was stretched to its limit. Modifications made to agree with one set of observations would upset agreement with another.

In 1543 the Polish astronomer Nicolaus Copernicus offered a way out.

He put the sun at the center of the universe, with earth orbiting it and revolving on its axis. The other planets likewise revolved around the sun. Unfortunately, Copernicus assumed that the orbits would be perfect circles (they are ellipses), and the predictions made from his model were actually less accurate that those from the highly evolved Ptolemaic model. Nonetheless, the fact that relatively accurate predictions could be made from such a simple model as Copernicus's was enough to convince many scientists that the correct path lay with this simpler, less accurate model. There was no clear-cut right answer, and a scientist's choice was likely to depend on such psychological factors as how religious he was (the Church insisted that earth was at the center of the universe), how willing he was to accept new ideas, and how greatly he valued simplicity in a model. The fact that the Copernican system was a step in the right direction was not completely clear until a few years later, when Johannes Kepler showed that by replacing the circular orbits with ellipses, the model provided nearly perfect agreement with observation. Afterward, astronomers could settle back down to normal science, making more and more accurate measurements and fiddling with the Keplerian model to bring it more closely in line with observation.

Kuhn refers to a change such as the Ptolemy-to-Copernicus revolution as a "paradigm shift." Scientists do their normal work within a given paradigm—a set of underlying assumptions and beliefs that define how to approach scientific questions and even which questions to ask. To fit their data, scientists will push and pull on the theories until the theories can no longer be stretched enough to cover all the observations. Anomalies start to pile up. Finally, someone offers an entirely new approach. If that approach seems better to enough scientists, it replaces the old, starting a new paradigm.

Philosophers of science have challenged Kuhn on a number of grounds. The weakest part of his argument is his contention that a paradigm shift marks a total break in how science is done. To Kuhn, scientists before and after a paradigm shift work from completely different assumptions, are trying to solve different problems, and, in effect, speak different languages. Yet this is clearly exaggerated. Einstein's theory of general relativity created a paradigm shift of the first order, changing the way that scientists think about space and time, yet physics students are still taught Newtonian gravity as part of their education. The old Newtonian paradigm is not some foreign, forgotten part of science, but instead remains as a special case of general relativity. It has been absorbed, not discarded.

Nonetheless, Kuhn's analysis offers some valuable insights for understanding technological revolutions. In particular, he argues that there is no completely rational way to make the decision to drop one paradigm in favor of another. Ultimately, although the reasons for the choice may be couched completely in logical arguments, scientists must rely to some extent on gut feelings. This is just as true—and probably more so—for a technological revolution. Because it is impossible to know a technology's true potential without developing it, decisions about going ahead with a technology must be made with incomplete information. Engineers can perform their calculations and do their tests, but in the end they must rely on their judgment.

And it is here, in the gray area where facts meet guesswork, that momentum becomes particularly important. When there is no clear rational basis for a decision, people fall back on their hunches, their biases, and their familiar patterns of thought. Normally this leads to caution and conservatism, and an engineering community gets behind a new technology only after the evidence is overwhelming that it's worth a try. Sometimes, however, the momentum works in the other direction, triggering a revolution before its time. Such was the case with nuclear power.

THE ATOMIC COUP

The devastation of Hiroshima and Nagasaki demonstrated the power of the atom in a most persuasive way, but there was no guarantee that this power would be good for anything besides destruction. Some of the Manhattan Project scientists had spent their spare time dreaming about the nuclear power plants they could build after the war, yet they were nowhere close to a blueprint, much less a test reactor or a prototype. They believed that such a power plant could be made to work, but they had no way of knowing how much it would cost, whether it could be made safe, or what to do with the radioactivity such a plant would generate. Nuclear technology was much further from reality than was the telephone in early 1876 or the jet engine in 1935. And there was no burning need for it—the United States had no shortage of other sources of power.

Yet, almost overnight, atomic energy went from a poorly known physical phenomenon to the next great thing. Enthusiasts urged putting nuclear energy to work not just in submarines and in electric power plants but in every way imaginable, from powering aircraft and rockets to digging canals and producing chemicals. Nor was it just talk. All of these ideas were seriously explored, some to the tune of hundreds of millions of dollars. By the mid-1950s, with none of the usual extended struggle and debate, nuclear energy had taken its place as an accepted technology. The decision had been made, the revolution had been won. Nuclear power was in.

How did it happen so quickly, with so few doubts or second thoughts? Going back to what was written about nuclear power in newspapers and magazines of the late 1940s and 1950s, one is struck by how little analysis of the atomic decision there was. Hard-headed businessmen might question how soon nuclear power would be competitive with electricity generated by coal or oil, but few had any doubts that its day was coming. And in the engineering community, the acceptance of nuclear power—a paradigm shift of the first order—was less a revolution than a coup. One looks in vain for any sign of resistance from conservative engineers who had doubts about the feasibility of nuclear power reactors.

The explanation for the unquestioning embrace of nuclear energy lies in a momentum built up from a number of sources. They included the generalized technological optimism of the time, longstanding dreams about releasing the power of the atom, a desire to find some counterbalance to the destructiveness of the atomic bomb, and Cold War political maneuverings. And once the intellectual commitment to nuclear power was made, it took on a life of its own.

The story begins at the turn of the century. In 1896, the French physicist Henri Becquerel had discovered that uranium spontaneously emitted some kind of radiation that could be detected with photographic plates. Marie Curie dubbed the phenomenon "radioactivity," but no one knew quite what to make of it. Then in 1900, while following up on Becquerel's discovery, Ernest Rutherford found that things were even more complicated than they had seemed. He discovered that thorium, a radioactive element closely related to uranium, gave off a gas that was itself radioactive. What was it? A previously unknown form of thorium? Or perhaps something else altogether? To find out, Rutherford enlisted chemist Frederick Soddy. Together they showed that the gas was argon, an element quite different from thorium, and they concluded that the thorium must be slowly disintegrating, transmuting itself into another element (or elements). The two scientists had observed, for the first time, the spontaneous decay of a radioactive element.

This in itself was a revolutionary finding. For centuries, scientists had tried—and failed—to transmute one element into another. Now they found it happened all by itself. Granted, thorium into argon doesn't have quite the appeal of lead into gold, but late-nineteenth-century physicists were just as excited by the one as fifteenth-century alchemists would have been by the other.

That discovery was soon overshadowed by an even more extraordinary and unexpected one. In 1903, while continuing their work on radioactive elements, Rutherford and Soddy uncovered a fact that would start the world on its march to the atomic revolution. When radioactive atoms such as radium or thorium disintegrated, they released a tremendous amount of energy—much more than did any process known up to that point. In Rutherford and Soddy's words:

> It may therefore be stated that the total energy of radiation during the disintegration of one gram of radium cannot be less than 10^8 gram-calories, and may be between 10^9 and 10^{10} gram-calories. . . . The

> union of hydrogen and oxygen liberates approximately $4x10^3$ gram-calories per gram of water produced, and this reaction sets free more energy for a given weight than any other chemical change known. The energy of radioactive change must therefore be at least twenty thousand times, and may be a million times, as great as the energy of any molecular change.

This was a startling, heart-stopping fact. Locked in the atoms of certain radioactive elements was an energy that dwarfed that of coal, oil, or gunpowder—if only it could be released in some controlled fashion. At the time, however, no one knew how that might occur. Scientists didn't even know that atoms consist of electrons orbiting a nucleus. Until Rutherford's 1911 discovery of the nucleus, it was assumed that an atom's negatively charged electrons were embedded in a diffuse, positively charged substance much like raisins scattered throughout a pudding. Nonetheless, Soddy was happy to speculate on what the energy of the atom might mean for mankind, and he became the first of a long line of people anticipating a technological revolution based on the atom.

In 1903, shortly after he and Rutherford calculated the energy released by the disintegration of radium, Soddy informed the readers of one popular magazine that radioactivity might prove to be a source of "inexhaustible" power. In a lecture the following year, he gave a concrete example: There was enough energy in a pint bottle of uranium, he said, to propel an ocean liner from London to Sydney and back. And in 1908 he published a book, *The Interpretation of Radium*, which assembled a number of his speculations, grown more grandiose over the years. "A race which could transmute matter," he wrote, "would have little need to earn its bread by the sweat of its brow. Such a race could transform a desert continent, thaw the frozen poles, and make the whole world one smiling Garden of Eden."

Soddy's book was well received and went to several editions. One person quite taken with Soddy's message was H.G. Wells, the British writer best known for science fiction tales such as *The War of the Worlds* and *The Time Machine.* Fired by the dream of a nuclear paradise, Wells wrote a novel of a future in which mankind has learned to liberate the energy locked within the nucleus of the atom. *The World Set Free*, published in 1914, envisioned a great war fought with atomic bombs, after which atomic energy was harnessed to create a world without want. In the prelude to the book, Wells has a professor (clearly modeled on Soddy) describe in a lecture what the rewards of harnessing atomic energy would be:

> It would mean a change in human conditions that I can only compare to the discovery of fire, that first discovery that lifted man above the brute. . . . Then that perpetual struggle for existence, that perpetual struggle to live on the bare surplus of Nature's energies will cease to be

> the lot of Man. Man will step from the pinnacle of this civilization to the beginning of the next. . . . I see the desert continents transformed, the poles no longer wildernesses of ice, the whole world once more Eden. I see the power of man reach out among the stars.

It was a powerful vision, this dream of an atomic Eden, and it lived and grew for thirty years after Rutherford and Soddy's discovery, even though no one could suggest a way that the energy of the atom might be tapped. In 1921, for instance, *Scientific American* discussed ways to meet humanity's growing energy needs. After mentioning coal, oil, natural gas, hydro, wind, solar, wave, and tidal power, it concluded: "But none of these possibilities is so attractive as that of atomic energy. It was Rutherford who said, 'The race may date its development from the day of the discovery of a method of utilizing atomic energy.' So enormous is this energy that it will confer on the man or the nation that learns to release and control it a power only less than that of the Omnipotent." Only one hurdle stood between mankind and near-omnipotence, but it was imposing: Although radioactive atoms spontaneously decay and release energy, that decay is a random, statistical process, and scientists knew of no way to trigger or to control it.

The trigger finally appeared in 1932 when James Chadwick, a British physicist working with Rutherford, discovered the neutron. This particle was approximately the same mass as the previously discovered proton, but it carried no electrical charge. This made it ideal for dealing with the nucleus. Because an atomic nucleus is positively charged, it repels any other object with a positive charge. A proton, for instance, must be accelerated to high speeds if it is to penetrate the nucleus. But a neutron can slip into a nucleus unimpeded. And once there, it can do several things. If it has enough energy, it may knock out one or more of the nucleus's protons and neutrons, transmuting the nucleus into a nucleus of some other element. A slower neutron may be swallowed, creating a nucleus with one more neutron than before. And, as Enrico Fermi showed in a series of experiments in 1934, some atoms that swallow neutrons become radioactive and spontaneously decay, just as radium, thorium, and other naturally radioactive elements do. This implied that neutrons can provoke a nucleus into releasing some of its energy. As Rutherford wrote in 1936, the practical implications were obvious: The neutron might be the key to tapping the energy of the atom "if only a method could be found of producing slow neutrons in quantity with little expenditure of energy."

That method turned up two years later, with Hahn and Strassmann's discovery of the fission of uranium atoms. Because the splitting uranium atoms released neutrons that would split other uranium atoms, a chain reaction could take place. The neutrons would not have to be added from the outside because, under the right conditions, the splitting atoms themselves would provide plenty of them. Thirty-five years after Rutherford and Soddy uncovered the energy of the nucleus, it seemed there was finally a way to release it.

Once it was clear that a chain reaction in uranium was possible, scientists didn't have to stop and think what this might mean. Decades of speculation and prediction had already created an image of a world with nuclear power. For example, the physicist Leo Szilard would later recall that when he heard of Hahn and Strassmann's discovery of fission, "[a]ll the things which H.G. Wells predicted appeared suddenly real to me." (This was the Szilard who later convinced Einstein to write his letter to President Roosevelt warning of the danger of the Nazis developing an atomic bomb.) For Szilard and others, it was taken for granted that atomic energy would be the source of unbounded power—for good or for evil, depending on the path mankind chose.

Unfortunately, this compelling image had been created from little more than the basic fact that the nucleus of the atom held a great deal of power. It did not take into account—because no one knew—what sorts of things might have to be done to release that power. There was no place in the vision for the problems of nuclear waste or the dangers of radioactivity, and so, at least at first, they were not taken as seriously as they deserved. Looking back on the early days, Alvin Weinberg recalls that even before the end of the war Enrico Fermi was warning his colleagues that the costs of nuclear power might be higher than people wanted to pay. "It was not clear," Fermi said, "that the public would accept an energy source that produced this much radioactivity, and that could be subject to diversion of material for bombs." But few paid much attention. The pure, uncluttered vision was too strong.

Thus the end of World War II saw the emergence of a small band of nuclear visionaries, mostly nuclear scientists from the Manhattan Project, who were convinced that atomic power would be a tremendous boon to mankind. By themselves they could have done little, but they found a receptive audience in the postwar United States, where faith in the power of technology seemed to take on an almost religious fervor.

This American faith in technology was nothing new, of course. For 150 years or more, Americans had embraced technology as the fount of material progress. But World War II brought this belief to new heights, particularly among the elite who formed public policy. This elite came mostly from industry, government, or the military, and they had just seen a war decided by superior technology—the radar that had won the Battle of Britain, the advanced airplanes and ships that turned the tide against the Axis, the sheer volume of weapons and materials produced by industrialized economies working at full tilt, and, more than anything else, the atomic bomb. To many, the Manhattan Project seemed the ultimate proof of man's ability to shape nature to his own ends. If we could conceive, design, and build such an incredible weapon in only a few years, could anything be beyond our grasp?

The success of the Manhattan Project also lent special authority to nuclear physicists, and when some of them spoke of the potential of nuclear energy, many others listened. These others may not have had the expertise to evaluate the nuclear claims on their own, but they had a faith in technology and they trusted the physicists. And so the nuclear momentum was transferred from a small group of scientists into the larger society.

David Dietz, science editor for Scripps-Howard newspapers, was among the earliest of those infected by enthusiasm for the coming atomic reign. In 1945, just months after the mushroom clouds appeared over Japan, he published *Atomic Energy in the Coming Era.* Among his predictions:

> All forms of transportation will be freed at once from the limits now put upon them by the weight of present fuels. The privately-owned airplane now suitable only for cross-country hopping will be equal to a flight across the Atlantic. There should be no difficulty in building passenger and cargo airplanes of any desired size. Planes carrying several thousand passengers, with as much cabin space as a luxury liner, will make non-stop flights from New York to India or Australia.
>
> Instead of filling the gasoline tank of your automobile two or three times a week, you will travel for a year on a pellet of atomic energy the size of a vitamin pill. . . . The day is gone when nations will fight for oil. . . . Larger pellets will be used to turn the wheels of industry and when they do that they will turn the Era of Atomic Energy into the Age of Plenty. . . .
>
> No baseball game will be called off on account of rain in the Era of Atomic Energy. No airplane will by-pass an airport because of fog. No city will experience a winter traffic jam because of heavy snow. Summer resorts will be able to guarantee the weather and artificial suns will make it as easy to grow corn and potatoes indoors as on the farm.

The only things he left out were that all children will eat their vegetables in the Era of Atomic Energy and that junk mail will be vaporized by smart, nuclear-powered mailboxes.

Although Dietz may have spiced things up a bit, as journalists sometimes do, his underlying message was echoed by many others, including those in a position to do something about atomic energy. James R. Newman, for example, served both as a science adviser to President Truman and as special counsel to the congressional committee that drafted the McMahon Bill, the basis for the Atomic Energy Act of 1946. His opinion of atomic energy's potential was not much different from Dietz's, minus the nuclear-powered cars and the artificial suns:

> This new force offers enormous possibilities for improving public welfare, for revamping our industrial methods and for increasing the standard of living. Properly developed and harnessed, atomic energy can achieve improvement in our lot equaling and perhaps exceeding the tremendous accomplishments made possible through the discovery and use of electricity.

Such enthusiasm might well have slackened as the difficulties in developing atomic energy became more apparent and as people were distracted by other, newer passions, but it never got the chance. In an unprecedented action, the U.S. Congress institutionalized the appetite for nuclear applications. A belief in the importance of nuclear energy would no longer be just a matter of opinion. It would be law.

It happened like this: With the war over, the federal government found itself in possession of a massive nuclear complex and a large body of scientific and technical know-how concerning nuclear energy. What should be done with it? The nuclear weapons program would continue, of course—with the wartime alliance with the Soviet Union unraveling and the postwar world looking increasingly unfriendly, that was a foregone conclusion. But what else?

One option was to throw open the Manhattan Project files, bequeath the new knowledge to whoever could use it, and let industry decide what to do with nuclear power. But Congress was not yet willing to share the country's hard-won nuclear expertise. With secrecy and luck, the United States might remain the only country with nuclear weapons for as long as a decade. On the other hand, the experts who advised Congress—mostly alumni of the Manhattan Project, still flush with the success of the most ambitious technological project in history—argued that the potential civilian applications of nuclear energy were too important to ignore. That left just one option: the federal government itself would have to get into the business of developing nuclear technology.

The Atomic Energy Act of 1946 codified this policy. Along with transferring control of the nuclear weapons complex from military to civilian hands—the major purpose of the act—it committed the United States to developing a technology that almost none of its citizens had heard of just a year earlier. The preamble to the act read as if its authors had been studying Soddy and Wells:

> Research and experimentation in the field of nuclear fission have attained the stage at which the release of atomic energy on a large scale is practical. The significance of the atomic bomb for military purposes is evident. The effect of the use of atomic energy for civilian purposes upon the social, economic, and political structures of today cannot now be determined. It is reasonable to anticipate, however, that

> tapping this new source of energy will cause profound changes in our present way of life. Accordingly, it is hereby declared to be the policy of the people of the United States that the development and utilization of atomic energy shall be directed toward improving the public welfare, increasing the standard of living, strengthening free competition among private enterprises so far as practicable, and cementing world peace.

Granted, it's hard to know how much weight to give the words of the preamble. Congress regularly serves up such high-sounding language to justify its measures, and no one takes it too seriously. But in this case, the actions spoke more loudly than the words. With the 1946 act Congress created the Atomic Energy Commission and charged it with devising and promoting peaceful uses of atomic power—even though no one knew just what those uses might be or whether they would prove practical or valuable.

According to the act, the AEC, whose five civilian members would be appointed by the President and approved by the Senate, would have a monopoly on atomic energy in the United States. It would own all fissionable materials and everything used to make fissionable materials, including patents and technical information; it would own and operate the nuclear weapons complex; it would be responsible for research into and development of atomic energy; and it would be entrusted with controlling and disseminating scientific and technical information relating to atomic energy.

The new law took the momentum toward nuclear power to a different plane. It aborted whatever debate might have taken place over the merits of nuclear energy and declared the United States to be officially committed to developing the technology. In effect, it institutionalized the belief that nuclear energy was destined for great things. In addition, the act set up a system to figure out what those great things were and then to develop them. The result, in the words of Cornell University sociologist Steven Del Sesto, was to create an "unrestrained and fertile developmental environment for the nuclear energy projects."

That environment was overseen by two powerful protectors which combined to push nuclear technology along on at as fast a pace as was technically feasible. The first of these was the AEC itself. It was never intended to be a neutral organization. One of its jobs was to promote the development of atomic energy, and throughout its history it would be dominated by true believers in the power of the atom to change the world for the better.

Lewis Strauss was a typical example of the type of technological enthusiast at the AEC who embraced the vision of nuclear scientists and took it even farther than the scientists themselves were willing to go. At age twenty, Strauss had helped organize Herbert Hoover's humanitarian Food Administration. He later earned business credentials during a career with an international banking firm. And he served in the naval reserve during World War II, retiring after the war as a rear admiral. When the Atomic Energy Commission was established in 1946, Truman picked Strauss to be one of its five commissioners, and in 1953 Eisenhower appointed him chairman of the commission.

Strauss's vision of a nuclear-powered future was remarkably similar to the one H.G. Wells had described forty years earlier. He outlined his views in a 1954 talk to the National Association of Science Writers. The United States, Strauss predicted, could expect to see electric power from the atom within five to fifteen years, depending on how hard the nation worked toward it. After that, the world would change forever. "Transmutation of the elements—unlimited power . . . these and a host of other results all in fifteen short years. It is not too much to expect that our children will enjoy in their homes electrical energy too cheap to meter, will know of great periodic regional famines in the world only as matters of history, will travel effortlessly over the seas and under them and through the air with a minimum of danger and at great speeds, and will experience a life span far longer than ours. . . . This is the forecast for an age of peace."

Decades later, when nuclear power was proving to be more expensive than electricity from coal, Strauss's "too cheap to meter" prediction would be recalled with scorn as hopelessly optimistic and naive. But his belief in the potential of nuclear technology was no stronger than that of many others; he simply phrased it more flamboyantly and said it for the record.

The AEC had a partner in promoting the nuclear vision, the second protector of nuclear fervor. Although it was not intended that way, the Atomic Energy Act had set up what was effectively a permanent nuclear advocacy group in Congress as well. Because of the special importance of atomic energy, the 1946 act came up with an innovative approach to handle it. The Joint Committee on Atomic Energy, consisting of eighteen members, nine from each branch of Congress, was created to be responsible for initiating all legislation dealing with atomic energy. This was unprecedented. In every other area, the Senate and House each had a committee working on legislation, so that the bills introduced in the two branches on defense, crime, and other matters were different, sometimes very different. The inconsistencies were worked out in conference, and the compromise sent back to the individual houses for approval. But with the Joint Committee, the legislation on atomic energy presented to the House and Senate was the same, giving the committee much greater influence than normal on legislation passed by Congress. Furthermore, since dealing with atomic energy demanded a great deal of specialized knowledge, the Joint Committee members gradually built up an expertise that gave even further weight to their recommendations. As a result, for years the Joint Committee had almost a free hand in controlling the general direction of nuclear power in the United States.

But the Joint Committee was not a disinterested watchdog over the AEC, as the act had intended. The committee naturally attracted members who were interested in atomic energy and believed it to be important. Even those members who were neutral when they joined got caught up in the enthusiasm after talking to their colleagues and sitting in on committee hearings that testified to the coming wonders of atomic energy. Furthermore, the members of the Joint Committee generally had none of the technical background that could help them understand the obstacles to developing atomic energy and therefore temper their enthusiasm. It was not surprising, then, that the Joint Committee often out-enthused the AEC.

The result was a government-directed technology with a built-in forward momentum. The only questions that got asked about nuclear energy were, How can it be put to work? and, How quickly can it be done? Sometimes this straight-ahead, no-looking-back approach worked out well. Rickover's successful development of the nuclear navy could not have moved nearly so fast otherwise. But sometimes the result was an extended foray into a blind alley. Two cases in particular—the nuclear plane and Project Plowshare—illustrate what can happen when ideas are allowed to run wild.

OUT OF CONTROL

The saga of the nuclear plane began in late 1945, with J. Carlton Ward testifying at a Senate committee hearing. Ward, president of the Fairchild Engine and Airplane Corporation, was pleading his case for continued government funding of aircraft development in peacetime, when the question came up of how best to deliver atomic bombs in future wars. Attacking Hiroshima and Nagasaki had been easy because the Japanese Air Force had been nearly destroyed, but future adversaries might be tougher. Assuming that the bombers would have to fly a long way to deliver their payloads, they'd be at a disadvantage to short-range fighters. Where could the Air Force turn? "An atomic plane," Ward said, "limited in range only by sandwiches and coffee for the crew."

Ward's suggestion made headlines (*Chicago Tribune*: "Predicts Atom Will End Limit on Plane Range") and got people talking. Although it wasn't really his idea—he had gotten it from Gordon Simmons, an engineer at Oak Ridge's gaseous diffusion plant—Ward soon became the head cheerleader for an atomic plane. He convinced the Army Air Force to look into its feasibility, and by 1946 Fairchild and nine other companies had signed a contract to study the concept.

The most practical approach to a nuclear aircraft engine appeared to be a modified turbojet, with a reactor taking the place of the combustion chamber. Air would enter the engine and go through a compressor, which would squeeze it to high pressure, and this high-pressure air would be heated by sending it through the reactor core. From there, the hot, expanding gas would stream through a turbine, turning the compressor, and then rush out the back of the engine, propelling the plane. It was a reasonable design—assuming that a reactor could be built that was powerful enough to propel a plane but light and compact enough that the plane could get off the ground. This, however, is not a simple task, and a number of difficulties quickly became apparent. For instance, air is not particularly efficient at absorbing heat, so any "direct cycle" engine in which air was passed through the reactor would have to be relatively large to provide enough thrust. An "indirect cycle" engine, in which liquid metal transferred heat from the reactor to the air, could be smaller, but it would be more complicated.

The worst problems centered on radioactivity. An atomic plane, said one early report, "would be dangerous on an airfield. When its reactor is running, all men in the vicinity will have to take cover, and the radioactive blasts roaring out of its tailpipes may poison the area permanently. . . . If atomplanes ever become common, the fields from which they fly will be thickly sown with radiation alarms and patrolled by safety personnel armed with radiation detectors." And if a nuclear plane crashed, it would likely spread radioactivity over a large area.

Furthermore, some way would have to be found to keep the radioactivity from killing the crew. Workers on ground-based reactors are protected by tons of concrete, lead, or other shielding materials, but that wouldn't work on a plane. The best hope lay in building a very compact reactor (which would reduce the amount of shielding needed), finding lightweight shielding materials, and putting the reactor as far from the crew as possible. To meet this last goal, some even suggested two-piece planes, with a nuclear-powered drone pulling a section that held the crew.

The Atomic Energy Commission, its hands full with more feasible projects, at first wanted little or nothing to do with the atomic plane. "[A] nuclear aircraft was an oxymoron. This was pretty much the view of all of us who had worked on reactors at Chicago and at Oak Ridge [during World War II]," recalls Alvin Weinberg, the long-time director of Oak Ridge National Laboratory. The Air Force was scarcely more enthusiastic than the AEC, once it had taken a close look at the plane. Various study groups found that, given enough money and time, a nuclear plane could probably be built, but it would never be particularly fast or high flying. Its one advantage was that, theoretically, it could stay in the air for months at a stretch. Unfortunately, the radiation made that a practical impossibility. No crew should be exposed that long to the engines' radioactivity, even with shielding. Secretary of Defense Charles Wilson summed up the defense department's attitude: "The atomic-powered aircraft reminds me of a shite-poke—a great big bird that flies over the marshes, that doesn't have too much body or speed to it, or anything, but it can fly."

Where Wilson saw a shite-poke, however, the Joint Committee saw an atomic-powered eagle with nuclear talons. The committee got involved in 1948, when it pushed the AEC and Air Force to perform a study on the feasibility of atomic-powered flight. The several dozen scientists in the so-called Lexington Project were initially skeptical but eventually decided that a nuclear-powered plane could fly, although the work would take at least fifteen years and cost $1 billion. That was promising enough that the AEC and Air Force, mindful of the Joint Committee's interest, began basic research toward an aircraft reactor.

The fundamental problem for reactor designers was that to provide enough thrust, the reactor had to run very hot—1500 to 2000 degrees Fahrenheit, or two to three times as hot as existing reactors. This meant that for such components as the fuel elements and reactor vessels, scientists would have to develop new materials that could stand up both to extreme temperatures and to an onslaught of neutrons. At the same time, researchers would have to devise new approaches to shielding and to conjure up a number of other innovations, all of which had to work together successfully, if the nuclear plane were to get off the ground.

Throughout the 1950s neither the AEC nor the Air Force could make up its mind on the project. In 1953, for instance, the Air Force recommended killing the nuclear plane altogether as a cost-cutting measure, but in 1957, after the Soviet Union's launch of the Sputnik satellite, the Air Force reversed itself and became eager to get an atomic-powered plane into the air as quickly as possible. The Joint Committee, on the other hand, held steady. It never lost faith in the plane, never believed that enough was being done to build it, and worried endlessly that the Russians might put a nuclear plane in the air first.

"If Russia beats us in the race for atomic bombers our security will be seriously endangered," said Senator Henry Jackson, a member of the Joint Committee, in 1956. "There have been estimates in the press and periodicals to the effect that an A-plane prototype may fly as early as 1958 or 1959. I do not agree with these predictions. It could fly at an earlier date. It probably will fly at a later date. The time schedules of our accomplishments in research and development are in proportion to the amount of effort which we apply to a given project."

In October 1957, Representative Melvin Price, another Joint Committee member, returned from a trip to the Soviet Union. The Russians, he reported, "are placing considerable emphasis on their own program to develop a nuclear-propelled aircraft." And if the Russians beat us to it, it would be our own fault, said Representative Carl Hinshaw, one of the project's most consistent supporters. "I can only estimate that the attitude of the people opposed to this project delayed it by as much as four years."

Meanwhile, the engineers trying to build the plane found it tough going. Oak Ridge National Laboratory, for instance, worked on an indirect-cycle reactor that used a liquid uranium solution as fuel. Continuously moving through the reactor, the uranium would chain react in a section containing a moderator, then move on to a heat exchanger where it could transfer its heat to air flowing through the jet engine. Oak Ridge researchers developed a workable fuel of uranium in molten salts of fluorine, such as sodium fluoride. Because these salts are exceptionally stable, they relieved the major problem of hot, circulating fuel—the corrosion it causes to the entire system. At the same time, other researchers at the lab were devising alloys and building valves and sensors that could stand up to the heat and radioactivity. In 1954 the Oak Ridge group put it all together to run a test reactor for a hundred hours at 1600 degrees Fahrenheit, hot enough for an aircraft engine. Next, the group designed a reactor of the same type that could fit into an aircraft. All told, several hundred Oak Ridge employees were on the project, and the nuclear airplane accounted for about a quarter of the lab's total budget. Yet by 1959 lab director Alvin Weinberg had concluded that the reactor "had little chance for success as a power plant for an airplane" and offered that opinion to the U.S. Department of Defense.

By 1961, both the AEC and the Air Force had lost their appetite for the program. After ten years and $1 billion spent, it looked as if it would take another decade and another $1 billion to get a finished product. The Joint Committee was still supportive, but it had few allies. Finally, newly elected President Kennedy, acting on the advice of his secretary of defense, killed the nuclear plane.

The pursuit of the nuclear-powered aircraft was certainly one of the more extreme products of the passion for the atom, but it was by no means the only one. There was the portable reactor, which the Army wanted for go-anywhere electricity and which scientists hoped to use in remote stations such as an Antarctic base. There was the nuclear cargo ship *Savannah*, which could travel much farther than conventional cargo ships without refueling—a trait that's unimportant to cargo ships since, unlike military vessels, they seldom stay at sea any longer than is necessary to get from port to port. There was the nuclear ramjet, never built, which was intended to propel low-altitude missiles at supersonic speeds. And there was the nuclear weapons program, which often got more support than even the Pentagon thought it needed. Time and again the Joint Committee insisted that defense department requests for bomb material were insufficient.

The twin ideas driving such programs were a generalized technological optimism and a sense that nuclear energy, because it was a million times more powerful than chemical energy, must have many important uses. It is difficult to comprehend today, but much of the Joint Committee's behavior through the 1950s and into the 1960s seems to have been driven by an almost childlike faith in the broad importance of nuclear power and the conviction that the United States, not the Soviet Union or some other country, should be the one to deliver nuclear technology to the world. In the case of civilian applications, a third element contributed to the momentum: a desire to compensate somehow for the violence and destructiveness of the atomic bomb. Everyone wanted to believe that the atom could be put to work peacefully.

Before the bombs were dropped on Hiroshima and Nagasaki, many of the Manhattan Project scientists had signed a petition asking that an atomic bomb first be exploded in an uninhabited area as a demonstration. The Japanese, they hoped, would be frightened into surrendering and so avoid the mass slaughter the bombs would wreak. The request was ignored, and the subsequent use of the bombs affected the scientists deeply. After the war, few stayed in weapons work. A majority returned to what they were doing before the war, and some, like Leo Szilard, left physics altogether. Szilard, greatly distressed by helping unleash this nuclear terror on the world, switched to biology.

But even the scientists who chose to study nuclear energy after the war felt the shadow cast by the mushroom cloud. Years later, David Lilienthal, the first chairman of the AEC, would reflect on the motivations of those people. They shared a conviction, he said, "that somehow or other the discovery that had produced so terrible a weapon simply *had* to have an important peaceful use. . . . Everyone, our leaders and laymen and scientists and military men, wanted to establish that there is a beneficial use of this great discovery. We were grimly determined to prove that this discovery was not just a weapon." This grim determination, combined with the conviction that nuclear power must have valuable applications waiting to be discovered, would spur efforts in a number of areas. Surely the most bizarre was Project Plowshare.

In 1956, Egypt was looking for money to build the Aswan Dam. When the United States and Great Britain said they wouldn't finance it, Egyptian president Gamal Abdel Nasser nationalized the Suez Canal. In the ensuing "Suez crisis," Israel invaded Egypt with the backing of England and France, and eventually UN troops came in and stopped the fighting. Egypt kept control of the canal and it was back to the status quo ante—except that the crisis had triggered an idea in Harold Brown, a chief scientist at the Livermore Radiation Laboratory in California.

Egypt's takeover had underscored just how dependent western countries had become on the canal, particularly on the Middle Eastern oil that passed through it. Perhaps, Brown thought, if an alternate canal could be cut through Israel, future crises in the Middle East might not be so menacing to the West. And Brown, who was a nuclear weapons designer, came up with a novel way to dig the canal. Forget conventional excavation methods, he said. Use nuclear explosives. A string of several hundred atomic bombs detonated in a line across Israel would open a path between the seas.

If this seems an outlandish, Strangelove-ian notion now, it did not seem so then. After all, the United States was still conducting open-air tests of its nuclear weapons at the Nevada Test Site, and the radioactivity from an atomic bomb used for excavation would be much less than from the above-ground explosions. If a nuclear weapon is exploded at the right depth underground, it forms a large crater, but the rocks and dirt that fall back into the hole trap most of the blast's radioactive debris. Thus a whole canal might be blasted out with no more radioactivity than was released from a few weapons tests. And most scientists at the time were generally unconcerned with the relatively minor health risks that might be caused by fallout from weapons tests. "People have got to learn to live with the facts of life, and part of the facts of life are fallout," said Willard Libby in a 1955 congressional hearing. Libby, a chemist who would receive the Nobel Prize in 1960 for the invention of carbon-14 dating of archaeological objects, was a commissioner of the AEC.

Brown's idea quickly captured the imagination of some of his fellows, and late in 1956, Livermore director Herbert York recommended to the AEC that it look into peaceful uses of the atomic bomb. And not just for digging canals. Scientists at Livermore as well as Los Alamos and Sandia, the other two nuclear weapons labs, had come up with a number of ideas for using the bombs that didn't involve killing people. The bombs might, for instance, be exploded deep underground and the resulting heat used to generate electricity. Such underground blasts could also be exploited to make plutonium for use in reactors (or more bombs). Or nuclear explosions might even be used to push rockets out of the pull of the earth's gravity.

The idea caught the fancy of the AEC, particularly Libby and AEC director Strauss, and the venture grew quickly. In 1959 the Commission approved $3 million in funding, then doubled that amount in 1960. In reference to Isaiah 2:4—"They shall beat their swords into plowshares, and their spears into pruninghooks"—the project was named Plowshare.

By early 1960 the AEC had mapped out three ambitious Plowshare schemes. Project Chariot would create a mile-long harbor at Cape Thompson in the northern part of Alaska by exploding five carefully placed bombs simultaneously. Project Gnome would set off a single ten-kiloton bomb in a salt formation 1,200 feet underground to test the feasibility of creating fissionable materials with atomic blasts. And Project Oil Sand would attempt to release petroleum from a formation of oil tar sands in Alberta, Canada. If nuclear explosions could heat the oil so that it flowed freely enough to be pumped out of the ground, the AEC said that Canada would then have more oil than the United States or the Middle East. "It could make the Canadians all sheiks," boasted Gerald Johnson, the Livermore scientist who was directing Project Plowshare at the time.

These plans faded with the realization that there were too many unanswered questions about the effects of atomic bombs exploded underground. How big a bomb was needed to create a hole of a given size? How deep should it be placed? How much radioactivity would be released? What would the force of the explosion do to the surrounding area? Would it, for instance, collapse shafts in nearby mines?

So the AEC decided to put Plowshare on firm footing with a series of tests. The first, in 1962, was code-named Project Sedan. A 100-kiloton bomb—eight times more powerful than the one dropped on Hiroshima—was set off 635 feet below the ground. The blast vaporized the surrounding rocks and created an expanding sphere of hot gases which pushed a mound of earth up from the desert floor. After the bubble had risen about 300 feet, it burst, showering dirt and rocks over a two-and-a-half-mile radius. When the dust settled, a crater 320 feet deep and nearly a quarter mile across had appeared, surrounded by a 100-foot-high lip of earth thrown from the hole.

The test was declared a grand success by the AEC, which distributed dramatic photos of the mound pushed up from the earth and of the crater left by the explosion. By this time the AEC had changed its focus from harbors and oil recovery back to canals, and Edward Teller, the new director of the Livermore laboratory, said Sedan had proved the feasibility of nuclear earth-moving projects. Atom blasting could do the job at 5 to 10 percent of the cost of conventional methods, he said. The press dutifully echoed the AEC, running stories with headlines such as "Digging With H-Bombs" and "Instant Canals."

For a year during 1963 and 1964, newspapers and magazines reported one proposed atom-blasting project after another. State officials in Tennessee, Alabama, Mississippi, and Kentucky wanted help from atom bombs in creating a 253-mile waterway connecting the Tennessee River with the Tombigbee River in Alabama. Nuclear explosives would blast out a thirty-nine-mile segment through a mountainous region and open a direct, if slightly radioactive, route to the Gulf of Mexico from Knoxville and other cities along the Tennessee. In California, state highway department officials worked with engineers of the Santa Fe railroad on a proposal to cut a two-mile channel through the Bristol Mountains. The bomb-made pass would offer a straight, convenient path for both a new interstate highway and a set of railroad tracks.

By far the most excitement was generated by the possibility of a new Panama Canal. The old canal, built in the early 1900s to accommodate ships of that time, was too small for many ocean-going ships in the 1960s. (The situation is worse now. Aircraft carriers, oil tankers, and many large cargo ships must go around South America to pass between the Atlantic and Pacific.) Furthermore, experts predicted that increasing traffic from those ships that could fit through the canal would overwhelm its capacity by 1980. And tensions between the United States and Panama exacerbated the situation. The United States, which had built the canal, owned and operated it, but Panama was pushing for more rights in the U.S.-run Canal Zone and for more financial assistance. As Harold Brown had decided several years earlier with the Suez Canal, the best solution might be to build a new canal, either through Panama or an adjoining country. The idea proved irresistible to atom-blasting aficionados.

"Our studies," said John S. Kelly, director of the Plowshare project in 1963, "show that a sea-level canal [through Panama] could be dug with nuclear explosives for less than one-third the cost and in about half the time conventional methods would require. And a canal excavated by nuclear blasts would be wider and deeper than a conventionally dug canal." A sea-level canal would avoid the problem of locks, whose size is the limiting factor on which ships can get through a canal. Instead, a clear path of water would connect the Atlantic and Pacific Ocean.

A U.S. Army study considered five possible routes for such a canal, two in Panama and one each in Mexico, Nicaragua, and Colombia. For each, a string of several hundred bombs would be set off at once, close enough to one another that their craters would overlap and create a continuous channel. In the excitement over such a tremendous engineering achievement, the dangers of radioactivity from the blasts were downplayed and the possibility that people might not want hundreds of atomic bombs exploded in a line across their country was ignored altogether. When *Time* discussed the atomic canal, for instance, it had only this to say about radioactivity: "Permanent population may have to keep away from the neighborhood of the new-dug canal for at least six months, but men under medical supervision may start working there in two weeks."

Despite the AEC's assurances that radioactivity was not a problem, the candidate countries for a nuclear canal all declined. The domestic blasting projects fared no better. Project Plowshare moved on.

Four years after the Canal commotion, atom blasting was being touted as the answer to recovering hard-to-get oil, gas, and minerals. In December 1967, the AEC, along with the Bureau of Mines and the El Paso Natural Gas Company, set off a twenty-kiloton hydrogen bomb 4,000 feet below the surface in a sandstone formation. The object was to release the large amount of natural gas trapped in the sandstone's pores, only 10 percent of which was ever likely to be extracted by normal means. The blast created a "chimney" filled with broken rocks and open spaces into which natural gas could collect as it filtered through fissures in the surrounding rock. By drilling into this chimney and others created by further atomic blasts, it should be possible to extract as much as 70 percent of the gas.

Within a year of this test, *Business Week* reported that several other companies were working with the AEC on their own gas-extraction projects. A consortium of oil companies was negotiating a project to use atom blasting to produce oil from shale rock. A copper company thought it could mine deep copper deposits by first creating a large underground cavity with a nuclear blast. And a natural gas company wanted to create an underground storage facility with a 200-kiloton blast a mile below the surface.

None of the ideas got past the planning stage, and few got past the talking stage. And eventually, Project Plowshare shriveled up and blew away. Its last hurrah was an experiment in 1973 aimed at loosening natural gas from sandstone in a Colorado gas field.

The dream of peaceful atom blasting had been done in by a combination of factors. The Partial Test Ban Treaty of 1963 had banned nuclear explosions that created "radioactive debris" that crossed international boundaries, and the Soviets complained that Plowshare tests were breaking that agreement. There were also some technical problems that were never overcome—engineers could not perfect the mechanics of forming a canal dozens of miles long in one stroke, and the experiments in loosening natural gas from sandstone were never completely successful. And the economics of peaceful atom blasts remained questionable. Uranium may be a million times as powerful as nitroglycerine, but atomic bombs—especially those carefully designed to minimize radioactivity or fit into a 10-inch drill hole—are still very expensive. Most importantly, the public was becoming increasingly worried about radioactivity. In 1974, for instance, Colorado voters banned any further atomic tests in the state. With the nation's new environmental consciousness, the possible benefits from Project Plowshare just didn't seem to be worth the risks.

In hindsight, the atom-blasting saga seems little more than a minor sideshow in the atomic energy circus. The $160 million spent on it over 18 years was hardly noticed by an Atomic Energy Commission whose annual budget exceeded $2 billion by the late 1950s. Although a certain group of scientists—many of them at the Livermore weapons lab—were quite committed to finding peaceful uses for nuclear bombs, the program never gained widespread support. But the Plowshare episode does offer an important lesson about the forces behind the development of nuclear technology.

As the focus of Plowshare moved from canals to oil recovery and harbor digging, back to canals, and finally on to mining for natural gas and minerals, it became obvious that the scientists involved were not really trying to answer such questions as: How can a canal best be built between the Atlantic and Pacific oceans? or How can natural gas best be recovered from sandstone formations? Instead, they had a solution—atomic bombs—and were casting around for problems that fit.

THE PUSH FOR POWER

The Plowshare and nuclear plane programs went on as long as they did because there were essentially no brakes to slow their momentum. The true believers at the AEC and on the Joint Committee had nearly complete discretion in developing atomic energy, and they tended to dismiss negative results as due to lack of effort or insufficient vision. It was not impossible, merely extremely difficult, to build a nuclear-powered aircraft or find a practical, peaceful use for atomic bombs, and so the supporters of these projects could—and did—argue that the proper response to problems was to try harder and spend more money.

In the beginning, the nuclear power program was in a similar position. It was all big plans and potential, with no actual information to provide a reality check. But as time went on, that began to change.

The AEC didn't start its first reactor development program until 1948. Before then, most of its energy had gone into making more and better bombs. Weapons designers were modifying and enhancing the Hiroshima and Nagasaki bombs and were developing the hydrogen bomb. Meanwhile, the uranium-enrichment plants and plutonium-production reactors were working nonstop to make enough fissile material for a nuclear arsenal. The peaceful uses of atomic energy would have to wait.

But after three years the weapons program was going well, and the commission turned to the civilian side. From the start, it took a rational, calculated approach to developing nuclear power, and it began by funding the design and construction of several types of reactors in order to learn the advantages and disadvantages of each. The designs included a light-water reactor that was the prototype for the reactor in the *Nautilus*; a breeder reactor, which would produce more fissionable material than it consumed; a homogeneous reactor, in which the uranium-containing fuel was in liquid form; a sodium-cooled, beryllium-moderated reactor, which was used for a while in the *Seawolf*, the world's second nuclear-powered submarine; and a sodium-cooled, graphite-moderated reactor.

The program quickly demonstrated that atomic energy could be harnessed to produce electrical power. On December 20, 1951, a breeder reactor designed at Argonne National Laboratory and built in Idaho at the AEC's reactor testing site produced the world's first atomic electricity. The amount produced was modest—about 100 kilowatts, enough to run the lights and equipment in the building that housed the reactor—but it was a psychological milestone. Fourteen months later, Oak Ridge's homogeneous reactor became the second to produce electricity, and it fed its 150 kilowatts into the TVA grid. The public had its first taste of atomic-generated electricity.

It was, of course, no different from electricity generated by coal-fired generators or hydroelectric plants. Electricity is electricity. And this fact threatened, sooner or later, to put the brakes on the nuclear power program. If a nuclear power plant could not produce electricity more cheaply than the alternatives, there was little reason to develop one. (In the 1960s, nuclear power would be touted as a nonpolluting alternative to coal, but that was mostly public relations; the bottom line was economic.) Over the next fifteen years, the fate of nuclear power would rest on the balance between enthusiasm about its potential and realism about its cost.

Early on, American business—which knew nothing about nuclear energy except what the AEC told it—reflected the official enthusiasm for the atom. In March 1947, for example, *Business Week* had reported, "Today no prudent businessman, no prudent engineer dares make plans or decisions reaching more than about five years into the future without at least weighing the possibility that the basis of his planning may be upset by the commercialization of discoveries about the atom. . . .Commercial production of electric power from atomic engines is only about five years away. That's the informed assumption today." Three years later, before the first atomic electricity had been generated, the magazine was already looking past it to even more intriguing uses of nuclear energy: "Right now, hardly anyone is interested in huge central-station atomic power plants. . . . America has plenty of fuel, and the prospect of saving a mill [tenth of a cent] or two per kilowatt-hour is not terribly exciting. What does look exciting today are the things atomic power promises to do that no other fuel is able to do at any price—particularly in the field of transportation. . . . In a decade or two, it may come to seem archaic and absurd that any big vehicle should devote carrying capacity to hauling its own fuel around, or should have to squat down every 500 or 1,000 miles to take on more fuel. It may seem natural and expected that a ship, a plane, a locomotive should run for months or years before anyone even has to glance at the fuel gauge."

Nor was it just talk. Businesses acted on these perceptions. In 1952, for instance, Goodyear Tire & Rubber Co. grabbed the contract to run the AEC's new uranium-enrichment plant in Pike County, Ohio, even though it had no previous experience in atomic energy. Its president explained why: "Atomic energy is the most significant development in our lifetime, so any aggressive company ought to jump at the chance to get in on the ground floor." Around that time, four separate industrial study groups released reports that extolled nuclear power's potential.

It was actually the AEC, which was discovering firsthand the obstacles to practical nuclear power, that was the most cautious player. Developing the test reactors had proved to be more complex and more expensive than originally thought, and the commission understood the importance of a slow, cautious approach. The Joint Committee did not.

For several years after the war, the Joint Committee had focused on military issues, but in 1951 it began to monitor the AEC's efforts to promote civilian nuclear power. First came a few staff reports, then in June and July of 1953 the Joint Committee held extensive open hearings on business and atomic energy. After some enthusiastic testimony from businessmen about how fast nuclear power could be developed, the Joint Committee decided that the AEC might not be moving quickly enough. Chairman Sterling Cole then wrote a letter to AEC chairman Lewis Strauss. The AEC's plans for developing nuclear power seemed rather vague, Cole wrote. Why didn't the commission offer a "3-to-5-year program of specific research and development projects" so that industry could participate more effectively?

The AEC responded in early 1954 by announcing an ambitious reactor program aimed at power production. The Navy-built reactor at Shippingport became part of this program, although the decision to build it had actually been made the previous year, and four other reactors were eventually included: a breeder reactor, a sodium-cooled reactor, a homogeneous reactor, and a boiling-water reactor at Argonne National Laboratory—a light-water design that was simpler than the pressurized-water reactor used at Shippingport. As in the earlier program, the commission was consciously trying out different reactor designs in order to determine which would be most suitable for civilian power plants.

In the meantime Dwight Eisenhower had been elected president, and he brought his own perspective to the atomic energy discussions. He worried that nuclear weapons had made atomic energy frightening to many people. To counter that fear, he wished to offer some redeeming use for the atom. Eisenhower laid out his vision in a December 1953 speech before the United Nations, in which he urged the countries of the world to turn away from nuclear weapons and to concentrate on "atoms for peace."

> It is not enough to take this weapon out of the hands of the soldiers. It must be put into the hands of those who will know how to strip its military casing and adapt it to the arts of peace.
>
> The United States knows that if the fearful trend of atomic military build-up can be reversed, this greatest of destructive forces can be developed into a great boon for the benefit of all mankind.
>
> The United States knows that peaceful power from atomic energy is no dream of the future. That capability, already proved, is here now—today. Who can doubt, if the entire body of the world's scientists and engineers had adequate amounts of fissionable material with which to test and develop their ideas, that this capability would rapidly be transformed into universal, efficient and economic usage?

In some ways, the speech was disingenuous. Eisenhower had no intention of stripping the United States of its nuclear arsenal. But his desire to counterbalance the destructiveness of nuclear weapons with peaceful uses of the atom was real enough, and it shaped his administration's atomic policies.

Throughout the 1950s, the AEC, the Joint Committee, and the president all pushed for nuclear power. If it had remained a purely government-run program, like the nuclear plane, it seems likely that this momentum would have quickly led to nuclear electric-generating plants. Indeed, some Democratic members of the Joint Committee had visions of a nuclear version of the Tennessee Valley Authority, the government-created corporation that generates and sells electricity from a number of hydroelectric plants. But Eisenhower and the Republicans on the Joint Committee were determined to keep government involvement to a minimum. The AEC could fund research and even provide subsidies to companies to build and operate nuclear plants, but the nuclear industry should be in private hands.

Eisenhower had come into office with small Republican majorities in both the House and Senate, and this allowed him to shift the public-versus-private debate decisively toward the private. In June 1953 he announced he would replace the retiring chairman of the AEC with Lewis Strauss, a businessman and conservative Republican. And in 1954 Congress passed a revised Atomic Energy Act that changed the ground rules for developing nuclear power. Where the Atomic Energy Act of 1946 had made atomic energy a government monopoly, the new act was designed to encourage businesses to get involved. It loosened up the flow of information about nuclear technology. It allowed the patenting of inventions relating to nuclear energy. And it allowed companies to own nuclear reactors—with the proper licenses from the AEC. The AEC would, however, still own the fuel and simply lease it out to operators of reactors. This gave the commission great influence over the economics of nuclear power, and for many years the AEC would subsidize the nuclear industry with fuel charges that represented only a fraction of the real cost of the fuel.

In the late 1950s and early 1960s, the private sector took on an increasingly large role in determining the path of nuclear power. Although the government would continue to push its development through various programs and subsidies, the most important decisions would be made in boardrooms, not government offices. The leaders of this emerging nuclear industry were generally just as enthusiastic about the future of the atom as their government counterparts, but their enthusiasm would be reined in by such practical matters as economic forecasts and cost-benefit analyses. And so an era came to an end. In the next phase of nuclear development, the vision that began with Soddy and Wells would continue to be important, but it would be modified by factors unique to the business world.

three

BUSINESS

In January 1975, the magazine *Popular Electronics* trumpeted the beginnings of a revolution. "Project Breakthrough," the cover said: "World's First Minicomputer Kit to Rival Commercial Models." Inside, a six-page article described the Altair, an unassembled computer that could be ordered from MITS, a company in Albuquerque originally founded to sell radio transmitters for controlling model airplanes.

To the uninitiated, it didn't look like much of a revolution. For $397 plus shipping, a hobbyist or computer buff could get a power supply, a metal case with lights and switches on the front panel, and a set of integrated circuit chips and other components that had to be soldered into place. When everything was assembled, a user gave the computer instructions by flipping the panel's seventeen switches one at a time in a carefully calculated order; loading a relatively simple program might involve thousands of flips. MITS had promised that the Altair could be hooked up to a Teletype machine for its input, but the circuit boards needed for the hookup wouldn't be available for a number of months. To read the computer's output, a user had to interpret the on/off pattern of flashing lights; it would be more than a year before MITS would offer an interface board to transform the output into text or figures on a television screen. And the computer had no software. A user had to write the programs himself in arcane computer code or else borrow the efforts of other enthusiasts. One observer of the early computer industry summed up the experience like this: "You buy the Altair, you have to build it, then you have to build other things to plug into it to make it work. You are a weird-type person. Because only weird-type people sit in kitchens and basements and places all hours of the night, soldering things to boards to make machines go flickety-flock."

But despite its shortcomings, several thousand weird-type people bought the Altair within a few months of its appearance. What inspired and intrigued them was the semiconductor chip at the heart of the computer. It was the Intel 8080, a sliver of silicon that contained five thousand transistors. In essence, the Intel 8080 was an entire computer etched onto a single chip—a microprocessor, in the industry jargon. When teamed with memory chips that stored operating instructions and kept track of calculations, the microprocessor could be programmed to perform an almost infinite number of tasks. Granted, it wasn't easy, and few people outside the community of computer enthusiasts wanted to make the effort. But the potential was there. The Altair could do on a small scale exactly the sorts of things that larger and much more expensive computers did.

So began the personal computer revolution. In less than a decade the hobbyist's toy would become a practical, even essential tool, one that would completely change the way people did their jobs and that would spawn entire new industries, from desktop publishing to electronic information services. And every personal computer today can trace its ancestry to that $397 computer kit sold by mail.

But despite its standing as the Adam of the personal computer tribe, the Altair was not the only personal computer in existence in January 1975. It wasn't even the best, not by a long shot. Although few knew about it at the time, the employees of one large corporation were already working with a personal computer that had many of the features of today's most modern machines. These workers entered data by typing on a keyboard, not by flipping switches. They saw the computer's output as letters and pictures on a screen, not as flashing lights, and that screen was rather sophisticated. Like the monitors of today's computers, it was "bit mapped"—every dot on the screen could be controlled independently, making it possible to create intricate and detailed images. The computer also included a "mouse" that could be used to point to something on the screen and manipulate it, and a user could open up "windows" on the computer screen to compare items from different parts of a document. Several dozen of the computers were connected on a network that allowed their users to pass information back and forth. The software for the computer included a powerful, easy-to-use word processor that functioned according to the principle of "what you see is what you get," or "wysiwyg" (pronounced "whiz-ee-wig")—the display on the screen showed exactly how the printed document would appear. And, when it was time to print, the personal computers sent their files to a laser printer that produced clean documents that looked as if they could have come from a typesetter.

The people using the ahead-of-its-time computer were workers at Xerox Corporation's Palo Alto Research Center (PARC), where the machine and its accessories had been under development since 1972. By the beginning of 1975, not only were many of PARC's scientists, administrators, and secretaries using the personal computers, but a similar system had been installed at Ginn & Co., a Xerox-owned textbook publisher. Employees there used the computers for editing books. The name of the Xerox computer was, by coincidence, very similar to that of the do-it-yourself computer hailed in the pages of *Popular Electronics.* It was the Alto.

With the Alto, Xerox engineers had created a personal computing technology that was years ahead of anything else available at the time. But this technology, the result of tens of millions of dollars spent on research by Xerox, would play only a peripheral role in the personal computer revolution unleashed by the Altair, and Xerox, despite having a sizable head start on the competition, would end up on the sidelines watching other companies split up a multi-billion-dollar market. What happened? As the creators of the Alto would discover, there is more to innovation than what goes on in the laboratory. The business and organizational environment in which a technology is created and nurtured plays a vital—and sometimes dominant—role in how that technology ends up.

PARC was a magical place in the mid-1970s. Xerox CEO Peter McColough had founded it in 1969 to provide Xerox with innovative products for dealing with information and had given it almost free rein. At the time, Xerox was essentially a one-product company—copiers made up the bulk of its business—and McColough wanted to diversify. Since he saw copiers as a way of handling information, McColough thought the natural direction for Xerox to head was to pioneer new ways of dealing with information. That meant computers, and McColough intended Xerox to become a major player. By offering its recruits an almost unlimited budget and the same sort of academic freedom they would normally find only in universities, PARC assembled the most impressive group of computer scientists anywhere in the world. A 1983 article in *Fortune* described the PARC computer crew this way: "Members were notorious for long hair and beards, and for working at all hours—sometimes shoeless and shirtless. They held raucous weekly meetings in the 'bean-bag room,' where people tossed around blue-sky concepts while reclining on huge pellet-filled hassocks. PARC's hotshots were not just playing at being geniuses. Before long, computer scientists recognized PARC as the leading source of research on how people interact with computers."

It was this atmosphere that produced the Alto, the laser printer, the Ethernet—a local-area network for personal computers—and the first word-processing program for a personal computer, as well as a set of tools to make interaction with a computer as easy as possible: the mouse, windows, icons, and more. Yet it was not enough for PARC to produce these wonders; Xerox had to recognize their value and act quickly. It would not.

Xerox was ill prepared to commercialize the products coming out of

PARC for a variety of reasons. One stumbling block was cultural: most of Xerox was extremely conventional and conservative, a blue suit/white shirt company that, like IBM, depended upon direct sales to businesses, while PARC was an outpost of casual dress and disdain for authority. The two groups did not communicate easily. Another obstacle was organizational. Xerox had no group designated to turn the research at PARC into salable products. Instead, the same people who made decisions about which types of copiers to develop were put in charge of finding uses for the ideas flowing from PARC, something they were poorly equipped to do. Despite the recognition by some of its top managers, including McColough, that the company needed to do something besides make copiers, when the time came to actually push into uncharted waters, the tendency was to hold back. Although managers at PARC, beginning as early as 1973, constantly urged the corporation to develop the Alto as a commercial product, Xerox decided to concentrate on less ambitious word-processing typewriters, which a number of companies were developing in the mid-1970s.

Eventually, after sustained prodding from Don Massaro, an outside entrepreneur appointed president of Xerox's Office Products Division, Xerox did bring out a product based on the Alto. Introduced in April 1981, the Star was an integrated office system aimed at business executives and managers. It consisted of a number of interlinked workstations connected with electronic filing cabinets and at least one laser printer. It was expensive—each workstation cost $16,595, and the laser printer a good deal more—but it had capabilities unavailable anywhere else. High-quality laser printing was one, ease of use another. And because everything was interconnected, the Star gave users an ability to communicate and transfer data back and forth that should have been extremely valuable for work in groups.

But the Star never caught on. Despite its virtuosity in many areas, it had several technical shortcomings. One was the lack of spreadsheet software, something that was fast becoming an important tool for business. And because the Star was a closed system—it would run only programs written by Xerox—customers could not use it to run any of the existing spreadsheet programs. Furthermore, since the designers had made Star do so many things, it was relatively slow compared to the less ambitious personal computers already on the market.

Still, it's hard to blame the engineers for the fall of the Star. If the PARC visionaries had had their way and made a Star-like system available in 1977 or 1978, things might have been quite different. Conceivably, Xerox could have become a major player in the emerging personal computing market and shaped people's ideas about what a personal computer should be. As it was, by the time Star came out, the influential Apple II had been around for more than three years, and the IBM PC—destined to become the industry's major standard—was about to be released. This world of Apples and

IBM PCs had little room for Xerox's vision of "personal distributed computing," in which personal computers were part of a larger system. Instead, the idea of a personal computer as an isolated, individual tool had become dominant.

But even as late as 1981 the Star might still have been able to carve out a place for itself if Xerox and PARC had been working together. Star was developed by the scientists and engineers at PARC with essentially no interest from the rest of the corporation, and while the PARC designers were tops at technical details, they could only guess at which features were most likely to make Star a commercial success. With no help or advice from the company's marketing experts, the Star's developers made what in hindsight looked to be obvious blunders. The lack of a spreadsheet program for the Star was one example. Even worse was the decision to make the Star a closed system. IBM had done this for years with its mainframes, selling a family of computers that ran software available only through IBM. The PARC strategy was similar: provide an entire suite of office products, from computers and software to printers and facsimile machines, all linked over the Ethernet. Since the technology was proprietary, any customer who wanted to expand or upgrade a system would have to go to Xerox. But the strategy wasn't suited for personal computing. Already hundreds of independent companies were writing software and designing hardware accessories for the Apple and other personal computers. No single firm—not Xerox, not even IBM—could produce enough applications to compete with this flood.

To make matters worse, when Xerox tried to market the Star system, the sales force had no idea of how to approach it. The company's sales staff consisted mainly of copier salespeople, who were unfamiliar with computers. To convince business executives to buy an expensive Star system, a salesman had to explain just how the Star would improve the performance of an office—and particularly why a network of interlinked computers was more valuable than a collection of less expensive, unlinked personal computers. To do this well, salespeople had to develop new skills and new approaches to their customers. This might have been possible if the Star sales force had remained a separate group, but in 1982, a year after the Star was introduced, the Star sales staff was merged with Xerox's much larger copier sales group, and any hopes of a truly effective selling effort vanished.

Ultimately, the Star's failure can be attributed to two very different miscalculations. The Xerox management did not see the potential of the Star until it was almost too late, and perhaps not even then. And the crew at PARC, who knew the value of what they had, did not understand that innovation is more than engineers chasing a dream. Modern technology is created not so much by individuals as by organizations, and the characteristics of those organizations—their histories, their leaders, their structures, their cultures, their financial well-being, their relationships with other organizations—will shape the sorts of technology that their engineers produce. It is well known, for instance, that large corporations have a difficult time developing truly innovative products. The case of the Alto is just one example among many. But the influence of business factors upon technology goes far beyond this sort of organizational inertia. To see just how far, let us return to the development of the personal computer and watch as two different companies set two very different standards for the personal computer industry.

THE APPLE AND THE GORILLA

The Apple I looked even less like a computer than did the first Altair. Designed by Steven Wozniak about a year after the Altair appeared, it was simply a circuit board with a microprocessor, memory chips, and interfaces that could be connected to a keyboard and television monitor. There was no case, no power supply. It did, however, have a programming language, BASIC, which allowed the user to instruct the computer much more easily than was possible with the first Altairs.

Wozniak, a young computer whiz working for Hewlett-Packard, hadn't thought of his machine as a commercial product. It was simply an intellectual challenge—a game, really—to test his programming and design skills. He had chosen the MOS Technology 6502 chip, a knockoff of Motorola's 6800 chip, as the microprocessor for his computer not because it was better than other available chips, but because it was much cheaper. The better-known 6800 cost $175, but Wozniak could get the 6502 for $25. In April 1976 he brought his first computer to a meeting of the Homebrew Computer Club, a band of San Francisco–area computer enthusiasts, and provided copies of the design to anyone who was interested.

But where Wozniak saw an intellectual challenge, his friend and fellow computer enthusiast Steven Jobs saw a chance to make some money. In the year since the introduction of the Altair, personal computers had become increasingly popular, although still just among a small group of hobbyists and computer buffs. Several companies had begun to offer personal computers for sale through the mail, clubs like Homebrew were springing up, and even a few retail stores were starting to sell personal computers. In January 1976, Jobs had suggested to Wozniak that they make and sell printed circuit boards based on Wozniak's design, and, after some hesitation, Wozniak agreed. Their initial idea was to sell only the circuit boards and let the customer get the chips and electronic components and install them on the boards in order to create a working computer. Jobs thought they might be able to sell about a hundred of the circuit boards for $50 apiece.

But plans changed quickly in July 1976. Paul Terrell, owner of one of the first computer stores and a frequenter of the Homebrew Club, liked Wozniak's machine—which Wozniak and Jobs had decided to call the Apple—and offered to buy fifty of them to sell in his shop. But, Terrell said, he wasn't interested in just the circuit boards. He wanted fully assembled computers. Suddenly the two Apple entrepreneurs had to shift gears. They had raised $1,300 to pay for developing the circuit boards, Wozniak by selling an HP-65 calculator and Jobs by selling his Volkswagen bus, but now they needed about $25,000. With Terrell's order in hand, they were able to get credit for the parts, and they set to work assembling the computers in Jobs's parents' garage. They hired Jobs's sister, Patty, and an old friend, Dan Kottke, to plug the components into the circuit boards. The embryonic company supplied Terrell the fifty computers he'd ordered and eventually was able to sell about 150 more, both by mail and through various computer stores in the San Francisco Bay area. The Apple I was priced at $666.66, a number Jobs chose because he liked the looks of it.

With the Apple I under his belt, Wozniak began work on what would become the Apple II. He had several improvements in mind. The new machine would include a keyboard, power supply, and the BASIC programming language, all things that had to be tacked onto the Apple I by the user. Most strikingly, the Apple II would display color, something that no other personal computer could do. At the time, most computer designers believed it would take dozens of chips to create a color circuit, but Wozniak came up with a clever approach that cut the number of extra chips down to a handful. When hooked up to a color television, the Apple II provided an unmatched show.

Although Wozniak thought that he and Jobs might sell a few hundred of the computers, Jobs saw the Apple II as a chance to create a truly successful business, and he sought advice from people in the computer industry about how to do that. Jobs soon hooked up with Mike Markkula, a thirty-three-year-old veteran of the semiconductor business who had worked for both Fairchild and Intel. Markkula had retired in 1975, after becoming a vice president at Intel. A millionaire from his Intel stock options, he had been relaxing at home, but by October 1976, when he first met with Jobs and Wozniak, he was ready to ease back into the industry. At first he agreed only to give the two advice on how to organize Apple, but after a few months he offered to join them. For a $91,000 investment, he took one-third interest in the company, with Jobs and Wozniak splitting the remaining two-thirds.

Now everything was in place. Apple had a brilliant designer in Steve Wozniak; a person experienced in computer marketing, distribution, and finance in Mike Markkula; and an indefatigable driving force in Steve Jobs. The combination gave the company a leg up on the competition at a critical point in the development of the personal computer. *Fire in the Valley*, a history of the personal computer, describes the situation like this:

> Dozens of companies had come and gone [in the two years since the introduction of the Altair]. Most notably, MITS, the industry pioneer, was thrashing about. IMSAI, Processor Technology, and a few other companies were jockeying for control of the market even as they wobbled. All of these companies failed. In some cases, their failure stemmed from technical problems with the computers. But more serious was the lack of expertise in marketing, distributing, and selling the products. The corporate leaders were primarily engineers, not managers. They alienated their customers and dealers. . . .
>
> At the same time, the market was changing. Hobbyists had organized into clubs and users' groups that met regularly in garages, basements, and school auditoriums around the country. The number of people who wanted to own computers was growing. And the ranks of knowledgeable hobbyists who wanted a better computer were also growing. The manufacturers wanted that "better computer" too. But they all faced one seemingly insurmountable problem. They didn't have the money. The manufacturers were garage enterprises, growing, like MITS had since January of 1975, on prepaid mail orders. They needed investment capital, and there were strong arguments against giving it to them: the high failure rate among microcomputer companies, the lack of managerial experience among their leaders, and—the ultimate puzzle—the absence of IBM from the field. Investors had to ask: If this area has any promise, why hasn't IBM preempted it?

Because of Jobs's entrepreneurial instincts, Apple now had what its competitors lacked. Not only did it have a solid product, thanks to Wozniak's engineering skills, but it also had experienced management in Markkula and Mike Scott, a Markkula protege from Fairchild who was hired as Apple president. Just as important, it had money—Markkula, as part of joining the company, had agreed to underwrite a $250,000 loan from a bank. And soon it would have one of the best advertising and public relations minds in the industry. Jobs had courted Regis McKenna, already well known for shaping Intel's image, and although McKenna had initially turned Jobs down, he eventually changed his mind. It was McKenna's agency that came up with the playful Apple logo, an apple with broad horizontal stripes in various colors and a bite taken out of it. The logo was designed to appeal to a broader public than hobbyists and computer enthusiasts. And it was McKenna who took out a color ad in *Playboy*, a move that brought national attention not just to Apple but to the emerging personal computer industry as well.

In 1977 Apple began to grow quickly. By the end of the year it was profitable and was doubling its production of Apple IIs every three or four months. To finance the rapid growth, Markkula had convinced a venture capital firm to put up money in return for a share of the company. And

Wozniak was creating one accessory after another: a printer card to hook up the Apple II to a printer, a communications card, and, most important, a disk drive. At the time, the only way to store relatively large amounts of data for a personal computer was on audio cassette tapes, which were slow and unreliable. Wozniak was the first to design a floppy disk drive for a personal computer. When it became available in June 1978, personal computer users for the first time had the option not only of storing large amounts of data but also of loading sophisticated software into their machines.

With that option, computer programmers began to write more useful software. The most important program for the Apple's success was VisiCalc, the first spreadsheet program. VisiCalc allowed a user to do financial calculations in a way never before possible, changing one or a few numbers and seeing immediately how all the other figures in a financial spreadsheet were affected. It appeared in October 1979 and was available for use only on an Apple for its first year, and its success drove much of the early popularity of the Apple II.

The defining moment for Apple came when Jobs visited Xerox PARC in the spring of 1979. Xerox's engineers showed Jobs the Alto with its brilliant graphics, mouse, icons, and windows, and Jobs was hooked. He quickly grasped the importance of the Alto's ease of use and how valuable that would be in a personal computer—the same lesson that the PARC engineers had failed to get across to the Xerox management for several years. Jobs decided that future Apple computers should have similar capabilities, and he hired away Larry Tesler, one of the scientists at Xerox who had worked on the Alto. It took several years to recreate Apple computers in the Alto's image, but by the early 1980s the company had done it. The Lisa, a relatively expensive computer aimed at business executives, was the first to offer the mouse-and-windows system that would become Apple's hallmark. It was released in 1983. Early the next year came the Macintosh, an inexpensive machine designed to attract a new group of users to personal computers. Although the effort to keep the price down had saddled the Mac with a small memory, and little software was available at first, the first inexpensive, user-friendly personal computer was an immediate hit. Later, Apple provided similar capabilities for the Apple II.

As Apple was developing its Lisa and Macintosh computers, it was awaiting the coming of the computer industry's 800-pound gorilla: IBM. For more than two decades IBM had been the dominant figure in the computer business, and no one doubted that, given the growing size of the personal computer market, IBM would soon weigh in.

Actually, the company had already weighed in, but in such an inept way that almost no one had noticed. In 1975, conscious of the hoopla surrounding the personal computer, IBM announced the 5100, its first personal computer. Unlike the Altair, however, the 5100 was much more "computer" than "personal." It was a full-featured machine with plenty of memory plus tape cartridges for data storage. It had a sophisticated operating system and could run programs written in either BASIC or APL, a programming language used primarily by scientists. And it weighed seventy pounds and cost many thousands of dollars. IBM had aimed the 5100 at scientists, a group the company had no experience in selling to and a group that studiously ignored IBM's offering. The 5100 was a major flop.

Undaunted, the same IBM development team that had produced the 5100 revamped it to appeal to the business market. But the new version, the 5110, was no more personal than the 5100. It sold for $8,000, far too much to be of interest to those who were intrigued by the Altair and the Apple, yet not enough to trigger the enthusiasm of IBM computer salesmen. Why try hard to sell a 5110, when traditional computers costing ten times as much could be sold with little more effort? The 5110 flopped too.

Despite the failures, IBM—and particularly its chairman, John Opel—remained interested in personal computers, and that interest grew with the steady expansion of the personal computer industry. So in 1980, when William Lowe, director of an IBM laboratory in Boca Raton, Florida, approached the company's senior management with a detailed proposal to develop an IBM personal computer, he quickly got a go-ahead. IBM would set up an independent business unit in Boca Raton, free of pressures from the rest of the company, to put the personal computer together—quickly. Lowe was given a year to bring it to market.

Because Apple and other firms had already defined what a personal computer should look like, and because IBM wanted to develop its own product fast, the IBM PC would be a computer unlike any the company had ever offered. Most of its components would be built not by IBM itself but by outside contractors. Its operating system—the language that gives instructions to the computer's processor—would be provided by someone outside the company, as would the software for the PC. More important, the computer would have an "open architecture" so that other companies could design and make components that would operate in it. And IBM would sell the personal computers through retail outlets instead of through its large in-house sales staff. All of these were major deviations from the normal IBM way of doing business.

The IBM Personal Computer, announced on August 12, 1981, made no attempt to be on the cutting edge. Its microprocessor, the Intel 8088 chip, was a fine choice for a personal computer and it made the IBM PC as good or better than any of the competition, but there were faster, more powerful chips available that IBM could have chosen. The computer used a cassette tape recorder for data storage (although a floppy disk drive was available as an option). The printer was a standard Epson model with an IBM label slapped on. Available software included VisiCalc and a word processor called EasyWriter, both originally developed for the Apple. The most surprising thing about the IBM computer was that there were no surprises—instead of special components, the machine was constructed from standard parts, and it ran standard software.

That was enough. IBM's reputation as the undisputed leader of the computer industry gave its personal computer instant credibility. Anyone who had been contemplating buying a personal computer was bound to at least consider the IBM PC. And for those who had questioned the value of personal computers, wondering if they were simply faddish toys, IBM's PC seemed proof that they were important and useful. IBM wouldn't sell something frivolous.

Almost overnight, the IBM PC became a standard for the personal computer industry. Companies rushed to develop components and peripheral devices for the PC. Software developers wrote programs to run on it. Some firms began to manufacture computers—IBM clones, they were called—that would perform similarly to the PC and be compatible with all its hardware and software. Meanwhile, many of the pioneering companies that had helped develop the personal computer industry found their sales drying up. All but a few went bankrupt. When the dust settled, only two companies remained with any real influence over the direction that the personal computer would take: Apple and IBM.

Today, those two standards define the personal computer. They have converged somewhat with time and may merge almost completely sometime in the future, but they still represent two very different visions of what a personal computer should be. The Apple is a people's computer, fashioned to be inviting and accessible even to those who have never used a computer before. Yet the technology underlying that friendly facade is aggressively state-of-the-art, with designs that win accolades and awards from the industry. The IBM is more businesslike, an accountant's machine instead of an artist's, and its structure and components are reliable but seldom inspiring.

The differences between the two computers reflect, more than anything else, the different business environments in which the two were developed. Apple began as a small start-up company with computers built by and for computer enthusiasts. "We didn't do three years of research and come up with this concept," Steve Jobs has said. "What we did was follow our own instincts and construct a computer that was what we wanted." That was possible only because Apple, as an embryonic, one-product firm, retained the flexibility and the recklessness of youth. Had Wozniak and Jobs been working for IBM, they would certainly have never produced anything like an Apple computer.

The same sort of difference between a small, new firm and a large, established one can be seen in how Apple and Xerox responded to the ideas coming out of PARC. Xerox, which had paid for the PARC research, could not appreciate its value. Despite the company's talk about developing an office of the future, Xerox found it almost impossible to step away from its established ways of seeing things and to envision radically new products. Even the PARC scientists were caught in a rut. Accustomed to thinking in terms of large, expensive office equipment and systems, they continued to focus on networks of linked computers even when the personal computer revolution was in full swing. At Apple, some resisted when Jobs came back from Xerox and pushed to put the Alto's features in the Lisa, which was already under development. But because Apple was still small and flexible enough to change direction, and because the aggressive Jobs wielded great influence, the Lisa—and later the Macintosh and Apple II—took on the trappings of the Alto. Years afterward, because of the popularity of the Apple operating system, IBM-type computers were provided with a similar operating system called Windows. Today, Windows is the dominant operating system for personal computers, and it can trace its ancestry back to those PARC scientists, whose company couldn't appreciate what they had done, and to Steve Jobs, who did.

Unlike Apple, IBM was already a huge company when it decided to become involved in personal computers, and the preexisting culture and concerns of the firm pushed its PC in a completely different direction from the Apple. IBM's customers were business people—always had been, and, as far as many in the company were concerned, always would be. So the IBM PC was intended to be a serious business machine. Let Apple and others speak of bringing computer power to the people and depict their machines as tools of popular insurgency; IBM was part of the establishment, and its computers would reflect that. Indeed, eyebrows shot up inside the company and out when, shortly before the PC's release, IBM decided to include a computer game as part of its optional software.

But of greater influence than IBM's corporate culture was the company's decision to develop its PC in an independent business unit (IBU) that was well isolated from the rest of the company. Over the years IBM has established a number of these IBUs to develop and market products new to the company, on the theory that real innovation demands freedom and flexibility. Although they're funded by IBM and staffed by IBM employees, the IBUs have little else to do with the corporation. They set their direction internally with only general oversight from top IBM management, and they're free of many of the rules and policies governing the rest of the company.

It was this freedom that allowed the IBM PC to break so many of the usual rules for IBM products. Only an independent business unit could look outside IBM for most of the computer's hardware and software or decide to sell the computer through retail outlets. And the choice of an open architecture for the PC went against every instinct of the corporate

IBM. For decades the company had protected its designs with patents and secrecy. Now it was, in essence, inviting other businesses to make components that could be added onto the IBM PC and to write software that would run on it—all without a license or permission.

Still, the independent business unit that developed the PC was a part of IBM, and it was constrained by the company's needs and goals if not by all its normal rules and policies. As industry observers James Chpolsky and Ted Leonsis have noted, IBM's standard strategy for emerging technologies was "noninvolvement until a market was established by the smaller and generally entrepreneurially driven companies in the industry—at which time IBM bullied its way in and, on the strength of its reputation for excellence and its superior manpower and resources, proceeded to dominate the market at the expense of the bit players who were often there at the beginning." This would be IBM's strategy for the personal computer: the independent business unit was intended to develop a product that would allow IBM to dominate the personal computing market.

Given this goal and the complexion of the existing market, the engineers and other specialists in the independent business unit found that many of their decisions were already made for them. IBM would not, for instance, be attempting to create a computer radically different from those already in the market. Instead the machine would be firmly in the mainstream of personal computer development. It would have a monitor, keyboard, and a box containing the processing unit. For output it could be hooked up to a separate printer, and for data storage and retrieval it would use a cassette recorder/player, with the option for a floppy disk drive. The IBM personal computer would adopt the best of what had already been developed, but it would not blaze new trails.

Because the personal computer business was growing so quickly and because other large companies were looking to jump in, IBM gave its independent business unit a year to get its product to market, and this, more than anything else, set the constraints that would shape the IBM PC. To get everything done in a year, the PC team had to subcontract much of the development work to companies outside IBM and to buy many of the computer's components off the shelf. Thus the main processor chip was the Intel 8088, the power supply was made by Zenith, the disk drives by the Tandon Corporation, and the printer came from Epson. The software, too, was farmed out, with Microsoft supplying both the basic operating system, PC-DOS, and a BASIC programming language, and other companies providing business software and the spreadsheet and word processing programs.

The most important consequence of the short development schedule was the decision to give the PC an open architecture. By doing this, the design team could lay out the basics of the computer and then have software development take place at the same time as the remaining hardware development. This saved time and allowed the PC to be created within a year, but it would also change the shape of the entire personal computing industry. Until that point, the Apple had been the closest thing to a standard that the industry had, but it had a closed architecture. No one but Apple and those companies licensed by Apple could make components or accessories for an Apple computer or write software for it, which gave Apple tight control over the direction its product would take. Other companies making personal computers did have open architectures, but they were generally incompatible with one another and none was accepted as a standard. The introduction of the IBM PC offered a standard that was open to everyone, and the industry blossomed. Because anyone with an idea and a little capital could create hardware or software that would work on an IBM—and also on the IBM clones that soon appeared—the personal computer began to evolve and improve far faster than would otherwise have been possible. Granted, that growth was often haphazard, but the rollicking, free-for-all market that IBM unleashed with its PC revolutionized the world. Since then, computing power and capabilities have risen dramatically sharply while costs have dropped even more significantly.

For its part, IBM seems to have had second thoughts about the wisdom of the open architecture. Yes, IBM-type computers now own more than 90 percent of the personal computer market, but IBM itself has only a small share of that. Indeed, IBM is no longer even the leading seller of IBM-type computers. In 1994, Compaq Computer Corporation claimed that title with 10.1 percent of all personal computer sales around the world, compared with only 8.7 percent for IBM. Since its introduction of the PC, IBM has offered other personal computer lines that aren't nearly so open in an attempt to reassert some control over its own products, but the industry is now far too large and diffuse for any one company, even IBM, to have much influence on it.

Ironically, following the success of the PC, IBM eventually decided against using independent business units to develop products. If IBM had made that decision before 1980, it might never have developed a successful personal computer, and it almost certainly would not have created a personal computer with an open architecture. In that case, the personal computer revolution would have proceeded much differently—and surely much more slowly.

The history of the personal computer reveals a different sort of momentum than we've seen in the past couple of chapters—a momentum with its roots in the business world. Technologies are pushed this way and that by the companies that develop them, market them, and buy and operate them, and the larger the company, the more influence business and organizational factors will exert over the development of a technology.

Nuclear power, as a complex, expensive technology that demands large organizations for its development and operation, was more prone to this sort of business momentum than most. And nowhere was it more obvious than during what came to be known as the Great Bandwagon Market.

In the early part of the 1960s, the nuclear industry found itself at a critical point. Nuclear power clearly worked—nuclear plants were already providing power for several electrical utilities—but the technology had not proven itself as a commercial product. Existing nuclear plants were not economically competitive with coal-fired plants, and although nuclear power costs were dropping, no one knew when, or even if, nuclear costs would catch up with coal. Nuclear power, which had seemed so promising in the 1950s, now looked as if it might turn out to be nothing special. But suddenly, within just a few years, much of the electric utility industry climbed onto a nuclear bandwagon. In 1966 and 1967, almost half of all new plants ordered by electrical utilities were nuclear. The reasons given at the time were purely economic: nuclear power, the utilities said, was now their cheapest option for generating electricity.

It wasn't that simple, however. Years later, when it became obvious that the nuclear power cost estimates were hopelessly optimistic, people took a second look at the industry decision to go nuclear. What they discovered was that the biggest turning point in the history of nuclear power was born not of some technical breakthrough but instead of a number of business factors, the most important of which was cutthroat competition between the two main U.S. manufacturers of nuclear reactors.

THE BANDWAGON

Build and sell. Sell and build. Build, build, build. Sell, sell, sell. That's the way it went throughout the 1950s and into the 1960s for VEPCO, the Virginia Electric and Power Company.

VEPCO, it seemed, could do no wrong. Covering most of Virginia, from the Washington, DC, suburbs down through Richmond, the state capital, and on into North Carolina, the company was constantly building new generating plants, but its supply of electricity just barely kept ahead of demand. VEPCO served an area growing faster than almost any in the country, and sales showed it. Kilowatt-hours of electricity were up 12.9 percent in 1959, 7.2 percent in 1960, 9.2 percent in 1961, and so on, year after year. And, as company executives were quick to point out, those ever-increasing numbers were more than just a byproduct of a booming economy. VEPCO was pushing them skyward with a series of aggressive campaigns to get its customers to use more electricity.

VEPCO consultants met constantly with architects, engineers, and building owners to encourage them to use plenty of electricity in offices and other commercial buildings: extra lights, air conditioning, electric heating. Then there was the agricultural sales program, "Farm Better Electrically." VEPCO representatives worked with the 4-H and Future Farmers of America, teaching the agriculturally minded youth of the state about electric motors and other equipment, and they traveled to individual farms to convince the older farmers of the value of modern electrical equipment. But the biggest push came in the home. Reddy Kilowatt, the electric industry mascot with a lighting-bolt body and light-bulb nose, assured consumers on television and in print that electricity was a friendly and helpful servant. A staff of home economists gave tours of all-electric kitchens and offered demonstrations to high school classes and various organizations. And year after year, VEPCO worked with dealers to push electric stoves and other appliances, sometimes offering to pay for part or all of the installation. The programs, particularly the residential one, had the desired effect. From 1955 to 1966, the average annual electricity use in residences nearly doubled, from 3,012 kilowatt-hours to 5,967, and sales to homes came to account for almost half of VEPCO's revenues from electricity.

With electricity demand setting new records each year, the utility had no choice but to continuously add new generating capacity. In 1959, VEPCO opened a 170-megawatt, coal-fired plant at Portsmouth Station. Before its stacks were dirty, the company had begun work on a 220-megawatt addition to it, which would open in 1962. Meanwhile, VEPCO was also building a 220-megawatt addition to its Possum Point coal-burning plant. (One hundred megawatts of electricity is enough for about 25,000 homes today, and was sufficient for about twice than many in the 1960s, when per capita usage was less.) In 1963, the company inaugurated a 200-megawatt hydroelectric plant. The next year, it put a new 330-megawatt unit into operation at its Chesterfield plant. Then in 1965 and 1966, VEPCO brought two massive 500-megawatt coal-burning units into service at Mt. Storm, West Virginia. At this point, the company had added nearly 3,000 megawatts of power within a decade, and the need for extra capacity was accelerating.

Life was good—for Virginia Electric and Power, its stockholders, and even its customers. As demand for electricity was going up, prices were going down. Throughout the 1950s and into the mid-1960s, the cost of coal was declining, with the savings passed along to consumers. And improvements in generating technology plus economies of scale from larger plants brought costs down further. Because of VEPCO's growing revenues and falling costs, state regulators allowed it to take steadily increasing profits, which translated into rising dividends for stockholders. As for company management, the job was rewarding, if not particularly challenging. Demand forecasts were easy—just assume a growth of 8 or 9 percent per year, and you wouldn't be too far off. The technology used in VEPCO's hydroelectric and coal-burning plants was mature, with few surprises. Because the company was healthy and growing steadily, financing for construction was relatively simple to arrange, either through bond sales or issuing new stock. And people in the company were happy. They were the "good guys." They provided an important service, helping the area grow and prosper, and they lowered their prices almost every year to boot.

To a large extent, the same was true for most U.S. electric companies during these years. From 1945 to 1965, electricity produced by utilities nearly quintupled, while prices dropped dramatically compared to other fuels. Utility stocks were safe investments offering a solid, if not spectacular, return—the ideal holding in a retirement account, for instance. The industry as a whole was adult, perhaps even a bit stodgy, and its path was clear: straight ahead.

By the mid-1960s, however, a few seemingly minor problems had begun to appear, both for the U.S. industry in general and VEPCO in particular. As the generating plants grew to 500 or 1,000 megawatts, their efficiencies—which had been increasing steadily with their size—stopped improving. These giant plants were complex and operated at high temperatures and pressures, all of which made them more difficult—and expensive—to operate. Coal costs, which had been declining for so long, flattened out and then began to increase. Growing concerns about pollution forced utilities to pay for extra equipment in their plants that did nothing to improve productivity.

VEPCO got a taste of the future with its 1,000-megawatt Mt. Storm plant. It was built in the coal fields of West Virginia to minimize the costs of transporting the fuel, a major portion of the operating costs of a coal-fired plant. But that meant that the electricity had to be carried to the utility's customers, most of them a hundred or more miles away. "Coal by wire," the company called it, and it set forth a plan for a 350-mile loop of high-voltage transmission wires running from West Virginia toward Washington, DC, down to Richmond, and back to the plant. The higher the voltage, the less electrical loss in transmission, so the company decided to use 500,000 volts. This would make it the highest-voltage transmission system in the western world, rivaled only by some in the Soviet Union. The towers carrying the transmission lines would be huge, from 100 to 150 feet tall, depending on the terrain. Since the construction costs would be about $80,000 a mile, the utility laid out a path for the system that was as direct as possible. It was a natural thing to do, but the company soon came to regret it.

The proposed path took the transmission lines directly through some beautiful, unspoiled Virginia countryside west of Washington that is home to many of the area's wealthiest people. Hundreds of large farms and estates are scattered about. More than a few have stables for the country folks' beloved horses, and it's not unusual to see people in town wearing riding boots and jodhpurs. Sometimes they've just returned from a fox hunt, a tradition still alive here.

So perhaps VEPCO shouldn't have been surprised when its plans ran into resistance. Residents of Clarke County pledged to spend $100,000 to fight the power line, while two hundred property owners in neighboring Fauquier and Loudoun counties signed a declaration against it. Two local men wrote an anti-power-line song and sang it at a meeting of the State Corporation Commission, which had the responsibility of deciding whether to approve the proposed line. (To the tune of "On Top of Old Smokey": "Don't give me no power line/Crossing my life/Don't give me no power line/No worry, no strife/Please keep your high towers/Away from my land/Don't cut my oak trees/Please let them stand.") Eventually VEPCO had to detour around the recalcitrant counties, staying in West Virginia as much as possible and even jumping into Maryland for part of the circuit. As a result, the planned 350-mile loop became 390 miles long—which, at $80,000 a mile, would cost the company an extra $3 million.

The coal-by-wire system had its technical difficulties as well. The first unit at the Mt. Storm station went on-line in 1965, but it didn't operate at 100 percent until the following year—not, however, because of problems with the station itself. Instead, the state-of-the-art transmission system slowed things down, as a series of equipment problems limited how much power it could carry. In its ambitious attempt to provide inexpensive power to its customers, VEPCO had tasted both technical frustration and public displeasure. It would not be the last time.

By the mid-1960s, the utility's management was considering new sources of power. Suitable sites for hydro plants were getting scarce, and coal costs were edging up, with the promise of further increases in the future. In 1964, the company announced plans for a different sort of power plant: a 510-megawatt pumped-storage hydro station that could "store" power in periods of low demand for use during peak demand. The station would consist of a series of reservoirs. When demand was low, some of VEPCO's excess power would be used to pump water uphill so that, when needed, it could be released to flow down through turbines, generating power as a normal hydroelectric plant would. Although this is a relatively inefficient way of producing electricity, it's cheaper than building a new generating plant that is used only during peak hours.

The pumped-storage plant would help meet the peak load, but VEPCO also had to contend with a growing "base load"—the minimum, around-the-clock load that a utility must provide. Base-load plants are a utility's workhorses, churning out power for eighteen, twenty, even twenty-four hours a day, every day. To do this most cheaply, a base-load plant's day-to-day operational costs, including fuel and maintenance, should be low. The construction costs are not so important since they can be spread out over the thirty- to fifty-year life of a plant that generates electricity almost continuously. A coal-fired plant, for instance, is typically a base-load plant: since coal is relatively cheap, the average generating cost goes down with each additional kilowatt-hour that the plant turns out. Peak-load plants, by contrast, should be inexpensive to build but can be expensive to operate. Since they are working only a few hours a day, the hourly costs are less important than the capital investment.

Sometime around 1965, VEPCO's executives began to consider nuclear power seriously. A few years earlier it would have been out of the question, since it cost several times as much as electricity from coal. But starting in 1963, prices for nuclear power plants had begun to look competitive, particularly with coal going up and everyone expecting that prices for uranium fuel would be coming down.

On June 14, 1966, VEPCO president Alfred H. McDowell offered an analysis of the situation to a meeting of the Richmond Society of Financial Analysts. The company had been weighing the nuclear option for some time, McDowell said, and the choice between nuclear and coal seemed to boil down to price—and coal was losing. "If the price [of coal] at the source continues to go up," he said, "if railroads are not able to put in new and visionary economics in delivering fuels, there is no question but what nuclear fuel will be used by half of the nation's utility companies by the year 2000." Advances in nuclear technology, he added, had brought the price of a nuclear plant down close to that of a coal-fired plant—between $105 and $110 per kilowatt versus $90 to $95 for coal—and nuclear fuel was cheaper than coal once transportation costs were taken into account.

What McDowell didn't tell the financial analysts was that VEPCO had already made its decision. Just eleven days later, the company announced it would build a nuclear station of about 750 megawatts along the James River. The location in rural Surry County was next to a waterfowl sanctuary and only a few miles upstream from the Newport News shipyard. Newport News was the birthplace of the *Enterprise*, the world's first nuclear-powered aircraft carrier, as well as a number of other nuclear ships.

The selection had been based mainly on economics, company officials said. Coal cost VEPCO about 28 cents per million Btu (British thermal units, a measure of heat generated), while nuclear fuel should run about 15 cents per million Btu. Coal cost only 14 cents per million Btu at the company's Mt. Storm mine-mouth plant, where transportation costs were negligible, but then VEPCO had to worry about transmitting the electricity on high-voltage lines—something that had proved to increase both expenses and management headaches. In weighing nuclear versus coal, McDowell told reporters, the company had asked the railroads if they could find a cheaper way to ship coal. But the railroads had already cut their price by 80 cents per ton earlier in the year, and they were unwilling to make further significant concessions. "What they offered was so minor," he said, "it hardly changed our economic evaluations." All things considered, VEPCO figured it would save about $3 million a year by building a nuclear plant instead of one that used coal.

Few other factors played a role in the decision. There had been some concern that the plant's cooling water—taken in from the James River and then returned much hotter—would harm the local oyster industry, but leaders of the state's marine industry didn't seem worried, so neither was VEPCO. Of course, there was always the possibility of opposition from local people who didn't want to live next door to a nuclear plant, but the company thought it could keep such worries to a minimum. For eighteen months VEPCO had been sending speakers to meetings of local civic organizations to talk about peaceful uses of atomic energy. And VEPCO had chosen its site carefully. Surry was a rural county with only 6,200 residents, most of them farmers and 65 percent of them black. Its only industry was a sawmill. It had no library, no movie theater, no dry cleaner, no barber, no beauty shop. One doctor, no dentist. The nuclear plant, a high-tech tenant paying high taxes, would be more than welcome. "If we had picked the whole world over, we couldn't have picked a better industry for Surry," said W.E. Seward Jr., chairman of the county's board of supervisors and owner of the sawmill.

On the surface, it appeared to be a carefully thought out, even conservative decision. A score of utilities across the country had already chosen to go nuclear, so the technology appeared to be well accepted. And VEPCO had made the effort to learn about nuclear power before making a major investment. In collaboration with three other utilities, it had taken advantage of one of the AEC reactor demonstration programs to build a heavy-water reactor in Parr, South Carolina, which began operations in 1962 and ran for several years. Most important, the cost estimates seemed to promise that a nuclear plant was the best option for VEPCO's customers, who, because VEPCO was a state-regulated monopoly, would pay according to what it cost the utility to generate electricity.

But there were a number of factors operating below the surface, factors that predisposed VEPCO and other utilities to take the nuclear plunge. These weren't discussed in the company's press releases about the new nuclear plant, but they were there nonetheless.

One factor was a steady government pressure on utilities to adopt nuclear power. The 1954 Atomic Energy Act had allowed private companies to own reactors, to obtain formerly classified information on nuclear power from the AEC, even to perform their own research and patent their innovations. Congress had assumed that these privileges—along with some financial incentives from the AEC—would be enough to convince the private sector to go nuclear.

But Congress was not very patient. Shortly after the passage of the Act, the AEC announced its Power Reactor Demonstration Program, designed to be a partnership between the commission and electric utilities. The companies would build the reactors; the AEC would waive its use charges for the nuclear fuel, bankroll some of the research and development for the reactors, and pay the utilities for some of the research they carried out themselves. Within three months of the program's start, the AEC had received proposals for three reactors, all of which were eventually built. The Yankee Atomic Electric Company, a consortium of New England utilities, constructed a pressurized-water reactor in Rowe, Massachusetts, similar to that at Shippingport, but larger—167 megawatts instead of 60. It was finished in 1961. Consumers Public Power District of Columbus, Nebraska, put up a 75-megawatt sodium-graphite reactor near Hallam, Nebraska, completed in 1963. Detroit Edison erected a 100-megawatt fast-breeder reactor at Monroe, Michigan, which went on-line in 1964.

And, in the excitement following the passage of the act, two giant utilities decided to construct nuclear plants without subsidies from the AEC. Commonwealth Edison contracted with General Electric to build a 180-megawatt plant in Dresden, Illinois, which used a boiling-water reactor. Consolidated Edison of New York bought a 163-kilowatt pressurized-water reactor from Babcock & Wilcox for a nuclear plant at Indian Point.

Still, the Joint Committee and other members of Congress felt that the drive to civilian nuclear power was going too slowly. In the first half of 1956 it had become clear that the British, not the Americans, would win the race to produce nuclear electricity for civilian use. Britain's Calder Hall plant, with a gas-cooled, graphite-moderated reactor, would go on-line October 17, 1956, thirteen and a half months ahead of Shippingport. More alarming was the possibility that the Soviet Union might also be moving ahead of the United States in the nuclear power race.

In June 1956, Senator Albert Gore and Representative Chet Holifield introduced a bill designed to speed up the American effort. It would direct the AEC to build large nuclear power plants to provide electricity for its uranium- and plutonium-production sites, instead of buying the electricity from nearby utilities. Since industry was not moving quickly enough on building commercial power plants, Gore and Holifield wanted the federal government to take the initiative. But their bill reopened the whole issue of public versus private power, which had first begun with the building of the government-owned Tennessee Valley Authority (TVA) electricity-supply system. Republicans in Congress were strongly opposed to government ownership of utilities or any other industry, and they saw the AEC's entry into the power-generation field as a step toward socialized electricity. On this basis the Republicans, allied with some Democrats, defeated the Gore-Holifield bill in the House after it had passed easily in the Senate.

The Republicans, however, were no less committed to nuclear power than the Democrats. They simply wanted to encourage it in different ways. In place of federal projects to build and operate power reactors, Republicans proposed giving private industry more incentives to build the nuclear plants. *Business Week* saw it this way in 1956: "Civilian atomic power is bound to benefit from an argument in the coming Congress. The basic question: Which side can do more for the atomic industry?" After the defeat of the Gore-Holifield bill, the Republicans were careful to show their hearts were in the right place: Lewis Strauss, the Eisenhower-appointed head of the AEC, announced a $160 million program to support private development of nuclear power.

The companies of the nascent nuclear industry appreciated this—but, they told Congress, it wasn't enough. Like a teen-aged driver with a red Corvette, they were having trouble getting sufficient liability insurance. Insurance companies would provide no more than about $65 million for a single plant, but the potential liability was much greater. According to a 1957 AEC safety study, a serious accident at a large reactor could conceivably kill thousands of people and cause billions of dollars of property damage. Even if the chances of such an accident were extremely small, no utility would be willing to risk bankruptcy by building a nuclear plant without enough insurance. In response, Congress passed the Price-Anderson Act of 1957. It required a nuclear plant operator to buy as much liability insurance as was available, but then committed the government to pay for any damages above that amount, up to a $560-million limit. In effect, the federal government was providing a $500-million supplemental insurance policy, free of charge, to the owners of any nuclear power plant built in the country. Furthermore, the act capped the amount of damages that would be compensated for any single nuclear accident at $560 million. If a major accident ever did cause billions of dollars in damage, the victims would be out of luck.

So Congress, by dangling a bunch of carrots and waving a stick, attempted to push the utility industry mule down the path toward a nuclear future. If the various incentives and subsidies didn't do the job, there was always the threat of a revived Gore-Holifield bill or something similar, which would expand the federal presence in the electricity-generating business. The creation of one or more nuclear TVAs was something no utility executive wanted to see.

To a large degree, however, Congress was pushing the utility industry in a direction it wanted to go. Executives in the electric power industry were—and to a large degree still are—technological enthusiasts. They have great faith in technology and its potential for making the world a better place. This can be seen, for instance, in VEPCO's 1962 annual report. The cover features the experimental nuclear plant VEPCO had built with three other utilities. In the foreground is a young boy, perhaps ten years old. He is looking at the plant's spherical containment vessel, which houses the reactor, and he is smiling. Inside the front cover, the accompanying text reads:

> A Vepco Shareowner inspects the Southeast's first nuclear plant. . . .
>
> In his lifetime, such projects will go far beyond the experimental stage. He will take atomic power for granted. In his lifetime, America will require an amount of power that may seem incredible today.
>
> Can these needs be met? They are being met. The nuclear plant is only one example of Vepco's policy of looking into the future—and building for future needs.

From here, the report goes on to describe the building of the Mt. Storm mine-mouth plant, the high-voltage lines that will carry its power, and a new hydroelectric project. It concludes:

> The list is long. For wherever our young Shareowner might travel throughout the Vepco system, he will find fresh evidence of Vepco's active planning for the future. Through far-sighted planning, far-reaching research, and consistent building and development, Vepco is insuring the power for generations yet to come.

In short, utility executives tended to see themselves as forward-looking folks, always on the lookout for ways to improve their customers' lives through technology. They would like nothing better than to assure a nearly limitless supply of cheap electricity by building nuclear power plants. There was just one problem: nuclear power wasn't cheap. At the beginning of the 1960s, even the most optimistic cost estimates had nuclear electricity as being significantly more expensive than electricity generated from burning coal. Most people assumed that eventually those costs would drop enough to make nuclear power the obvious choice, but it was impossible to say when.

The combination of technological enthusiasm and government pressure was enough to convince some utilities to invest in nuclear power even when they knew they would lose money. When the Duquesne Light Company of Pittsburgh made its successful bid for the Shippingport reactor, it offered to pay the AEC much more than the company would earn by selling electricity from the plant, but Duquesne's management judged that getting in on the ground floor of the nuclear age was worth the initial multimillion-dollar loss. Similar reasoning motivated VEPCO to build the experimental reactor in South Carolina. VEPCO and its three partners expected to spend $20 million for construction and another $9 million to operate the plant for five years. (The AEC would put in up to $15 million in research and development, but nothing for construction or operating expenses.) It was a reasonable cost, VEPCO thought, for gaining experience in operating a nuclear reactor.

A number of other companies followed suit in the late 1950s and early 1960s. Northern States Power Company of Minneapolis constructed a 62-megawatt boiling-water reactor at Sioux Falls, South Dakota; Consumers Power Company of Jackson, Michigan, put together a 50-megawatt boiling-water reactor at Big Rock, Michigan; and Philadelphia Electric Company erected a 45-megawatt graphite-moderated, gas-cooled reactor at Peach Bottom, Pennsylvania. One company, Southern California Edison, announced an ambitious plant that would be several times larger than any built or planned at the time—a 370-megawatt pressurized-water reactor at Camp Pendleton, California. These were the risk-takers and trailblazers among the electric utilities. Their faith in the promise of nuclear power was such that they were willing to lose money now in the expectation of greater returns in the future.

This faith could push nuclear power only so far, however, and by the early 1960s the limit seemed to have been reached. Some utilities had decided to build reactors with help from the AEC, and a handful had done it on their own, but almost all were small prototype reactors. Meanwhile, the bulk of the utility industry remained on the sidelines, waiting for more certainty.

James Jasper, a sociologist at New York University, explains the situation as a standoff between the two types of people who predominated in the utility industry: the technological enthusiasts and what Jasper calls the "cost-benefiters." In the late 1950s and early 1960s, there was simply not enough information to say when—or even if—nuclear power would be important to electric utilities in the United States and other countries. And, Jasper notes, "In situations where there are no clear data, everyone must rely on intuitions filtered heavily through worldviews." The intuition of the technological enthusiasts told them that nuclear power would be an important—perhaps even revolutionary—advance in producing electricity, and that progressive utilities should get on board. The cost-benefiters, on the other hand, approached the nuclear decision as an exercise in economic analysis, looking to allocate resources in the most efficient way. If the dollars didn't add up, the nuclear option didn't make sense. Neither group could sway the other. The technological enthusiasts saw the cost-benefiters as excessively cautious; the cost-benefiters weren't willing to take the value of nuclear power on faith—they wanted some numbers.

The standoff was broken in dramatic fashion in December 1963, when the utility industry's cost-benefiters got the numbers they'd been asking for.

Jersey Central Power & Light announced it had signed a contract with General Electric (GE) to build a 515-megawatt nuclear plant at Oyster Creek, forty miles north of Atlantic City. For just $68 million, GE would provide everything: the reactor and steam generator and the rest of the equipment, the building to house them, the land, employee training, licensing costs. That worked out to an astounding $132 per kilowatt, little more than the construction costs of a coal-fired plant of similar size. And when fuel and operating costs were taken into account, the utility calculated that producing a kilowatt-hour of nuclear electricity would cost less than the 4.5 mills (0.45 cents) it took to generate a kilowatt-hour from coal. Best of all, GE guaranteed the price for the entire plant. Jersey Central would not have to worry about cost overruns or unforeseen expenses.

It seemed too good to be true, and the utility industry took a while to appraise the new situation. Was this the long-predicted nuclear breakthrough, or was it wishful thinking? To counteract the second-guessing, Jersey Central took the unprecedented step of making public its financial analysis comparing the nuclear plant with a coal-fired plant at the same location and a mine-mouth plant in the Pennsylvania coal country. The study, published a month after the initial announcement, made a variety of assumptions about the nuclear station's operating characteristics, but none of them seemed unreasonable in light of the performance of earlier nuclear plants. No matter which assumptions it used, the company figured the nuclear plant to be more economical than a coal plant in the same location; a mine-mouth plant might be cheaper in some scenarios, but only if the nuclear plant did more poorly than the company expected. All in all, nuclear seemed the best bet.

Even Philip Sporn was convinced. The former president of American Electric Power Company had been asked by the Joint Committee on Atomic Energy to assess the Jersey Central report. Sporn, an old coal man, had been one of the few people in the utility industry to openly criticize nuclear proponents for being too optimistic, and the previous year he had thrown cold water on an AEC report that found nuclear power to be "on the threshold of economic competitiveness." Sporn ran through the numbers that Jersey Central had provided and replaced some that he thought were unrealistic, but his conclusions were not much different from the utility's. The power wouldn't be as cheap as the company claimed, but it would still be less expensive than coal power in many places around the country.

Then, to clinch its case, GE published a "price list" in September 1964. An interested utility could buy a 50-megawatt nuclear plant for $15 million, or $300 per kilowatt of capacity. This wasn't cheap—generating costs would be about 10.4 mills per kilowatt-hour—but utilities could get much less expensive nuclear power, GE said, by buying larger plants. A 300-megawatt plant would cost only $152 per kilowatt of capacity and produce power at an estimated 5.2 mills per kilowatt-hour. A 600-megawatt plant, about the size of Oyster Creek, would be $117 per kilowatt to build and 4.2 mills per kilowatt-hour to run. And for the really ambitious, GE would construct a 1,000-megawatt plant (five times larger than anything the company had actually built) for a mere $103 per kilowatt; it should produce power at a tidy 3.8 mills.

These figures quickly shifted utility thinking on nuclear power, especially when Westinghouse began offering a similar deal on its nuclear reactors. Now the cost-benefiters calculated nuclear power to be a more efficient investment than building more coal-fired plants, which put them in agreement with the technological optimists on the question of whether to go nuclear. The result was what Philip Sporn would later dub "the Great Bandwagon Market." No nuclear plants were ordered in 1964, as the industry debated the significance of Jersey Central's and GE's claims, but in 1965 U.S. utilities wrote contracts for eight nuclear reactors. Then utilities ordered forty-eight more nuclear plants in 1966-67, or nearly half of all electrical-generating capacity ordered in the United States.

Neither the technological enthusiasts nor the cost-benefiters inquired too closely as to what had happened in 1963 to suddenly allow GE to build nuclear plants much more cheaply than in the past. The cost-benefiters left technical questions like that to the engineers, and the technological enthusiasts chalked it up to the normal course of technological advance. In reality, however, the price breakthrough at Oyster Creek had little to do with technical advances by GE engineers and everything to do with the company's ongoing competition with Westinghouse.

LOSS LEADERS

The rivalry between GE and Westinghouse goes back a hundred years, to the conflict between their founders, Thomas Edison and George Westinghouse. Over the years, the two companies have sold many of the same product lines, from commercial generators and motors to light bulbs and refrigerators. Westinghouse has always been the smaller of the two, but, as the 1960s dawned, it didn't intend to accept second place in the emerging nuclear market.

Both companies had become involved with the naval reactors program early on. Westinghouse built the pressurized-water reactor for the *Nautilus*, and GE the liquid-metal reactor for the *Seawolf*. Because of Rickover's decision to go with the pressurized-water type, Westinghouse not only took an early lead over GE in building reactors for the Navy but it also got a head start in developing nuclear power for the civilian market. Westinghouse built the reactor for Shippingport, and even before construction began on that plant, the company had agreed to provide the reactor for Yankee

Atomic plant in Rowe, Massachusetts. In 1960 it announced a contract for Southern California Edison's giant 370-megawatt plant at Camp Pendleton.

But GE was not far behind. Leaving the pressurized-water reactor to Westinghouse, it gambled on a boiling-water design, in which the water coolant was allowed to boil. This had the advantage of simplicity: instead of two systems of pipes connected by a steam generator, the boiling-water reactor has just one—a circuit that leads away from the reactor core to a steam turbine, through a condenser, and back to the core. Furthermore, the boiling-water reactor has no steam generators, the tricky components that have been a constant source of trouble for pressurized-water reactors. On the minus side, the water and steam in a boiling-water reactor's single loop is radioactive, so workers doing maintenance and repair on the turbine must be protected. Furthermore, a break in a steam pipe is a more serious problem than in a pressurized-water plant. At about the same time that Westinghouse landed the Yankee-Rowe contract, GE agreed to provide a 180-megawatt boiling-water reactor for Commonwealth Edison's Dresden plant. A few years later, Pacific Gas & Electric chose GE for a 325-megawatt boiling-water facility to be built at Bodega Bay.

Throughout the late 1950s and early 1960s, a number of other companies took a stab at the nuclear reactor business—North American Aviation, Combustion Engineering, Babcock & Wilcox, ACF Industries, Allis-Chalmers. Of all these, only Babcock & Wilcox could boast a project comparable with those of GE and Westinghouse—a 163-kilowatt pressurized-water reactor for Consolidated Edison's Indian Point plant. That gave Babcock & Wilcox a good claim for third place in the nuclear power sweepstakes, but everyone knew that spots one and two were reserved for GE and Westinghouse. The two old foes dominated because of their size, their background in making electrical equipment, their experience in the naval program, and, particularly, their commitment to nuclear power.

"The atom is the power of the future—and power is the business of General Electric," said GE's chairman, Ralph Cordiner, in January 1959, and he meant it. The company had 14,000 employees working on atomic energy projects of one sort or another and could boast $1.5 billion in government atomic contracts and another $100 million in private deals. It had spent $20 million of its own money on nuclear power, including $4 million to build a prototype of the Dresden plant. In all, General Electric expected to lose $15 million to $20 million on the Dresden deal alone, but counted the money as an R&D expense. Westinghouse had not spent quite as much of its own funds on nuclear power but could point to even more contracts. It was just as committed to nuclear power—financially and psychologically—as GE.

That commitment sprang from several sources. Both firms had been leaders in the field of electrical generation, so they saw the move to nuclear power as a logical next step. Both had learned the trade by building reactors for nuclear submarines, and each wanted to leverage its experience with military technology into an advantage in the commercial sphere. And underlying everything else, policy at both companies was shaped by a technological optimism and a conviction that nuclear power would someday be a technology of major importance. It was a conviction with roots not in any financial analysis but rather in basic beliefs about mankind and progress. GE's Cordiner put it like this: "Civilization is moved forward by restless people, not by those who are satisfied by things as they are." Developing nuclear power was more than a job. It was a mission, a manifest destiny.

As the two companies entered the 1960s, it seemed that the era they dreamed of was tantalizingly near. The cost of nuclear-generated electricity had been coming down as the firms gained experience designing and building nuclear plants. With just a little more practice, they thought, it should be possible to build plants that could hold their own against coal, at least in parts of the country where coal was expensive.

This was a critical time. As both GE and Westinghouse recognized, a small advantage early on might be parlayed into dominance later on. If one of the two could win a few reactor orders, the experience building them could be used to bring costs down further and thus capture more orders, thereby gaining more experience and increasing the company's advantage even further, until it was well in front of its rival.

As it happened, Westinghouse suddenly seemed ready to grab that lead. By coincidence, in December 1962 and January 1963, Westinghouse had four major reactor contracts appear almost simultaneously. One of them, for Southern California Edison's 395-megawatt San Onofre reactor, had actually been announced in 1960, but it had taken the utility two and a half years to negotiate a lease with the Marine Corps for its land at Camp Pendleton. A second order was for a giant, 1,000-megawatt reactor Consolidated Edison wanted to put just across the East River from Manhattan. That project would eventually be scuttled due to resistance from New York City residents worried about a reactor accident. The third was a 490-megawatt plant in Malibu Beach ordered by Los Angeles Water & Power. It was later canceled because of worries over an inactive earthquake fault. The fourth was the only one of the bunch that both was a new order and would actually be built: a 500-megawatt plant for Connecticut Yankee Atomic Power. In retrospect, it hardly seems as if Westinghouse were about to corner the market, but that's how it looked to GE at the time.

Moreover, GE was worried that European competitors might cut into its market. The French and British were developing a radically different technology—gas-cooled reactors—that was widely seen as having greater long-term potential than the light-water reactors of Westinghouse and GE. In Britain, the Advanced Gas-Cooled Reactor had just begun operations at

Windscale. If that plant allowed the British to open up a substantial lead, GE might never catch up.

"We had a problem like a lump of butter sitting in the sun," recalled GE's vice president of planning, John McKitterick, in a 1970 interview. "If we couldn't get orders out of the utility industry, with every tick of the clock it became progressively more likely that some competing technology would be developed that would supersede the economic viability of our own. Our people understood this was a game of massive stakes, and that if we didn't force the utility industry to put those stations on line, we'd end up with nothing."

So GE took a bold, calculated risk. It would offer to build an entire nuclear plant for a utility at a fixed price. Such a contract, which was unlike anything the utility industry had seen, came to be known as a "turnkey." To start up such a plant, the pitch went, all a utility had to do was walk in and turn a key. The turnkey approach was calculated to appeal to utilities in a number of ways. At the time it was standard industry practice for generating plants to be built on a cost-plus basis, with the utilities responsible for any overruns, but GE agreed to pay for any cost overruns on the turnkey plants. Furthermore, few electric utilities knew anything about building a nuclear plant. GE's offer saved them from having to learn. But the make-or-break feature of the turnkey contracts was the price. If GE could supply a nuclear plant at a price that made it competitive with coal, it thought it could sell several of them quickly and steal a march on Westinghouse. But could it really build plants for such a price?

No one at GE knew. At the time, none of the costs for nuclear-generated electricity being offered by the AEC and the reactor manufacturers was anything more than an educated guess. There was very little experience in actually building nuclear plants—the Dresden unit, which opened in 1960, and Yankee Rowe, which opened in 1961, were the only completed nuclear plants of more than 100 megawatts. How much would a 400- or 500-megawatt plant cost? It was impossible to say. The cost estimates depended on a number of subjective factors, such as how important economies of scale were likely to be in building big nuclear plants.

Everyone in the industry knew that larger plants should be more efficient and operate at a lower per-unit cost. This is the case for other types of plants, from chemical refineries to coal-fired power plants, and there were obvious reasons to expect it from nuclear plants as well. The site for a 500-megawatt plant need not be much larger than the site for a 250-megawatt plant, for instance, so the proportional land costs—measured in acres per megawatt or some such unit—would be lower for the larger plant. The reactor vessel surrounding a 500-megawatt reactor is not much more expensive than the vessel for a 250-megawatt reactor. The instrumentation and controls are not very different for the two sizes, nor is the number of people needed to run the plant. Of course, building larger plants posed certain problems, too. It is not possible to double a plant's size simply by doubling everything in it—putting twice as many fuel elements in the reactor core, for instance—so enlarging a plant demands a lot of extra design work. The conventional wisdom was that the cost savings in building larger plants would outweigh the extra expenses, but without actually building nuclear plants of various sizes, it was impossible to predict how it would all add up.

Besides economies of scale, GE and other reactor manufacturers also assumed that the costs of nuclear plants would come down as more of them were put up. The design costs could be spread over several plants, and the practice gained from the early plants would make construction faster and more efficient later on. But again, the actual amount of these savings was mere guesswork until the plants had been built.

These uncertainties left plenty of wiggle room for the GE management, and, facing the threat of Westinghouse getting a big lead, the company made what in hindsight were extremely optimistic cost estimates. It would not lose too much money, the company thought, if it offered to build the Oyster Creek plant for $68 million, a price that made it cheaper than a coal plant. Furthermore, the company figured, it would break even or perhaps make a small profit if it could sell two more plants identical to Oyster Creek and spread out the design costs.

Once GE had settled on the assumptions it would use in estimating the cost of Oyster Creek, those assumptions became a baseline for all other cost calculations. Nine months after the Oyster Creek announcement, GE published its list of prices for turnkey plants of various sizes—prices calculated according to the same guesses about economies of scale and costs of construction that went into the Oyster Creek price. But the acceptance of the Oyster Creek assumptions went far past GE. The AEC, the Joint Committee, most of the nuclear industry, and nearly all of the media accepted the figures uncritically. Even Philip Sporn, normally a skeptic, swallowed the basic postulates underlying GE's estimates—such as the ability to predict construction costs with essentially no experience in building such plants—and quibbled only with a few details. Less than two years earlier, Sporn had calculated that nuclear power should cost 7.2 mills per kilowatt-hour. After examining the Oyster Creek numbers and toning down some of the more optimistic assumptions, he calculated that the proposed plant would produce electricity at 4.34 mills per kilowatt-hour.

In retrospect, there is something almost surrealistic about the arguments that went on over Oyster Creek. Would it generate electricity at 3.68 mills per kilowatt-hour over thirty years, as the final Jersey Central report claimed? Or would it be 4.34 mills, as Sporn said? No one had built a nuclear plant larger than a third the size of Oyster Creek, much less operated one for thirty years (Shippingport was approaching seven years of operation at the time), yet serious people were calculating what the cost would be, accurate to two decimal places. What could they have been thinking?

Alvin Weinberg, director of Oak Ridge National Laboratory at the time, offers one answer. In his memoirs, *The First Nuclear Era: Life and Times of a Technological Fixer*, he recalls being swept along by a nuclear fervor:

> I find it hard to convey to the reader the extraordinary psychological impact the GE economic breakthrough had on us. We had created this new source of energy, this horrible weapon: we had hoped that it would become a boon, not a burden. But *economical* power—something that would vindicate our hopes—this seemed unlikely. . . . I personally had concluded that the commercial success of nuclear power would have to await the development of the breeder. I was therefore astonished that there was a more direct, easier path to economical nuclear power, as evidenced by the GE price list. And because we all wanted to believe that our bomb-tainted technology really provided humankind with practical, cheap, and inexhaustible power, we were more than willing to take the GE price list at face value.

Later, Oak Ridge engineers would make their own calculations—calculations that assumed large economies of scale and that resulted in numbers very close to those of GE. As a result, Weinberg and his Oak Ridge colleagues became some of the biggest cheerleaders for the light-water plants, telling anyone who would listen that cheap nuclear power had arrived.

Meanwhile Westinghouse, perhaps the only organization with the expertise and experience to recognize GE's numbers as the gamble they were, found itself in a bind. Although it doubted GE's ability to make a profit on such deals, Westinghouse had little choice but to match them. "The competition was rather desperate in those days," said Westinghouse executive vice president Charles Weaver. "To meet it, we had to go a route we didn't necessarily feel was desirable but that we could stand up under." Like GE, Westinghouse offered to build entire nuclear plants for a single, guaranteed price, one that was competitive with GE's. Later, Babcock & Wilcox and Combustion Engineering would make similar offers.

Only years afterward would it be evident how badly the reactor manufacturers had misjudged—or misrepresented—the real costs of building nuclear power plants. On average, nuclear plants ordered in the mid- to late-1960s cost twice as much to build as estimated. GE and Westinghouse quickly realized what was happening and stopped offering the turnkey contracts in 1966, but by then they had agreed to build a dozen nuclear plants for a fixed price. Although the two companies have never released exact figures, analysts estimate that on those twelve plants GE and Westinghouse together lost as much as a billion dollars.

Not surprisingly, most of the losses came in areas that the companies had little or no experience with. The cost estimates for the reactors themselves were relatively accurate, recalls Bertram Wolfe, formerly a vice president and general manager in charge of GE's nuclear energy division. But the reactor is a relatively small part of the total plant, and the companies greatly underestimated how expensive it would be to build the rest of it. In particular, Wolfe says, "We lost money on the turnkeys largely because we didn't know how to control site construction costs." On site, GE subcontracted most of the work building the plants, and many of those contracts were not on a fixed-price basis. For example, the architect-engineers who designed the plants worked on a cost-plus basis, Wolfe says, so "they had no incentive to do it on schedule and on cost." And the labor costs were much higher than expected, for several reasons. The buildup of the Vietnam War pushed hourly labor costs up as much as 20 percent in some parts of the country. The amount of labor spent building the plants was much more—sometimes as much as two times more—than what the architect-engineers had projected would be needed. And low productivity and labor disputes stretched out the construction schedules. In one three-month period at Oyster Creek, for instance, work was stopped for fifty-three days because of jurisdictional disputes among the unions.

But despite their cost, the turnkeys did what they were intended to do. As the most expensive loss leaders in history, they succeeded in pulling the electric utilities into the nuclear store. And once there, the utilities continued to shop. In 1965, while three utilities were signing contracts for turnkey plants, five others were announcing agreements to build nonturnkey stations. Instead of having GE or Westinghouse do everything, these utilities were to act as general contractors, buying the reactor from one company, hiring a second to design the plant, a third to build it, and so on. The utilities took the turnkey prices at face value and assumed that they could build nuclear plants at similar costs without price guarantees.

The most dramatic example of this came in June 1966, when the Tennessee Valley Authority said it would built a two-unit, 2,196-megawatt nuclear plant at Browns Ferry on the Tennessee River, 30 miles west of Huntsville, Alabama. General Electric would supply the reactor and most of the other major equipment at a guaranteed price, but this was not a turnkey deal. TVA itself would be in charge of designing and building the plant. The deal thrilled the nuclear industry, not just because of its size—equal to the combined capacity of all the nuclear plants then operating—but also because of its location. TVA's territory included plenty of low-cost coal, which it was turning its back on. "An atomic bomb in the land of coal," *Fortune* called it.

At about the same time, VEPCO was announcing its decision to build a nuclear plant. To VEPCO's south, Duke Power said it would enter the nuclear era with a two-unit, 1,400-megawatt plant that would be even cheaper to build than the TVA station. Babcock & Wilcox, with its first large reactor order, would help them put in the plant for less than $96 per kilowatt. Across the country, utilities were looking at the numbers and deciding that nuclear power made sense.

Even after GE and Westinghouse stopped offering turnkey contracts, the utility industry continued to order nuclear plants, one after another. It was not for another couple of years that the utilities began to realize what the reactor manufacturers had already figured out: that building nuclear power plants was much more expensive than originally advertised. In 1968, orders for new plants plummeted, and by 1969 the number of sales was back to where it had been in 1965.

The commercialization of nuclear power in the mid- to late-1960s was, more than anything else, a product of the business environment: the culture of technological optimism at the utilities and the reactor manufacturers, the fierce rivalry between GE and Westinghouse, and the arrogance of the electric utilities, whose two decades of constant growth had convinced them that they were capable of tackling anything. These factors combined to create the Great Bandwagon Market, which established nuclear power as a serious alternative to coal years—or perhaps even decades—before it would otherwise have happened.

And once nuclear power was established in this way, it didn't matter that the original reasons for building the plants were faulty. The nuclear industry had gained valuable experience with the technology, learning its strengths and weaknesses in ways that weren't possible without building dozens of nuclear plants. And this in turn made possible a second bandwagon market in the early 1970s.

Throughout the 1960s and into the 1970s, demand for electricity in the United States continued to grow at about 7 percent each year. The power had to come from somewhere—if not nuclear plants, then fossil-fuel plants or hydro power. But utilities were finding fewer and fewer places to put dams for new hydroelectric plants, and fossil-fuel plants were facing increasing pressures. Because of growing environmental concerns and the resultant creation of the Environmental Protection Agency in 1970, utilities were being forced to cut back on air pollution. For some, this meant a switch from high-sulfur to more expensive low-sulfur coal, while others converted their coal plants to burn oil and then had to find a source of low-sulfur oil. The increased demand drove up prices for both low-sulfur coal and oil, and the oil embargo of 1973 pushed oil prices up even further. Meanwhile, the price for natural gas was also increasing sharply, as supplies could not keep up with demand. It got so bad that in 1975 the Texas Railroad Commission —which regulated utilities in that state—prohibited utilities building any new power plants that burned natural gas.

Meanwhile, the nuclear industry was assuring its customers that previous problems were under control. It had been surprised once by cost overruns and construction delays, but the industry was sure it wouldn't happen again. Nuclear electricity wouldn't be as cheap as promised in the mid-1960s—instead of $120 per kilowatt of capacity, it might be as much as $200 a kilowatt—but at least no one would have to worry about too-optimistic cost estimates, and with the increased prices of coal and oil, nuclear power was still competitive with fossil fuels. Furthermore, while fossil-fuel costs were expected to keep rising, nuclear energy was expected to become cheaper, as the architect-engineers and other members of the construction teams gained experience in building nuclear plants.

Thus the years 1970 to 1974 saw a second bandwagon market even larger than the first. It reached its peak in 1972 and 1973, when sixty-five reactors were sold. By 1974, the United States had two hundred reactors either operating, under construction, or on order.

But suddenly the bottom dropped out of the market. Provoked by the Arab oil embargo and skyrocketing prices, Americans began to look for ways to conserve energy. The utilities, accustomed to seeing demand for electricity double every ten years, now projected a doubling only every thirty to forty years. Realizing they wouldn't need all the plants they had ordered in the first half of the 1970s, they began to cancel orders, both nuclear and non-nuclear. Other factors also came into play that made nuclear power less and less attractive to utilities: increasing regulatory demands; court challenges and delaying tactics by well-organized antinuclear groups; growing public concerns about nuclear safety, particularly after the Three Mile Island and Chernobyl accidents; and cost overruns that had not, after all, been gotten under control after the first bandwagon market.

Today the United States has just over one hundred operating nuclear plants, all of them ordered before 1975. Those plants provide more than 20 percent of the electricity generated in the country. In all likelihood, very little of this would exist if GE and Westinghouse had not both been convinced that nuclear power would become a valuable technology and had not each been committed to grabbing a big chunk of that market for itself.

four

COMPLEXITY

Things used to be so simple. In the old days, a thousand generations ago or so, human technology wasn't much more complicated than the twigs stripped of leaves that some chimpanzees use to fish in anthills. A large bone for a club, a pointed stick for digging, a sharp rock to scrape animal skins—such were mankind's only tools for most of its history. Even after the appearance of more sophisticated, multipiece devices—the bow and arrow, the potter's wheel, the ox-drawn cart—nothing was difficult to understand or decipher. The logic of a tool was clear upon inspection, or perhaps after a little experimentation.

No longer. No single person can comprehend the entire workings of, say, a Boeing 747. Not its pilot, not its maintenance chief, not any of the thousands of engineers who worked upon its design. The aircraft contains six million individual parts assembled into hundreds of components and systems, each with a role to play in getting the 165-ton behemoth from Singapore to San Francisco or Sidney to Saskatoon. There are structural components such as the wings and the six sections that are joined together to form the fuselage. There are the four 21,000-horsepower Pratt & Whitney engines. The landing gear. The radar and navigation systems. The instrumentation and controls. The maintenance computers. The fire-fighting system. The emergency oxygen in case the cabin loses pressure. Understanding how and why just one subassembly works demands years of study, and even so, the comprehension never seems as palpable, as tangible, as real as the feel for flight one gets by building a few hundred paper airplanes and launching them across the schoolyard.

Such complexity makes modern technology fundamentally different from anything that has gone before. Large, complex systems such as commercial airliners and nuclear power plants require large, complex organizations for their design, construction, and operation. This opens up the technology to a variety of social and organizational influences, such as the business factors described in chapter 3. More importantly, complex systems are not completely predictable. No one knows exactly how a given design will work until it has been built and tested—and the greater the complexity, the more testing it needs. It is particularly difficult to anticipate all the different ways that something may go wrong. As anyone who has bought an early version of a computer program knows, a few bugs always slip through the design and testing phases. And while this is only an inconvenience when the product is software for a personal computer, it can be disastrous when it's a rocket boosting a space shuttle with seven crew members into orbit or a genetic engineering experiment that accidentally creates a new and dangerous organism. In truly complex systems, no amount of testing or experience will ever uncover all the possibilities, so decisions about risky technologies become a matter of how much uncertainty one is willing to put up with and how much one trusts the designers.

The complexity that marks modern technology is not merely a matter of how many pieces are part of the system. Indeed, that is not even the most important factor. Instead, the defining feature of a complex system is how its parts interact. If the parts interact in a simple, linear way, even a system with many parts is not very complex. Consider a Rube Goldberg device: Flipping a switch turns on a hair dryer, which melts a block of ice sitting in a bowl, the resulting water floats a toy boat, which rises up and knocks loose a marble, which rolls down a chute and sets off a mousetrap, which does something else, and forty-seven steps later the process ends by turning on a light bulb or opening a door. Is this complex? It's certainly complicated, but it is not complex in the way that scientists and engineers speak of complexity. Each piece in the crazy gadget has a single, well-defined role to play which doesn't vary from one enactment to the next, and the performance of the entire system is the sum of the performances of its parts.

By contrast, the components of a complex system interact with one another, and the actions of any one component may depend upon what others in the system are doing. The more interaction among the components, the more complex the system, and the harder it is to predict what the system will do from knowing how any given component will perform.

Furthermore, a complex technology generally demands a complex organization to develop, build, and operate it, and these complex organizations create yet more difficulties and uncertainty. As we'll see in chapter 8, organizational failures often underlie what at first seem to be failures of a technology.

Why has technology gotten so complex? Part of it is simply the accumulation of additions and enhancements over time. Inventors have always come up with ways to improve a device or to give it an extra capability, and often the price is adding extra pieces. The simple bow becomes the less simple crossbow. The Model T becomes the Ford Taurus. But a deeper reason for the increasing complexity today is that the nature of invention has changed. Throughout most of history, innovation has been a trial-and-error affair. A ship designer would try a different hull shape or a different set of sails and see what happened. A paper maker would experiment with various ingredients in varying proportions in an attempt to get a more durable page. This incremental, experience-driven approach to invention kept progress at a snail's pace.

That changed with the birth of modern science. Equipped with a theoretical understanding of the natural world, inventors devised imaginative new ways to put nature to work. Discoveries about how gases behave led to the invention of the steam engine. Studies of electricity and magnetism opened the door to telephone, radio, and much else. The discovery of nuclear fission resulted in both the atomic bomb and nuclear power. Years of research into semiconductor physics made possible the creation of the transistor. And not only did scientific theory lead to these initial discoveries, but it also guided much of the subsequent improvement in the devices. Knowing how such things as gases or electric currents work makes it possible to come up with complicated designs and predict how they should perform.

And, inspired by the success of the physical sciences, over the past century engineering too has become a highly mathematical, theory-laden discipline. Engineers have a good idea how an electronic circuit or a chemical reactor will perform before it leaves the drawing board.

More than anything else, it is this theoretical understanding of scientific and engineering principles that both enables and causes technological complexity. It would be impossible to develop a plane with 6 million parts by trial and error. There aren't enough people in the world to perform all the trials, nor enough time left before the sun dies. At the same time, no one would ever attempt to design a plane with 6 million parts unless there was some a priori reason to think it would fly. The complexity of a Boeing 747 or any other modern technology arises because scientific and engineering theory guides the design effort, indicating what should be possible and how to achieve it.

It is possible, but difficult, for a layman to peek in on that design process and watch complexity appear. Modern devices are so complicated and the theories on which they are based so highly evolved that nonspecialists have trouble understanding what's going on and why. Still, it's useful to have a sense of just how complexity arises in the natural course of developing a technology. To that end, let us step back a few hundred years to a time when modern science was still pretty new and when it was still possible for a single person to conceive and develop one of the world's most sophisticated devices.

COMPLEXIFYING STEAM

Credit for the first practical steam engine goes not to James Watt, whose well known work on that machine came several decades after its invention, but to Thomas Newcomen, a man not nearly as famous as he should be. Newcomen was an ironmonger—a seller of iron and iron products—in Dartmouth, twenty miles from Plymouth in the southeastern corner of England. He was born in 1663 and died in 1729, but aside from these dates and a few other basic facts about his life, little is known of him. No image of him survives, and his burial spot is unknown. We have no definite details about his education or training, although historians believe he was apprenticed to an ironmonger in Exeter for a while before returning to Dartmouth to start up his own business. There he sold metal goods and hardware, and likely practiced blacksmithing to produce some of what he sold. And at some point during this obscure career, probably when he was in his early forties, Newcomen invented the engine that would do more than any other device to set England on the path to the Industrial Revolution.

Newcomen lived at a time when coal was in growing demand. Centuries of cutting trees both for fuel and for the charcoal used in ironmaking had devastated forests in England and the Continent, creating a serious shortage of wood. Coal was a natural substitute, and England had plenty of it, yet extracting it was made difficult by the water that collected in the mines. If a mine was close to a suitable river or stream, then a water wheel could supply the power to pump out the mine. Otherwise the choice was animal power or man power—neither of which was cheap or particularly reliable. Newcomen's contribution was to devise a (relatively) simple machine that could drain the mines of water, that didn't need to be near a river, and that would work longer and harder than any animal or man could. It would open up England's coal mines and, more importantly, give the world an entirely new source of power.

Newcomen's engine consisted mainly of a large, vertical piston and a beam that rocked back and forth on a central support like a giant seesaw. The piston sat several feet below one end of the beam, attached to it by a chain. Each time the piston moved downward, it would pull down on that end of the rocking beam, forcing the other end up. The opposite end was attached to a suction pump, similar to the hand-operated pumps you still see on some old water wells, and each downstroke of the piston would bring gallons of water gushing up through a pipe from the mine below.

The piston was the heart of the engine—and Newcomen's contribution to modern technology. A foot or more across, it moved up and down inside a hollow cylinder that was closed at the bottom. The cylinder was set above a steam boiler and joined to the boiler by a valve. As the piston rose in the cylinder, pulled upward by a counterweight attached to the other end of the beam, the valve was opened so that steam filled the expanding space beneath the piston. Once the piston reached the top of its stroke, the valve was closed and a jet of cold water was injected into the cylinder. The cold water condensed the steam and created a vacuum in the cylinder, causing the piston to shoot back down into the cylinder with great force. When the piston reached the bottom of the stroke, the cycle would begin anew. Only the downward stroke, with the piston being sucked to the bottom of the cylinder, provided any power. The upward stroke was merely a way to get the piston back into place for the next downstroke.

Newcomen's first successful engine, built in 1712 at Dudley Castle in Staffordshire, could pump ten gallons per stroke at twelve strokes per minute. Later ones were larger and pumped more.

Although Newcomen's machine is called a steam engine, no work was actually done by steam. On each downstroke, as the steam condensed and the piston was "pulled" downward by the vacuum, the real force at work was the weight of the atmosphere above the piston. At sea level, atmospheric pressure is 14.7 pounds per square inch, so a 12-inch-diameter piston has more than 1,600 pounds of atmosphere pushing down on it. Normally there would be an equivalent amount of atmospheric force pushing up on the piston from below, but in Newcomen's engine the cylinder below the piston was evacuated on each downstroke. The role of the steam was simply to replace the air under the piston with water vapor, which could then be condensed into water to create a vacuum beneath the piston. For that reason, it's more accurate to call Newcomen's machine an "atmospheric engine." But because it led the way to other machines, slightly modified, that did use steam as a motive force, historians of technology generally refer to it as the first steam engine.

Newcomen's invention was the result of a concerted, decade-long effort to put recently discovered scientific principles to work. ("Recent," that is, by the standards of the time—within the past generation or two. "Recent" today is more likely to mean the past year or two, if not the past few months.) In 1643 the Italian physicist Evangelista Torricelli had shown that at sea level the atmosphere exerts a pressure equal to that of a column of mercury 30 inches high. Later, the English physicist Robert Boyle had performed a series of experiments with gases, showing that their pressure varies directly with their temperature and indirectly with their volume. At around the same time, Otto von Guericke was demonstrating the power of a vacuum by using an air pump to remove some of the air from a cylinder. When Newcomen went to work on his engine, probably shortly after 1700, he had all the scientific information he needed to make it work. His genius was in designing a machine that could take advantage of atmospheric pressure by creating a vacuum under the piston quickly and repeatedly. Others had condensed steam to create a vacuum, but never in a way that produced useful power.

In the decades after Newcomen's initial success, his engine spread across England. It was used mostly in mines, but sometimes also to lift water for water wheels (for a mill, say) or for public water systems. Europe was not far behind England in putting the engine to work, and the first Newcomen engine in North America began operating at a New Jersey copper mine in 1755. After Newcomen's patent expired in 1733, anyone could build the engines, and a number of manufacturers did get into the business, adding a variety of improvements based on what had become a large amount of operating experience. Still, the design remained basically unchanged until James Watt reworked the engine starting in the 1760s. Even after that, the earlier Newcomen machines—less efficient but also less expensive than Watt's—stayed popular. Many of them were working well into the nineteenth century, and one at Sheffield was still pumping out a coal mine in 1923, a hundred years after it had gone into service.

As useful as Newcomen's engines were, they had two major limitations. They were very wasteful of fuel, which wasn't a particular hardship when they were operated in coal mines, but which made them costly to use elsewhere. And, because of their design, their only real use was in pumping water. The single-stroke action of pulling down on a beam over and over again was fine for pumping, but it could not be readily adapted to other applications, such as turning a mill wheel. Watt's contribution would be to rework the steam engine into a machine that could power the Industrial Revolution.

Watt, a Scotsman, had been trained as a maker of scientific instruments. After a year's apprenticeship in London, he found a job as an instrument maker at the University of Glasgow, not far from his home town of Greenock. It was there he first encountered a Newcomen engine. Sometime in 1763 or 1764, when he was twenty-seven and had been at the university for seven years, Watt was asked to repair a working model of a Newcomen engine used in the natural philosophy class. He quickly figured out how it worked, and, as often happens when someone approaches a problem fresh, he also spotted a major design flaw of the machine—one that no one else had recognized or, at least, that no one had ever fixed. It is not unbelievable that Watt was the first to notice, because catching the flaw required not just a mechanical understanding of how the engine worked but also an appreciation of the laws of thermodynamics and how they applied to the machine.

When the steam flowed in under the rising piston, Watt realized, all but a fraction of it condensed immediately because the surrounding cylinder—having just been cooled by a jet of water—was at a relatively low temperature. This meant that several times as much steam was used—and several times as much fuel was consumed—as was theoretically sufficient to fill the piston on each stroke. In essence, a great deal of energy was wasted reheating the cylinder each time a new burst of steam came in. Ideally, Watt recognized, for maximum fuel efficiency the temperature of the piston should be kept at a constant 212 degrees. This conflicted, however, with another requirement of the machine: in order to condense steam effectively, the piston needed to be at a relatively low temperature, perhaps 50 or 60 degrees. This contradiction hobbled Newcomen's invention.

In May 1765, while out for a walk on a Sunday afternoon, Watt suddenly realized how to fix the problem: build a machine with a condensing chamber separate from the cylinder and keep the two at different temperatures. He quickly came up with a way to do this. The cylinder would be isolated and kept hot by circulating steam around it, while off to one side of the cylinder, attached to it by a valve, would be a condensing chamber kept cool by flowing water. With each stroke, as steam was filling the cylinder and the piston was moving up, an air pump would be creating a vacuum in the condensing chamber. Then, when the piston reached its full extent, the valve between the cylinder and the condenser would open. The steam would rush out of the cylinder into the vacuum of the condenser, where it would be condensed into water and drawn away. This would create a vacuum beneath the piston, just as with the Newcomen engine, but with much less waste of steam.

By making the steam engine a bit more complicated and expensive to build, Watt could increase its efficiency considerably. "I have now made an engine that shall not waste a particle of steam," he told one friend in the first flush of excitement after his brainstorm. What he didn't realize was just how difficult it would be to modify the Newcomen engine in this one, seemingly simple way.

Watt applied for a patent on his separate condenser in 1765, and it was granted in 1769, but it was not until 1776—eleven years after the patent application—that he actually sold and built a working steam engine using the new design. Part of the delay had nothing to do with the machine itself. Watt had to support himself during these years, which limited the time he could devote to developing his invention. And his original partner in the undertaking, John Roebuck, could not provide all the resources Watt needed. It wasn't until Roebuck went bankrupt in 1773 and his interest in Watt's invention was passed on to the industrialist Matthew Boulton that Watt was given free rein to perfect the steam engine.

Nonetheless, it was engineering details that took up much of the decade of development. Watt worked and reworked the condensing chamber, for instance. He originally planned a single large cell flooded with cool water, then later switched back to a cooling jet of water as in the Newcomen engine, and still later designed a condenser made of a series of narrow tubes with water circulating around them.

The piston and its cylinder proved even trickier to perfect. To keep both at 212 degrees, Watt had to circulate steam not only around the cylinder's walls but also in the part of the cylinder that lay behind the piston, and this complicated things considerably. First, Watt had to design a "stuffing box" for the piston rod to pass through so that none of the steam heating the back of the piston would escape. Second, and more challenging, Watt had to make sure that the steam on the back of the piston didn't leak around the piston and into the cylinder, spoiling the vacuum on that side of the piston. In the Newcomen engine, the problem had been solved by keeping a thin layer of water between the piston and the sides of the cylinder, which kept air from leaking around the piston and into the cylinder. Watt, however, had to find another way, since spraying water onto the piston and cylinder would cool them off and defeat the purpose of the separate condenser. For years he would try one material and then another. He experimented with wood and various metals for the piston, and, for the packing, woolen cloth, felt, "English pasteboard made of old ropes instead of paper," even paper pulp and horse muck. Nothing worked. In the end, it would take an improvement in cylinder-making technique. A new boring method invented by John Wilkinson, a friend of Boulton's, produced cylinders that were truly cylindrical their entire length, without the small imperfections that marked those made with other processes. Using these perfectly shaped cylinders, Watt found that the packing demanded less flexibility and that lead backed with oakum worked nicely.

Watt and Boulton put the first two of their engines in service around March 1776. Neither functioned smoothly, and both needed constant attention and repair, but they did work as advertised, providing the same power as a Newcomen engine with a fraction of the fuel. And, more importantly, they offered Watt a chance to learn where the problems with his invention lay. His next machine would be that much better for the experience, and the next one even more so. Meanwhile, Watt was coming up with a series of design innovations that would revolutionize the steam engine and make it valuable to a much larger market. For this, Watt would come to be seen as the father of the steam engine.

At first, the Watt engine was used just as Newcomen's was—to pump water—but there was a growing need for power for other applications. The English Industrial Revolution was under way, driven mostly by water. Mills and factories were set up along rivers and streams, the only places they could get the necessary power. Once Watt had developed his fuel-efficient steam engine, Boulton encouraged him to adapt it to turn axles like a water wheel. "The people in London, Manchester and Birmingham are steam mill mad," he wrote to Watt in 1781. "I don't mean to hurry you, but I think in the course of a month or two, we should determine to take out a patent for certain methods of producing rotative motion from . . . the fire engine." In response, Watt came up with several ways to use the up-and-down motion of the piston to turn a wheel on an axle, and suddenly he had a whole new market for his machine. Flour mills, iron works, cotton spinning, and more began to rely on the power of Watt's steam engine. Later, steam-powered ships, railroads, and even automobiles would appear, all taking advantage of the newfound ability to turn a wheel with coal power.

Watt's innovations didn't stop with the separate condenser and a method for transforming the stroke of a piston into rotation. Since the piston was now completely enclosed, the power to move it could no longer come from the pressure of the atmosphere on one side of the piston pushing against a vacuum on the other. Instead, Watt replaced the atmosphere with steam at atmospheric pressure, and the result was the same. But now his machine was a true steam engine—the push on the piston was provided by steam instead of the atmosphere—and this opened up an entire series of improvements that had not been possible with Newcomen's engine. For example, Watt realized that since his engine drew steam into the cylinder on both sides of the piston, there was no reason to put a vacuum on just one side. It made more sense to create a vacuum alternately on one side, then the other, so that both the upstroke and the downstroke were power strokes. The piston would both push and pull, doubling the power on each stroke.

That improvement forced Watt to rework the rocking beam, which transformed the downward pull of the steam engine into an upward pull on a water pump. In his early machines, the beam was usually a simple oak log. Since it had basically the same function in Watt's engine as it had in the Newcomen engine, people knew from decades of experience what sort of beam to choose for what sort of job. But the invention of the double-acting engine changed everything. Now the engine was not just pulling down on the beam on its downstroke but was also pushing up on the beam on the upstroke. The repeated reversals of stress on the log put much more strain on the wood and made it much more likely to break. In response, Watt used various methods of supporting and adding strength to the beam, and he tested the different designs by hanging heavy weights from them. Ultimately the solution came in replacing wooden beams with cast-iron ones.

In the double-acting engine, Watt was using the pressure of steam, not of the atmosphere, to move the piston, and this made it possible to push on the piston with more than the normal atmospheric pressure of 14.7 pounds per square inch. All one had to do was increase the pressure of the steam. Watt himself never built engines with steam pressures much greater than atmospheric pressure, fearing the danger of boiler explosions if he were to use high-pressure steam, but his successors had no such qualms.

Throughout the nineteenth century, engine pressures rose steadily, until they reached 180-200 pounds per square inch by 1900. These higher pressures not only allowed more power from a smaller engine, but they also simplified the design somewhat. Since the push came from the pressure of the steam on the piston, there was no longer any need for a vacuum on the other side, and so the engines no longer needed a condenser. It was the relative compactness of these high-pressure engines that allowed them to be used for railroads and, later, automobiles.

Another of Watt's major innovations was a governor that would prevent a steam engine from running too fast. It contained two heavy balls attached to a rod mounted on a vertical shaft. At rest, the rod and balls hung down, but as the shaft spun around, the assembly rose up, moving higher as the shaft went faster. At a certain speed, the rod holding the balls would push on a butterfly valve in the pipe carrying steam to the engine, closing the valve somewhat and slowing the engine down.

By the time Watt retired in 1800, the steam engine had gone from a simple and inefficient but workable method of pumping water, to a powerful, efficient, but increasingly complex machine that could power a variety of tasks. The new engines were more expensive to build, but they made up for that by costing less to operate. (Indeed, Watt initially provided his engines for a fee that amounted to one-third the cost savings in coal compared to a Newcomen engine of the same power, paid annually for twenty-five years. This way, even someone who already owned a Newcomen engine would save money from the very start by replacing it with a Watt engine.)

The complexity of the Watt engines increasingly made its presence felt. Besides being more expensive to make, for example, the engine required a much higher level of manufacturing skill than the Newcomen machine. Engineering tolerances were smaller and demands on materials more stringent. Accurate boring of the cylinder was essential, the piston had to fit tightly in the cylinder, the timing of the valves letting steam in and out had to be exquisitely precise, and the entire steam system had to have as few leaks as possible. Furthermore, the proper lubricant proved hard to find. The close fit between piston and cylinder and the high heat of those two pieces strained the lubricating abilities of many common oils, and, worse, the oil in the cylinder tended to be dissolved and carried away by the steam. Watt eventually used mineral oil with lead dissolved in it to keep it from mixing so easily with the steam.

All this meant that the new steam engine demanded much more from the "infrastructure" (although this was not a term that would have been used at the time) than did Newcomen's machine. Each Watt engine was built at the spot where it would operate, with local workers fabricating most of the engine according to plans provided by Watt. Only a few small, critical pieces, such as valves, were made at one of Boulton's factories and shipped out. On site, it was a constant struggle to locate materials of the proper composition and purity, and it was difficult to find local craftsmen whose work met the stringent requirements of the more efficient machine. In one letter to Boulton, Watt complained about the work of a foreman named Dudley who had been assigned to get a large steam engine running in September 1778:

> On Thursday they had attempted to sett ye Engine to work before they had got it ready. I had told them that they had not [enough cold water] for above 7 or 8 strokes pr minute. However as there was a great number of spectators, Dudley thought he would show them some what, and accordingly sett off at the rate of 24 sts pr minute, he soon got all his water boiling hot and then they seemed to be at a loss why the Engine would not go. . . . I also found many air leaks [which] convinced me he had never sought after them, he had neglected to put oakum about the necks of ye regular spindles, and had used no means to prevent over opening of exhaustion regulator.

Eventually, in 1795, Boulton and Watt opened a factory devoted to manufacturing most of the parts for their engines. The engines were then shipped by canal and assembled on site.

Watt also found maintenance and repair to be a problem. A Newcomen machine was a rugged thing, relatively insensitive to changes in operating conditions. Feed it coal and water, and it would run. If its boiler rusted out, a sheet of lead could be hammered over the hole; leaks around the piston could be fixed by replacing the disk of leather used as a seal. But a Watt engine needed more careful handling if it was to retain the efficiency that was its selling point. Air leaks, for instance, had not been a major worry in the older machines but could greatly weaken the newer ones. They had to be plugged whenever they appeared. And the valves, which were greater in number and operated more quickly in the Watt engine, demanded much more careful maintenance.

This pattern of greater demands on everyone—manufacturers, suppliers, operators—only increased over the next century and beyond as inventors continued to make improvements to the steam engine. As with Watt, their principal motivation was to increase its efficiency and thus provide more power with less fuel and less cost. And, as with Watt, an unintended consequence was an increase in the complexity of the machine.

In earlier Watt engines, the steam valve was kept open until the steam pushed the piston all the way to the end of its stroke, but engineers quickly realized that they could improve efficiency by keeping the valve open for only the first part of the stroke. Even after the valve was shut and no more steam entered the cylinder, the steam that was already there would expand and push the piston the rest of the way through the stroke. Watt himself recognized that this expansion of steam was a way to increase the efficiency of his engines, and he even tried the idea on one experimental machine, but he never pursued it because there was little improvement in performance when working with steam at normal pressure. At higher pressures, however, this trick can greatly boost efficiency, so after 1800 there was a steady increase in the pressure—and, accordingly, the efficiency—of the steam engine.

But this way of getting the high-pressure steam to do extra work carries its own inefficiency. As steam expands, its temperature drops, so by the time the piston has reached the end of its stroke, both it and the cylinder have been cooled somewhat by the steam. This creates the same problem—on a smaller scale—that Watt recognized with the Newcomen engine: on each cycle, the incoming steam wastes some of its heat warming the piston and cylinder back up from the final temperature of the previous cycle.

Engineers devised a number of stratagems to get around the problem. They connected several cylinders in a row, with the steam traveling to each in turn, so that the temperature drop in each was only a fraction of what it would be if all the work were done in one cylinder. They installed "steam jackets" around the cylinders and circulated steam through them to keep the cylinders at a constant temperature. They also used the same type of steam jackets to "superheat" steam before it went into the cylinder. Each improvement increased the efficiency of the steam engine—but usually at the cost of increasing its complexity.

This process would probably still be going on had it not been interrupted by a new technological revolution: the invention of the steam turbine. The reciprocating engines of Newcomen, Watt, and their successors all had an unavoidable flaw. They depended on pistons that moved up and down, or back and forth, and thus were constantly reversing their motion. The resultant speeding up and slowing down of the piston with each stroke wasted energy. In contrast, steam turbines use a simple circular motion much like that of a water wheel or a child's pinwheel, which gives them the potential to operate at higher efficiencies than the steam engine.

Since the 1880s, when the first practical versions appeared, steam turbines have kept the same basic design. Inside a closed cylinder, one or more wheels revolve on a central shaft. On the outer rim of each wheel is a series of closely spaced vanes or shallow cups that extend around the entire diameter. A high-speed jet of steam is sprayed at the outer edges of the wheels, and as this steam passes through the vanes, it pushes on them and turns the wheels and the central shaft. The ultimate effect is very similar to the way that a gust of wind turns a toy pinwheel, except that all the action takes place in a tightly enclosed space so that the energy tapped from the steam is maximized.

Because turbines avoid the problems of friction, leaks, and direct mechanical contact that plagued the reciprocating steam engines of

Newcomen, Watt, and their successors, the turbines can exploit steam of much higher temperatures and pressures. This in turn has allowed turbines to achieve much higher efficiencies than were possible with the piston machines. (Turbines are also much quieter because they don't have the intense vibrations generated by the back-and-forth movements in the piston engines.) Within a few decades of their development, turbines had taken over most applications of steam engines. They propelled ships such as the *Titanic* and the British battleship *Dreadnought,* they powered locomotives, and they produced electricity. Today's nuclear and fossil-fuel electrical generating plants use giant turbines that operate at pressures of several thousand pounds per square inch and put out hundreds of thousands of horsepower—a far cry from the atmospheric-pressure, 50-horsepower Watt engines that were state-of-the-art two centuries ago. But of course, this power and efficiency has come at a price. Like the steam engine before it, the steam turbine has gradually become more complicated, more difficult to build, and more sensitive to flaws and imperfections.

THE COSTS OF COMPLEXITY

Complexity is not, in itself, a bad thing. Granted, engineers prefer simple designs to complex ones when the two will do the same job equally well—after all, simple designs are generally less expensive to make and easier to repair and maintain. But if a device should not be more complex than it needs to be, neither should it be simpler. By adding extra pieces, engineers are able to create machines that do more things and do them better than the less complex alternatives. The challenge then becomes deciding how far to go. When do the costs of the complexity outweigh its benefits?

It is easy enough to calculate part of the cost. From long experience, engineers know approximately how much a new component will add to a price tag and how likely the component is to break down. But adding extra pieces takes another toll, one that is more subtle and difficult to measure. Complexity creates unpredictability. The more complex a system, the more difficult it is to understand all the different ways the system may behave—and, in particular, to anticipate all the different ways it may fail. Interdependence among parts creates entirely new ways that things can go wrong, ways that engineers often overlook or ignore. Thus many technological failures chalked up to mechanical breakdown or design flaws are more accurately described as the children of complexity. Consider, for example, the car.

Automobiles are a mature technology. The auto industry is 100 years old, and much of automotive engineering is a matter of incremental modifications to existing, well-tested designs. Yet it seems nearly impossible to bring out a new line of cars without some flaw in it.

If you were unlucky enough to buy an early 1994 Chrysler Neon, your car may have been recalled as many as three times in its first few months of operation—even though Chrysler had been boasting of the strict quality controls on its new line. There was a difficulty with water leaking onto some computer chips and causing starting problems. There was a snag with the seals on the master cylinder of the antilock brakes allowing air into the brakes and hurting their stopping power. And there was a stumble with the car's brake brackets. Understandably, Neon owners were not happy. But neither were they alone. General Motors' Quad 4 engine, used on many of the company's cars, was recalled four times between 1987 and 1993. Honda has a reputation for quality, but in 1993 the company called back 1.8 million of its Accords and Preludes to repair a problem with the fuel filler pipe that could lead to gasoline leaks or fires. All told, auto manufacturers recalled a total of 11 million vehicles in the United States in 1993, with some companies fixing more cars than they made.

Why is it so hard to get everything right? Clearly, part of the answer is that the technology is constantly changing as the manufacturers try to make their cars better than, or at least different from, last year's models. If automakers were to freeze their designs, a few years' experience with millions of customers would allow them to iron out almost all the bugs and produce cars with few or no unforeseen problems. There would still be the expected problems, of course, such as brakes needing repair and the inevitable oil leaks that appear in older engines, but such wear and tear is a normal part of operating any machine. It can be predicted and, for the most part, handled with regular inspections and maintenance. But why should those few years of experience be necessary in the first place? Can't the carmakers simply design and build a car without bugs?

From the accounts in the press, it would seem the answer should be yes. Each time a recall is announced, the flaw is described as the result of carelessness or undue haste, with the implication that if everyone would just be a little more careful, none of this would happen. GM's problems with the Quad 4 engine, for instance, were blamed on rushing it into production without enough testing. In the case of the seals on the Neon's brake cylinders, the culprit was the rubber compound used in the seals. The company that supplied the compound had changed to a new formulation without verifying that it would work as well as the original.

Perhaps in a perfect world, where no one made mistakes and carmakers had unlimited amounts of time to spend on design and testing, there would be no recalls. And it's certainly true that the companies that put more time and effort into their cars generally have fewer problems. But to blame the flaws solely on human failings is to ignore the role that complexity plays. A car has thousands of parts, and each part must be designed and tested independently. The more parts, the more things that can fail, so increasing the number of parts increases the odds that something will be overlooked. Furthermore, the parts are interconnected, with the performance of one being affected by the performance of others. Thus a new rubber compound that seems perfectly equivalent to the old stuff in a laboratory may behave quite differently when put to work in a brake, where there are heat and vibrations and brake fluid.

Because of these interconnections, in any system as complex as a car it's virtually impossible to foresee and test for all the ways that something can go wrong. In hindsight, perhaps, it might seem that a car manufacturer should be able to anticipate a water leak affecting a computer chip and to do its best to prevent such leaks. But there are so many possible combinations that can lead to problems—most of which never happen—that it's inevitable a few will be missed. The best an automaker can do is to test, test, and retest, driving the cars as much as possible under as many situations as possible, in hopes that the bugs will surface during the tests instead of after the car is on the market. In that sense, perhaps, auto recalls can always be blamed on insufficient testing, but it's more accurate—and ultimately more useful—to see them as a manifestation of the machine's complexity.

A similar case can be made that many airliner crashes are due at least partly to the complexity of the machine. Take the case of the 1974 crash of a Turkish Airlines DC-10 which killed 346 people. The plane had taken off from Paris and reached an altitude of 12,000 feet when its cargo doors flew open. The sudden shift in pressure caused the floor above the cargo hold to collapse, breaking the hydraulic control lines for the rear engine and rear control surfaces, which ran along the floor between the passenger compartment and the cargo hold. The plane plunged groundward and crashed in a forest. There were no survivors.

In the most literal sense, the crash was caused by the unfortunate union of two design flaws: the decision to run the hydraulic lines above the cargo hold, and poorly engineered latches on the cargo doors. The latches were designed so that it was possible for the baggage handlers to believe they had locked the doors even when the lock pins were not fully engaged. Apparently the ground crew for the Turkish Airlines plane made that fatal mistake, and when the difference in pressure between the cargo hold and the outside atmosphere became too great, the partially engaged lock pins could not hold the doors closed any longer.

On a deeper level, however, it was the plane's complexity that lay at the root of the disaster. That complexity had prevented the plane's designers from seeing that a number of seemingly unrelated decisions could interact to produce the tragedy. One such decision was putting the hydraulic lines above the cargo hold. Another was the choice of latch design. A third was the choice not to reinforce the floor between the hold and the passenger compartment. A stronger floor could have withstood the sudden depressurization, but it would have also added to the weight of the plane. And a fourth was designing the cargo doors to open outward. This was done for both structural reasons and for convenience in loading and unloading. But if the doors had opened inward, they could not have burst open when the pressure inside was greater than the pressure outside. None of these design decisions was, by itself, an obvious mistake. Only when they came together in a way the designers never anticipated did they cause an accident. One of the hallmarks of a complex system is that events can combine in unexpected ways and make the entire system behave unpredictably.

Ironically, by the time of the DC-10 crash, the accident was completely predictable. Four years previously, a pressure test of the first DC-10 fuselage caused a cargo door to fly open and the floor to collapse. Not wanting to completely redesign the door, the plane's manufacturer, McDonnell Douglas, tried a series of "Band-Aid" fixes—minor alterations that did little or nothing to make the system safe. Then, two years before the Turkish Airlines crash, a nearly identical accident took place on an American Airlines flight from Detroit. At 12,000 feet, a cargo door blew out and caused the floor to buckle, taking out the hydraulic controls for the rear of the plane. But because the plane was lightly loaded, the pilot was able to maintain control and bring it to a safe landing. By chance, the pilot had trained himself on a flight simulator to handle just such a situation, flying a plane without being able to use the rear engine and control surfaces, and he knew exactly what to do when the accident hit.

Even after the 1972 incident, McDonnell Douglas resisted major changes to the aircraft's design. Instead, with the approval of the Federal Aviation Administration, the company applied more Band-Aids, such as a 1-inch peephole through which it was possible to examine the lock pins. That wasn't enough.

It's possible to blame the whole accident on corporate greed and indifference to human safety—and, indeed, some observers have done just that—but the complexity of the DC-10 certainly played a key role. It led the plane's engineers to settle on a design that allowed such an accident to occur.

Nowhere are the difficulties created by complexity more obvious than in the bugs that infest computer programs. There are no physical parts that can fail unexpectedly in software. Any errors occur because the programmers who wrote the software made a mistake. And because of the incredible complexity of many of today's computer programs, mistakes appear with troubling regularity. One famous mistake appeared some years ago in the avionics software written to control the F-16 jet fighter: when the plane crossed the equator, the software commanded it to flip upside down. Fortunately, the bug was discovered during flight simulation instead of during a real mission.

AT&T wasn't so lucky with a software bug that was written into a program for the computers that control its long-distance telephone network. In January 1990, the bug triggered a series of shutdowns of AT&T's switching computers which crippled the long-distance system for nine hours. During that time, only about half the 100 million phone calls placed with AT&T got through.

The details of that shutdown are somewhat complicated, but are worth understanding. The switches that AT&T and other telephone companies use to route telephone calls are actually large computers designed specifically for the task and controlled by switching software. The system is fantastically complex. In early 1990, AT&T relied on 114 switching centers, each with a main computer and backup computer, to dispatch long-distance calls around the country. Many of the centers used the state-of-the-art Signaling System 7, whose software contained up to ten million lines of code—a hundred times as many as a business word-processing program might have. Because its developers knew that the software could not be tested completely and because the results of a breakdown in the switching computers can be very costly—both to AT&T and to businesses left without long-distance service—the software was designed to be tolerant of faults. If one switch failed, there was always a backup to step in, and the system was designed to route calls around a malfunctioning center if necessary. If one center went on the blink, the other 113 should keep working, and AT&T customers would never know the difference. Ironically, it was this fault-tolerance structure that set the stage for the cascading series of failures that paralyzed much of the system for a large part of the business day.

Just before 2:25 P.M. on January 15, one of the switching centers in New York City went down. It was a minor mechanical problem, the sort of thing the software was designed to handle, and, as planned, the switching center took itself out of service, sending out messages to the other centers that they should not send any calls to it until the problem was solved. The switching center's maintenance software fixed the problem within six seconds and the center started sending calls back out to the other centers, signaling that it was rejoining the loop. Upon receiving these calls, the other centers began resetting their programs to once again route calls to the New York center. But then the unexpected hit. Something about the pattern of the calls arriving from the New York center set off a flaw in the software, disrupting data at some of the other centers. These centers responded to this glitch by taking themselves temporarily off-line while the problem was fixed, and then coming back on-line. But when they came back on-line, the pattern repeated itself. Soon, all across the country, switching centers were taking themselves out of the loop and then coming back in, with their reentrances causing other centers to fail. It was six and a half hours before AT&T scientists finally figured out how to fix the problem, and it was nearly midnight before the last switching center was put back into working order.

The postmortem revealed a tiny flaw—a single error in one line of the program that controlled the switching computers. The flaw had been introduced the month before, when AT&T programmers had modified the switching software in order to cure what had been an annoying but minor problem in the telephone network. Over the previous two years, the company had been replacing its Signaling System 6, in use since 1976, with the Signaling System 7, a faster computer system that could handle twice as many calls as its predecessor. By the end of 1989, the new version had been installed in 70 percent of the network, and a glitch in its software had become apparent. Sometimes when a switching center went off the network to fix a problem, the center lost a few calls as it came back on-line. So in December 1989, AT&T engineers modified the software slightly to fix this glitch. But in doing so they introduced another problem into the software, one that was not immediately apparent. The program worked fine for several weeks, but then on January 15, the right sequence of events triggered the day of chaos.

Even after the accident, it took several days for AT&T engineers to figure out exactly what had gone wrong. They tried various things until they were able to make a test computer fail in the same way that the phone system had, and then, knowing what events triggered the failure, they were able to find the flaw in the software.

The accident could be blamed on carelessness by the programmers, but it's more accurate to point to the (necessary) complexity of the software. It's possible to go through the program line by line and make sure that each individual instruction makes sense, but it is not possible to guarantee that the program as a whole has no flaws. Even a program with only a few hundred lines of code may have dozens of lines that are "conditional"—their actions vary according to the circumstance. If X=0, do this; if X=1, do that. Such a simple program can have thousands of alternate paths of computation, and which one it follows depends on the initial input. With millions of lines of code, or even just thousands, it is simply not possible to check every path the program might take to make sure that each performs as planned. The best that the programmers can do is to be careful in the preparation of the program and then to test the software by applying it in ways that users seem likely to.

This is a problem not just in telephone switching software but in every sort of large computer program. When a commercial program initially comes to market, its first users are essentially guinea pigs. A number of them will discover—sometimes painfully—that the software doesn't perform exactly right in every situation. It freezes up, often wiping out whatever data had not been saved, or it refuses to perform a certain function. The users complain to the company that sold the software, and its programmers go back and figure out what went wrong. With enough experience, they can eventually offer a product with almost no bugs.

This way of catching mistakes is unacceptable for certain types of software, however—computer programs used in a nuclear plant, for instance, or software that controls the launch of ballistic missiles. In those cases, even one bug can be too many. Computer scientists don't have good answers for this dilemma. One approach is to check the software extensively before it is put to actual use. When the Canadian electric utility Ontario Hydro decided to use computers to run the emergency shutdown systems for the reactors at its Darlington nuclear plant, Canadian regulators forced the utility to prove that the computers would work as planned—a software-verification process that took nearly three years. The software engineers tested the program in a number of ways, including trials that simulated expected circumstances in the plant, trials that fed random data into the computers, and direct mathematical proofs that the software did what it was supposed to. After many man-years of intense work, the engineers found no major flaws in the software.

No matter how closely one checks, however, the possibility remains in such complex programs that there are hidden errors that could lead to a breakdown. The best approach may be to design the overall system so that a failure in the software will not be disastrous—for instance, by having a separate set of safety devices to protect the system in case of a software crash.

The same sort of thinking can be applied to physical systems where complexity creates uncertainty—commercial aircraft, the space shuttle, nuclear power plants, chemical plants, and the like. Because of the uncertainty, the designers usually provide a margin of error by including safety features, backups, or redundancies. But protection costs money—for extra equipment, for engineering design work and testing, for losses in efficiency, and so on—and the makers of complex devices must draw the line somewhere. Unfortunately, the same complexity that necessitates the margin of error also makes it impossible to know how big the margin should be. So, ultimately, the decision must come down to educated guesses and subjective opinions about how conservative one should be. And sometimes, as the crashes of the AT&T system and the DC-10 show, people guess wrong.

It seems to be virtually a law of nature: If people are going to misjudge the effects of complexity, they will underestimate rather than overestimate them. Perhaps the most spectacular example is the 1974 Rasmussen report, which attempted to gauge the likelihood of a major accident at a nuclear power plant. As is described in chapter 6, the Atomic Energy Commission hired MIT nuclear engineering professor Norman Rasmussen to prepare the study in the hopes that it would calm rising fears about nuclear safety. And, indeed, the work was quite encouraging about the safety of nuclear plants. As the AEC presented it to the public, a person was no more likely to die from an accident at a nuclear plant than from being hit by a meteor, and a core meltdown could be expected at a nuclear plant only once in every million years of operation. Less than five years later, the reactor core at Unit 2 of Three Mile Island melted down. Postaccident analyses blamed the accident on a combination of factors, each of which was relatively minor by itself but, when added together in an unexpected way, led to the partial meltdown. Ironically, Rasmussen had identified just this sort of chain of events as a possible cause of accidents in nuclear plants, but his report had seriously underestimated how likely such accidents were.

Not all the effects of complexity are as dramatic as the meltdown at Three Mile Island or the crash of a long-distance telephone system. Indeed, most of the effects are subtle and low-key. Nonetheless, their cumulative effects are greater—in a financial, if not a psychological, sense—than the occasional spectacular failure.

With every new technological product, whether it is a word-processing program or telephone-switching software, a steam engine or a nuclear power plant, there is a learning curve. The designers, the manufacturers, and the users must all put time into learning about the product's capabilities and quirks. This wasn't a big problem when technological innovation was mostly trial and error. Change came so slowly that learning easily kept up. But today, because engineers can conceive of and design devices unlike anything that has ever been built, people end up using things they don't really understand. That's where the learning comes in. In manufacturing a new device, for instance, the maker must figure out which techniques do the best job at the least cost. The manufacturing process may uncover features of the product's design that looked fine on paper but just don't work in practice. And no matter how careful the designers have been in their estimates, the real manufacturing costs only become clear with experience. Similarly, the users of a new machine discover what the capabilities of the product are in practice, instead of in theory. They gradually learn the best way to operate the new machine to get the best performance out of it. And they, too, find things that the designers will need to rework. Generally, the developers of a new product try to do much of the learning themselves, with computer modeling, laboratory tests, and the building and operating of prototypes. But inevitably there is much that remains to be learned when the product is released.

The commercial jet airliner offers a good example. As Stanford economist Nathan Rosenberg has pointed out, the steady improvement of these machines over the past few decades is a product not just of improved aeronautical design but also of the experience gained from countless hours of flight. For instance, when airlines first began flying jets in the 1950s, they had no way of knowing how often the jet engines would need to be serviced, so they scheduled overhauls at the same intervals that were used for the piston engines—about every 2,000 to 2,500 hours. With experience, the airlines learned they could safely stretch that out to every 8,000 hours or so, saving money and helping make commercial air flight more economical. Not just maintenance requirements but also the aircraft's performance in the air had to be gradually discovered, Rosenberg writes. "With the advent of the jet engine, aircraft moved into a new high-altitude, high-speed environment where they encountered unanticipated aerodynamic effects, including lethal stalling. There was an extensive learning experience before the behavior of the new aircraft in their new environment was reasonably well understood." The airframe, too, was something that could be designed and predicted only up to a point. The aircraft manufacturers were generally quite conservative with their designs initially, but as they gained experience and knowledge about how a given airframe performed, they became willing to make a number of modifications, such as stretching the fuselage.

In practice, the fruits of this learning are incorporated into the body of knowledge that engineers draw upon in future designs. Indeed, despite the general perception of engineering as a calculational science with explicit procedures and mathematical formulas, engineering has historically been mostly experimental—particularly for new technologies. Guided by experience with similar devices, by general principles, and by their own intuition, engineers come up with designs they think will work, but then they must build them and see how close reality comes to expectations. Eventually, the experience they gain is systematized into lists of facts and rules that tell future engineers what works and what doesn't. In the early part of the twentieth century, for instance, there was no body of knowledge that could be called "aeronautical engineering." The early airplane designers were flying by the seats of their pants as they tried different shapes and placements of wings, different engines and propellers, and different types of controls. As the body of knowledge grew and was connected with such theoretical sciences as fluid dynamics, airplane design became a true engineering discipline, but even today it remains as much an art as a science.

Not surprisingly, the more complex the technology, the longer and more extensive the learning process. And, also not surprisingly, people often underestimate just how much learning will be needed as a new technology moves from the drafting table to the commercial market. It can be an expensive mistake.

LEARNING ABOUT NUCLEAR POWER

As we saw in the last chapter, reactor manufacturers and utilities lost billions of dollars in the 1960s and 1970s by trusting optimistic predictions of how much nuclear plants would cost to build and run. The first victims were GE and Westinghouse, whose turnkey projects lost them a combined $1 billion. They were followed by dozens of utilities which typically found their nuclear plants costing two or three times as much as expected. Study after study has looked at what happened and why, and a variety of culprits have been fingered: incompetent utility managers, regulators who kept changing the rules, greedy corporations that pushed the commercialization of nuclear power before it was ready. No doubt all of these played a part, but the root cause was deeper and more impersonal. No one in the nuclear industry—not the reactor manufacturers, not the architect/engineers who designed the plants, not the construction companies that built them, and not the utilities—realized just how much learning this new and complex technology demanded. There was a great deal of essential knowledge that could only be gained by experience—by actually building nuclear plants and seeing what went wrong. But, blinded by technological hubris, the industry didn't think it had much to learn.

Rickover, who oversaw the construction of more than a hundred reactors, probably understood the importance of practical experience better than anyone else. In a paper he wrote for a technical journal, he described the difference between "paper reactors"—those that existed only as engineering plans—and real reactors. A paper reactor, he wrote, generally has the following features:

- It is simple.
- It is small.
- It is cheap.
- It is lightweight.
- It can be built very quickly.
- Very little development is required; it will use off-the-shelf components.
- It is in the study phase; it is not being built now.

A real reactor, he wrote, is somewhat different:

- It is complicated.
- It is large.
- It is heavy.
- It is being built now.
- It is behind schedule.

- It requires an immense amount of development on apparently trivial items.
- It takes a long time to build because of its engineering development problems.

In the early days, when nuclear power was still new and mysterious, people took for granted the importance of gaining direct experience with the reactors, and the early AEC programs were careful to allow for this "learning by doing." The commission started with experimental reactors, moved to small demonstration units, then proceeded with a series of successively larger and more complex reactors. At each step the reactor's performance was evaluated, and the AEC made sure there was enough time between steps to apply whatever lessons were learned in the design, engineering, and construction of the reactors.

But by 1963 nuclear power no longer seemed so daunting, and the nuclear industry decided it knew enough to jump to much larger plants without the cautious sequential learning of the previous decade. There would still be learning by doing, of course, but the nuclear establishment believed the learning costs would now be so minor that they would be outweighed by economies of scale. The only important role for learning would be to bring down the cost of nuclear electricity even more, as experience taught companies how to build and operate plants more efficiently. So reactor manufacturers began offering—and utilities began buying—plants two, three, and more times as large as any yet constructed, secure in the belief that nuclear power was a mature technology and that whatever learning took place would merely make it more economical than it already was.

It was, in retrospect, an exceptionally optimistic stance. Centuries of engineering experience had shown that scaling up is never as easy as it sounds. Consider, for example, doubling the size of a modern automobile. With a body twice as long, twice as wide, and twice as high, it would weigh eight times as much. But the frame, whose strength is proportional to the cross-sectional area, would be only four times as sturdy. It would have to be beefed up. The tires, to carry eight times as much weight, would have to be inflated to a much higher pressure, demanding thicker—and less resilient—rubber. This would make the ride much bouncier, so the springs and shock absorbers—already strained by accommodating the extra weight—would have to be modified still more. To stop the car would demand eight times the braking force, but the brake pads and drums would have only four times the surface area. They'd have to be made bigger. And so on.

In any technology, from cars to coffee makers, scaling up is more than a matter of making things bigger. As the size increases, the design must change. If the size increase is small, the changes can be minor. But bigger steps in scale demand increasingly radical redesign. For that reason, the standard engineering practice in scaling up complex devices is to take it bit by bit, testing and redesigning incrementally.

Ironically, in the mid-1960s the electric industry had just finished a refresher course in the importance of not scaling up too quickly. Beginning in the previous decade, in a quest for economies of scale, U.S. utilities had sharply boosted the sizes of their coal- and oil-fired electric plants. The largest units went from less than 250 megawatts in 1957 to about 700 megawatts in the mid-1960s and more than 1,000 megawatts by 1970. But it soon became apparent that the new, large plants were less reliable than the older, smaller ones. As Commonwealth Edison chairman J. Harris Ward explained in a 1967 interview, the bigger the plant, the harder it was to keep running: "We have one coal-fired unit with 800 miles of pipe. If there's one small clot somewhere along those lines, we might have to shut it down for repairs. And it's not easy to get a man down there to fix it."

But despite the experience with fossil-fuel plants, executives like Ward were convinced that nuclear power would somehow be different. In the same 1967 interview, he argued that nuclear plants should be able to avoid the difficulties that had plagued their coal-fired cousins. A nuclear plant, he said, "is more susceptible to technical improvement. After all, coal is coal."

Ward was not alone in his conviction. The people of the nuclear industry were heirs to an impressive string of successes: the Manhattan Project, Rickover's naval reactors program, and a series of increasingly ambitious reactor projects that had begun with Shippingport. This success had bred a confidence, almost a cockiness, among the nuclear brotherhood. With hard work and good engineering they could, they were sure, overcome all obstacles to putting the atom to work.

In one sense they were right. Since its birth on a squash court at the University of Chicago, the nuclear reactor had grown from a primitive pile of graphite bricks and uranium spheres to a sleek, slick machine containing precisely designed arrays of tube-shaped, Zircaloy-clad fuel elements interspersed with control rods and sunk in a pool of cooling water, all of it monitored by sophisticated instruments and controlled to exquisite accuracy. After twenty years and numerous design-build-test cycles, the reactor had become a relatively mature technology, and throughout the 1960s and 1970s nuclear reactors would cause few problems.

But the reactor is only a small part of an entire nuclear plant, and in 1963 the rest of the plant was far from a mature technology. This would become increasingly obvious over the next decade as the reactor manufacturers found themselves in a predicament very similar to the one that James Watt had faced two centuries earlier in building the first pumping stations powered by his new engine.

Suppliers, for instance, were a constant concern. Companies that agreed to provide crucial parts often either were late with delivery or provided materials that were not up to standard. The resulting delays and reworkings stretched out the construction and jacked up the costs of the plants. The worst of these fiascoes was the nondelivery of reactor vessels by Babcock & Wilcox.

B&W, as it was known, was a well-respected company with a long and successful record of making boilers and other large steel vessels. It had catered to the electric utility industry since 1881, when it provided the boilers for one of the first American central-station power plants, in Philadelphia. And when the peaceful atom began to stir in the 1950s, B&W decided that nuclear energy would become as important to the company as electricity had been. Jumping into the atomic race, B&W built the reactors for Consolidated Edison's Indian Point 1 plant and for the nuclear ship *Savannah*, but its executives thought the company could make more money faster in a different niche: it would supply the pressure vessels for reactors that other firms built. The pressure vessel is the large, thick, steel container that surrounds the reactor and holds the coolant—either pressurized water or boiling water, depending on the type of reactor. It is difficult and demanding to manufacture, and the giant pressure vessels for the larger nuclear plants would demand extra care, but B &W had had decades of experience with just this sort of structure. It seemed like a perfect fit for the company.

Anticipating a large market for nuclear power, Babcock & Wilcox built a plant devoted solely to making pressure vessels, and shortly after the plant opened in 1965, the orders started to pour in. Soon B&W had sold its projected output of one pressure vessel a month for several years in advance. But the company fell behind schedule from the start. It had badly underestimated how difficult it would be to get a new plant to quickly produce pressure vessels to the exacting requirements of the AEC, and each unit was taking much longer than expected. Furthermore, its own suppliers were slow in providing the state-of-the-art equipment that B&W needed for its plant. By mid-1969, the company admitted to its customers that every one of the twenty-eight pressure vessels under construction at the plant would be late—some by as much as seventeen months. Eventually, GE and Westinghouse yanked many of the half-finished vessels they had ordered from the plant and shipped them to other firms to finish the job. Those firms, which got an easy entree in the pressure-vessel market, were the only winners. Babcock & Wilcox lost money, market share, and reputation. GE and Westinghouse saw the delivery of their reactors slowed. And the utilities, forced to stretch out the construction schedules for their nuclear plants, had to absorb both the extra interest payments during construction and also the cost of backup energy sources while the nuclear plants were being finished.

On site, the problems for the nuclear plants were as bad or worse. Tradesmen accustomed to building fossil-fuel plants were asked to do things differently for nuclear plants, and they responded no better than did the craftsmen of the 1770s, who thought that if something was good enough for the Newcomen engine, it should be good enough for that new-fangled Watt engine, too. Bertram Wolfe, formerly a vice president and general manager in charge of GE's nuclear energy division, recalls that many of the difficulties GE encountered with their turnkey plants stemmed from "building the plants to nuclear standards." The concrete of the containment vessel, for instance, had to be certified to meet various requirements, such as standing up to certain pressures. In the end, the concrete might not be much different from that poured for a fossil-fuel plant, but the need for constant checking and certification meant the workers had to learn new habits in working with it. Welding was a similar story. The pipes in a nuclear plant faced pressures no higher than those in a coal-fired plant, Wolfe notes, but the consequences of a leak were much greater, so the welds had to be made to much more rigid standards. "We trained the welders in special classes," he remembers. It wasn't that the new ways of doing things were particularly difficult, though. "It was more of a change in attitude and approach."

And indeed, more than anything else, it was the old attitudes and approaches on which the new nuclear industry stumbled. Over the course of several decades the electric utility industry had developed ways of doing things that worked for the fossil-fuel technology—construction methods and operating techniques that were well suited for a plant with coal or oil burners at its heart. When nuclear power arrived, few in the industry saw any reason to change those old habits. Nuclear energy was, in a common phrase, "just another way to boil water." Sure, there was a nuclear reactor in place of the old coal burner, but everything else was the same.

But everything else wasn't the same. Although it took years to sink in, the nuclear industry came to realize that there is a critical difference between the non-nuclear parts of a nuclear plant and the corresponding sections of a fossil-fuel plant: In a nuclear plant, those parts are hooked up to a nuclear reactor. That changes the rules.

If some item in a coal-fired plant fails—even a major item—it's not a disaster. Suppose a steam pipe ruptures. Then the turbine will be shut down and the generator will quit producing electricity. The flow of coal into the burner will be stopped and the burner fire will go out. That's it. That's the extent of the accident. As soon as possible, workers will repair or replace the pipe, and the plant will be restarted. The utility loses money because its plant has stopped producing electricity for a while and because it has to pay for repairs, but this is the only damage.

Suppose, however, that the same steam pipe rupture occurs in a nuclear plant powered by a boiling water reactor. (The situation for a pressurized water reactor is similar.) As before, the turbine and generator halt, but this time the troubles are only starting. Because the steam in a light-water reactor serves two purposes—to drive the turbine and also to carry heat away from the reactor core—the break in the pipe shuts down the cooling of the fuel in the core. This is a serious problem. Automatic devices will have scrammed the reactor, pushing control rods into the core and shutting down the chain reaction, but, unlike the situation with a coal burner, it's not possible to extinguish the "fire" completely. Once a reactor has operated for a while, radioactive byproducts of the fission build up in the fuel, and the decay of these elements provides a second source of heat. In a reactor that has been in operation a month or more, the decay heat shortly after shutdown is almost 10 percent of the total heat the reactor produces at full throttle—and this decay heat continues even after the control rods are in place. If the core is not kept covered with water, the fuel will eventually get so hot that it becomes liquid and melts its way through the floor of the containment vessel and into the ground below. In the worst case, such an accident could kill thousands of people and cause billions of dollars of damage.

To guard against this, nuclear plants are equipped with backup cooling systems to flood the core with water, but even if these systems work as planned, the rupture of a steam pipe is much more serious than it would be in a coal plant. The steam in the pipe, having passed through the core, is itself slightly radioactive, so the rupture releases radioactivity into the plant, threatening the workers. And even after the accident is over, the radioactivity of the pipe and surrounding areas makes the job of repairing the damage much more difficult, dangerous, and expensive.

In short, although it wasn't obvious to most utilities at first, even the seemingly familiar section of a nuclear plant—the part that turned steam into electrical power—was in effect a new technology. It's hard to imagine any accident in a coal-fired plant that could endanger people outside the plant boundaries, but a large-scale nuclear accident threatens people dozens or even hundreds of miles away.

The need to guard against this risk had two effects. First, the AEC held nuclear power plants to much higher standards than coal or oil plants were expected to meet. Nuclear plants had to be designed more carefully, built with higher-quality materials and more exacting workmanship, and operated with greater care and vigilance. This caused few problems with the nuclear sections of a plant, which had always been recognized as demanding particular care, but the utilities and their contractors had to learn—painfully and expensively—to treat the rest of the plant according to nuclear, rather than fossil-fuel, standards. It was remarkably like the situation that Watt had encountered in the 1770s, when he struggled to change the old Newcomen-engine mentality. And, like Watt, the utilities lost time and money in the learning process.

Besides demanding higher standards, the AEC also required that nuclear plants be equipped with an array of backup systems and safety equipment designed to prevent accidents or, at least, to keep an accident from endangering people outside a plant. The safety devices, which were connected to both the reactor and the rest of the plant, added a layer of complexity to an already complex system.

This extra complexity meant that practical experience would be particularly important in learning to design and use nuclear plants with these safety systems, but when the rush to large plants started in 1963, the systems were still in the formative stages. Consider, for example, the emergency core-cooling system. When the first reactors were built, they had no need for such a system—their cores were so small that, even in a loss-of-coolant accident, they could not generate enough heat to melt the fuel and allow it to escape from containment. It was not until the reactors got bigger—more than about 100 megawatts—that core meltdown became a worry. In response, the AEC demanded that the larger reactors have emergency core-cooling systems which would inject water into the reactor core in case anything interrupted the primary supply of cooling water. But these systems, at first little more than a collection of pipes and pumps grafted onto the rest of the plant, had not gone through the same rigorous series of design-build-test cycles that the reactors had. Throughout the 1960s and into the 1970s they would have to be revised constantly as calculations, tests, and experience pointed to flaws or areas that needed improvement.

The story was the same for safety regulations in general. In the early- to mid-1960s, the AEC set reactor safety criteria not by experience but according to the judgment and guesswork of its staff. But as the utilities built and operated reactors and the AEC gathered and analyzed data on their performance, the commission learned that its original assumptions had not been as conservative as believed. In response, it strengthened regulations, forcing utilities to rework their plant designs and change their operating procedures. These constant changes in safety requirements added greatly to the cost of the plants.

The problem was made worse because the AEC was trying to hit a moving target. At one point, its data came from the operations of plants no larger than 200 megawatts while utilities were making plans to build 800- to 1,000-megawatt plants. The result, according to a study by the RAND Corporation, was that "the regulatory division could never 'catch up'; backfitting became an inevitable ingredient of the regulatory program." And, of course, backfitting is much more expensive than doing it right the first time.

Of all the mistakes the U.S. nuclear industry made in the 1960s and

1970s, the single most damaging was the failure to settle on one or a few standard designs for nuclear plants. Standardization maximizes learning. It allows people to learn from others' experience as well as from their own. But the utilities never saw any need for it.

Why not? For one thing, utilities were accustomed to custom-building their fossil-fuel plants, and it had never been a problem. Also, no part of the utility industry was really conscious of how much something cost. The utilities themselves, as regulated monopolies, got paid according to their expenses—as long as the cost of a generating plant wasn't unreasonable, the utility was allowed to set its rates to get a reasonable return on its investment. And the companies that built plants for the utilities—including nuclear plants—generally were paid on a cost-plus basis. Perhaps if the utility industry had been more cost conscious, it would have been more interested in standardization as a way to keep down costs. Finally, since nuclear technology was changing so rapidly during the 1960s and 1970s, with the designs steadily improving, there never seemed to be a good point at which to stop and standardize.

As it was, the utilities ordered each nuclear plant as a custom unit, and the sizes of those units increased yearly throughout the 1960s and into the 1970s—300 megawatts, 600, 900, and eventually as large as 1,200. It was as if American, Delta, United, and the other U.S. airlines each demanded jets customized to their individual specifications and bought bigger and bigger planes each year. The jets might not crash, but the costs to build them, fuel them, and maintain them would go through the roof.

By contrast, the French nuclear industry did standardize, and it avoided many of the problems that plagued the U.S. industry. Through the 1960s, France had begun construction on only a few units, all of them gas-graphite reactors developed by the French nuclear industry. In 1969, President Georges Pompidou decided to switch the technology from the home-grown reactor to U.S.-designed light-water reactors. Five years later, after the Arab oil embargo, the French national program took off, ordering sixteen reactors in 1974 and another twelve in 1976. By this point, the French had a decade of U.S. experience to learn from, and they took advantage of it.

France's one utility, the state-owned Electricité de France (EdF), decided that all its reactors would be based on a single design licensed from Westinghouse. One company, Framatome, would build the pressurized-water reactors, while another, Alsthom-Atlantique, would have a monopoly on the steam turbines paired with the reactors. EdF itself would design and build the remainder of each nuclear plant. Under this arrangement, the French nuclear program maximized learning. Not only did it benefit from many of the mistakes made by the U.S. program in the 1960s and early 1970s, but it also focused its own experience by standardizing everything it could. This included the reactors and turbines, the plant designs, the control rooms, and even the construction procedures and contracts. Framatome did increase the size of its reactors, but, unlike its American counterparts, it did so slowly, taking the time to learn how its reactors performed before changing them significantly. Its earliest reactors, first sold in 1970, were 900 megawatts. From there Framatome moved to 1,300 megawatts in 1976 and finally to 1,450 megawatts in 1987.

The care and caution paid off. The French nuclear program has been one of the world's most economical. Its plants cost $1 billion to $1.5 billion apiece—about $1,000 per kilowatt, or about what the most economical American plants cost and half the average cost of U.S. nuclear plants. EdF has consistently built plants in six years or less, much more quickly than most American utilities, which has also helped keep costs down. And once the plants open, they are very reliable, ranking among the world's best in the percentage of time they're available to generate electricity.

In the United States, a few utilities have taken a similar approach—with similar success. Duke Power is one of them. Acting as its own architect-engineer and construction contractor, Duke developed an in-house expertise in designing and building nuclear plants. It has standardized its plants to a degree unusual in the United States. Its first, the Oconee plant at Seneca, South Carolina, has three identical Babcock & Wilcox pressurized-water reactors. Its second and third, McGuire at Cornelius, North Carolina, and Catawba at Clover, South Carolina, switched to reactors from Westinghouse, but they are still pressurized-water reactors, and all four units in the two plants are the same. As a result, Duke's construction costs have been comparable to EdF's, and its capacity factors—a measure of operating performance—have been among the best in the United States.

If nothing else, the nuclear industry's lessons in the importance of learning have dispelled the technological hubris with which it began the nuclear era. Time after time, the industry believed that it understood the technology very well, only to be surprised again. One of those surprises has arisen in the steam generator, which in pressurized-water reactors uses the heat from the high-temperature, high-pressure cooling water to boil water in a second system of pipes leading to the steam turbine. The steam generators worked well for years, but then, after a decade or so of operation, a number of them were hit by unexpected buildup of "green grunge," which choked off the flow of heated, pressurized water and forced expensive repairs. Other steam generators have developed leaks. At the Trojan plant operated by Portland General Electric, for instance, leaks in the steam generator led the company to shut down the reactor permanently in 1992, after less than 17 years of operation. The humbling lesson is summed up by Bertram Wolfe, the man formerly in charge of GE's nuclear division: "If you've got a plant that's supposed to last 40 years, it takes you 30 to 40 years to understand what to expect."

five

CHOICES

During the 1940s and early 1950s, when atomic energy was new, it was common to hear reactors described as nuclear "furnaces." They "burned" their nuclear fuel and left behind nuclear "ash." Technically, of course, none of these terms made sense, since burning is a chemical process and a reactor gets its energy from fission, but journalists liked the terminology because it was easy and quick. One loaded fuel into the reactor, flipped a switch, and things got very hot. If that wasn't exactly a furnace, it was close enough.

And actually, the metaphor was pretty good—up to a point. The basement furnace burns one of several different fuels: natural gas or fuel oil or even, in some ancient models, coal. Nuclear reactors can be built to use plutonium, natural uranium, or uranium that has been enriched to varying degrees. Home furnaces have a "coolant"—the air that is circulated through the furnace and out through the rest of the house, carrying heat away from the fire. Reactors have a coolant, too—the liquid or gas that carries heat away from the reactor core to another part of the plant, where heat energy is transformed into electrical energy. There, however, the metaphor sputters out. In a nuclear reactor, the coolant not only transfers heat to a steam generator or a turbine, but it also keeps the fuel from overheating. The coolant in a furnace does nothing of the sort. And most reactors use a moderator to speed up the fission reaction. The basement burner has nothing similar. But the most importance weakness of the furnace metaphor is that it obscured just how many varieties of reactors were possible—and, consequently, obscured the difficult choice facing the early nuclear industry: Which reactor type should become the basis for commercial nuclear power?

The possibilities were practically unlimited. The fuel selection was wide. The coolant could be nearly anything that has good heat-transfer properties: air, carbon dioxide, helium, water, liquid metals, organic liquids, and so on. And the options for the moderator were equally broad: carbon, light water, heavy water, beryllium and other metals, organic liquids, and even, in some reactors, no moderator at all. With all these choices, selecting a reactor type became like ordering Chinese food from a fixed-price menu: take one item from column A, one from column B, one from column C. Some combinations might work better than others, and some might not work at all, but the number of combinations was huge.

Although few laymen realized just how many options existed, the members of the nuclear community did. Even before the end of World War II, workers in the Manhattan Project had dreamed up a variety of reactors they thought would be feasible. By 1955, at the Peaceful Uses of the Atom Conference held in Geneva, Alvin Weinberg reported that several hundred types of reactors were conceivable; of these, many were obviously impractical, but a hundred or so might actually do the job.

Which would be best for generating electricity? No one knew. Each reactor type was likely to have its own characteristic set of advantages and disadvantages. Without designing, building, and operating the reactors, choosing among them would be nothing more than educated guesswork. Yet it would cost far too much to develop a number of competing lines independently and see which was most effective as a mature technology. How, then, should the choice be made?

Many new technologies face a similar predicament. When personal computers first came onto the market beginning in the 1970s, a large number of companies offered mostly incompatible machines: MITS, IMSAI, Cromemco, Apple, Processor Technology, Tandy, Osborne, IBM, and many, many others over the first decade of the personal computer industry. When videocassette recorders were initially offered to consumers, people had to choose between Betamax and VHS formats. More recently, crystalline-silicon and amorphous-silicon technologies have been dueling over the market for solar cells. In each case, two or more versions of a technology have competed for adoption as each is evolving and improving.

How do we—and how should we—make choices between such competing technologies? Traditionally, economists have paid relatively little attention to this question, simply assuming that the market will do the right thing. According to classical economic theory, when a choice exists among various alternatives, the market will take the one that offers the most efficient use and allocation of resources—i.e., the one that gives the biggest bang for the buck. This is a nice, comforting notion: As long as the various technologies are competing in a free and open market, we can expect to arrive at the best of all possible outcomes.

Unfortunately, it ain't so. To make a rational decision, anyone—even a vague, sweeping entity like the "market"—needs information, and for nascent technologies, information is in short supply. How difficult and expensive will it be to perfect this version or that? What problems will appear? Will any of them prove to be show stoppers? What is the long-term potential of each of the options? Early in the development process, no one knows the answers to any of these questions. People can make forecasts and predictions, of course, basing them on experience from earlier development efforts, but the future seldom behaves as expected. And when the technology is complex, the forecasters might as well be reading tea leaves.

In the face of this uncertainty, decisions about a technology hinge on the sorts of nontechnical factors we've already encountered. The gut feelings of engineers. Political forces. Business considerations. The historical coincidences that put this person in that place at just the right time to have a deciding influence on the path a technology takes. For all practical purposes, most of these factors can be considered to be random, since they are impossible to predict. The development of a technology proceeds not as a well planned-out campaign (although parts of it clearly are well thought out) but rather as a disjointed effort that only in hindsight seems to have been moving toward a particular conclusion.

Nonetheless, economists and others have traditionally assumed that in some vague way this haphazard process does—with the help of the market—zero in on the best choice. The collective result of millions of individual decisions, each of which might be only partially rational or even completely irrational, will itself be rational. As in Darwin's theory of evolution, natural selection will ultimately result in the survival of the fittest. And certainly it often works out this way. The long competition between AC and DC described in the first chapter did end with the better technology taking over the market, though it was more difficult than it should have been.

But the best technology doesn't always win. Early decisions, made with relatively little information, can strongly affect the evolution of a technology. If, for instance, one approach attracts more investment early on, that could speed its development and allow it to open up a lead over its rivals, giving it an advantage in attracting even more investment. Or an early lead in sales when the market is small can snowball into a dominant position later on, even if the product is not obviously superior. Such influences make technological choice path dependent—the ultimate outcome can depend on a myriad of small, seemingly insignificant factors. Economists have discovered—to their surprise—that far from being completely rational, markets can sometimes settle on a technology that is clearly inferior to other choices. Indeed, even if the individual participants in a market all behave in a completely rational way, making their decisions logically according to the best available information, a weaker technology may get "locked in," shutting out other, potentially better choices from the market.

Consider, for example, the automobile. Although the internal-combustion engine has been the king of the road for more than half a century, some still wonder if we didn't take a wrong turn near the beginning.

STEAM VERSUS INTERNAL COMBUSTION

At the turn of the century, when horseless carriages were first becoming common in Europe and the United States, three different types were running neck and neck in popularity: electric, steam, and gasoline. The electric cars were particularly appealing because of their clean, quiet operation and smooth starting and acceleration. And they were fast—at the time, they held the world's land speed record. But they had what proved to be a crucial technical flaw: the batteries that powered them were heavy and had to be recharged frequently, making the cars unsuitable for sustained use. (Indeed, it is only now, a century later, that decades of hard work in battery technology are starting to make the electric car practical.) This left two technologies—steam power and internal combustion—competing to propel the coming automotive revolution, and although we know the outcome now, at the time it looked like a horse race.

The great advantage of internal combustion was that it enabled a lot of power to be generated by a relatively small, lightweight engine. Its great disadvantage was that it wasn't well suited to operate at the speeds most convenient for an automobile. The standard four-stroke internal combustion engine, invented by Nikolaus Otto in 1876, goes through a cycle of four steps, only one of which provides power: a fuel-air mixture is drawn into a cylinder; a piston compresses this mixture; the mixture is ignited, pushing the piston out and generating power; and the exhaust gases are pushed out of the cylinder, emptying it for the next cycle. This is inherently a very jerky, uneven process, so the piston was attached to a large flywheel which smoothed the in-and-out movement of the piston. But the presence of the flywheel limited the engine's range of speeds; the larger the flywheel, the smoother the operation but the narrower the limits. And the mechanics of the gasoline engine imposed another set of limits. Since each cycle in a gasoline engine produces only a small amount of power, the engine must be run at several thousand cycles per minute to be most efficient. On the other hand, the engines would stall if slowed much below a thousand revolutions per minute. The result of all these limits was that the internal combustion engine had an effective operating range of perhaps eight or nine hundred revolutions per minute up to several thousand.

This range matched up poorly with the demands of the automobile, whose wheels needed to turn at anywhere from a few dozen revolutions per minute to a thousand or more at high speeds. To build a gasoline-powered car, inventors thus had to develop a transmission system that could gear down from several thousand revolutions per minute to three hundred, or six hundred, or nine hundred, depending on the speed a driver wanted to go. It was a difficult, complicated task, and although engineers did eventually manage to develop effective transmission systems, they remain a trouble spot for automobiles even today—a reminder that the internal combustion engine is not well suited, at least in this one respect, to the task of powering an automobile.

The steam engine had its own set of pluses and minuses. Like the electric cars, steam-powered vehicles could be much smoother and quieter than their gas-powered counterparts. And they had plenty of power—in 1906, a Stanley Steamer set a land speed record of more than 127 miles per hour. Because the engine turned at hundreds of revolutions per minute instead of thousands, there was no need for a transmission, and wear and tear on internal parts was much less. A more mature technology than internal combustion, steam engines posed fewer engineering problems to solve before they could operate smoothly and dependably. And they were much simpler—the engine in one later Stanley model had only twenty-two moving parts, no more than were contained in the self-starter alone on a gas-powered car. Finally, steam-powered cars were much easier to drive. Not only did the operator of a gas-powered automobile have to worry about working the clutch and shifting gears simultaneously, but starting the car was a major production. A driver or brave assistant had to prime the cylinder, advance the spark, and then turn the crank, being careful to let go quickly. The crank could spin around and break an arm if one wasn't careful. As one automotive author wrote in 1904, "Every steam carriage which passes along the street justifies the confidence placed in it; and unless the objectionable features of the petrol carriage can be removed, it is bound to be driven from the road, to give place to its less objectionable rival, the steam-driven vehicle of the day."

The steamer's weaknesses sprang not from its engine but from its burner, in which a variable fire was kept burning, and from its boiler, in which water was turned into steam to push a piston and power the engine. Until automatic controls were perfected in the 1920s, the boiler demanded a great deal of attention from the driver. Furthermore, early models weren't equipped with condensers to capture and reuse the water, so the boilers needed to be refilled periodically. The signature of the old steamers was a cloud of water vapor trailing the car.

Taking everything into account, the steamers and gas-powered cars seemed evenly matched. Both offered advantages and disadvantages to their operators, and each had attributes that advocates could point to as evidence of natural superiority—compactness for the internal combustion engine, and simplicity for the steam engine. Even after thirty years of development, it still wasn't clear which was the superior choice. The gas-powered car was already taking a commanding lead in the market, but as Harvard historian Charles McLaughlin writes, "The Doble steam car, built in the late 1920s and early 1930s, was unsurpassed in trouble-free operation and performance by any car at that time." That Doble automobile, however, was steam's last hurrah—one final, failing attempt to bring the steam engine to an industry that had already decided internal combustion was the way to go.

If the two technologies were so well matched, why did gas-powered cars come to monopolize the market? The classical economic explanation would be that it was a rational market decision, that in the final analysis internal combustion must have been the superior technical choice. But many people who know more about automobiles than economics have challenged that. What really happened, they say, was that much more development went into the internal combustion engine than into its steam counterpart. S.O. White, an engineer who worked on automatic transmissions in the 1940s, put it this way:

> If the money and effort that have been put into the gasoline engine had been concentrated on the steam engine, its boiler and controls, we would not now be discussing automatic transmissions. The gasoline engine is so firmly entrenched that we must have some kind of multispeed transmission to make up for its deficiencies. We can only use the steam car performance as an ideal we would like to attain.

Even now, engineers calculate optimal efficiencies for steam- and gas-powered cars and wonder if steam might have proved to be a better option.

The reality is that the automotive competition between internal combustion and steam power was not decided according to which was the better technology. Instead, the victory of internal combustion was the result of a confluence of factors that were mostly independent of the technical issues.

One factor was that the developers of the gas-powered car, people like Ransom Olds and Henry Ford, were not just good engineers but also savvy businessmen. Many of the early producers of the gasoline cars set up shop in and around Detroit, and there they developed an entirely new way of doing business, mass-producing one or a few models and then selling them at low prices through large distribution networks.

By contrast, the Stanley Company—the leader and longest-lived of the steam-powered car manufacturers—was a relic of an earlier time of careful craftsmanship and local markets. The Stanley twins were inventors who were much more interested in designing and building excellent cars than in selling them by the millions. Their Massachusetts factory employed skilled workmen who could customize a car, even changing the wheel base according to the request of the buyer, and who might attach the boilers differently depending on who was doing the assembly. This not only made the cars much more expensive to build than, say, Ford's Model T (which was available in any color you wanted, so long as it was black), but it made them more difficult to service. That didn't bother the Stanleys, however. They were content to sell a relatively small number of cars to aficionados who were willing to pay a premium price. Before the start of World War I, the Stanley Company's entire yearly production—about 650 cars—was about what Henry Ford turned out in one day.

This approach worked for more than a decade, but it couldn't last. The Stanley Company was a relatively small business with little capital reserve, and a series of setbacks combined to sink it. Because the Stanleys tried to keep their cars as simple as possible, they had never installed condensers to recycle the water. This was not a major problem at first, since most of their cars were operated in New England, where everything was close together and people seldom drove long distances, and, besides, there were plenty of horse troughs along the sides of roads where the cars could take on more water. But in 1914 an epidemic of hoof-and-mouth disease broke out, and most of the public horse troughs were removed to prevent its spread. Suddenly the Stanley brothers were forced to develop a condensing system for their engines, and they cut back production drastically until they had finished it. By 1916 they had come up with a condenser and started to revive production, but they were blindsided once again. In 1917, the United States entered World War I, and the government restricted consumer manufacturing output. By law, a company could produce only half of its average annual output over the previous three years, which left the Stanley company able to manufacture only a very few cars. After the war, the weakened company seemed about to get back on its feet when hit by a third stroke of bad luck: the 1920 recession. It drove prices down and left the company—which had bought its inventory of parts at high prices just before the recession hit—unable to make a profit. It never recovered.

Other firms making steam-powered cars either went out of business or switched to the internal combustion engine, but McLaughlin concludes that the gasoline car's takeover of the market was not due to any technological superiority: "The principal factor responsible for the demise of the steam car was neither technical drawbacks nor a conspiracy of hostile interests, but rather the fact that its fate was left in the hands of small manufacturers. Even White and Stanley, the most prominent makers, failed to introduce the innovations in production and selling which would have enabled them to survive in the face of Detroit competition."

POSITIVE FEEDBACKS, QWERTY, AND TECHNOLOGICAL LOCK-IN

It's not important to us whether the steam automobile was potentially superior to the gas-powered car—some automotive engineers argue that it was, but no one really knows. What is important is that the fate of the automobile was not foreordained. It depended on a sequence of actions and events that, had they been different, could easily have yielded another outcome. If, for instance, the backers of the steam-powered car had congregated in Detroit and developed a product that could be mass-produced and mass-marketed, while the proponents of the internal-combustion engine had been content to make expensive cars for a small market, perhaps we'd all be driving steamers today.

This is the same path dependence we encountered in chapter 1. There's nothing particularly surprising about the idea that different paths should lead to different outcomes—unless you've taken a few economics courses. Classical economics describes a world determined by the balancing of various forces—supply versus demand, cost versus benefit, and so on. In this view, the fate of a new technology is settled by such things as its cost, how well it does its job, how well it meets a demand, and how it compares with other technologies that do the same or similar jobs. There is no place in the equations of classical economics for an outbreak of hoof-and-mouth disease.

In recent years, however, a new breed of economist/historian has begun studying the question of how technologies evolve, and these researchers find that it's not possible to understand the introduction of a new technology without looking closely at its history. Technological innovation depends upon a complex and subtle interplay of forces: the timing, content, and dissemination of scientific discoveries; the existing technological infrastructure; market judgments about the potential of and demand for a new technology; decisions made by organizations, whether in business or government; and the actions of a few key individuals, such as inventors, influential industrialists, and powerful government officials.

This new approach to understanding technological development differs from that of classical economics in a number of ways, but the most important is probably its emphasis on increasing returns, or "positive feedback." Conventional economics concentrates on situations in which diminishing returns push the market to an equilibrium. When oil prices skyrocketed in the 1970s, people began conserving energy by buying smaller cars, driving less, and lowering their thermostats, thus decreasing the demand for oil and pushing its price back down. Or in 1994, when Congress voted a ban on "assault weapons," people rushed to gun stores to buy them before the law went into effect; this increased demand created a sharp increase in prices for the guns, pushing the demand back down, since fewer customers were willing to pay the inflated charges. This is standard freshman economics: when something pushes on the market, the market pushes back, reestablishing a balance. In the language of the physical sciences, which economists have adopted, this push is called "negative feedback." It acts the same way in a market that it does in a machine or electronic device, nudging everything toward an equilibrium—an equilibrium that, according to classical economics, marks the most efficient use of resources.

In modern economies, however, there are many situations in which increasing returns, not diminishing returns, hold sway. One of the most familiar examples is the Betamax-versus-VHS competition for the videocassette recorder market. When the two formats came on the market, consumers had to choose one or the other, since a Beta video recorder would not play VHS tapes and vice versa. Many experts thought the Beta format was technically superior to VHS, but the advantage was not so large that other factors could not cancel it out—which indeed is what happened. At first, the two formats had roughly equal market shares, but then VHS got a small advantage, thanks perhaps to its marketing campaign, or some quirk of customer psychology, or just plain luck. With more VHS customers, the stores that rented and sold videotape movies made more titles available on VHS. This in turn led still more people to choose VHS, which guaranteed even more VHS titles in the stores, and so on in a self-reinforcing cycle, until VHS became the de facto standard for the industry.

Brian Arthur, an economist at Stanford and the Santa Fe Institute who has studied the effects of such positive feedbacks, notes that they create uncertainty in the market. Because small, random events that happen early can be magnified to have great importance later, the eventual outcome can depend quite sensitively on circumstances—it is path dependent. A chance laboratory discovery may push a developing technology in an entirely different direction than it would have taken otherwise. The unwillingness of venture capitalists to provide funds to a struggling company aborts what could have been a major new market a decade later. A computer language developed for early computers, which had relatively little power or memory, becomes a standard that shapes the programs written for much more powerful computers a decade later. In particular, Arthur notes, such path dependence implies that the outcome cannot be predicted with any certainty ahead of time. When VCRs were first offered for sale, no one knew which format would predominate.

Negative and positive feedbacks create markets with very different properties. The usual way of visualizing a classical, negative-feedback market is to think of a marble inside a bowl. The force of gravity pulls the marble down toward the bottom of the bowl, which is its equilibrium point. No matter where the marble starts out or what path it takes, its eventual position is easy to predict. Positive feedback turns this picture upside down—literally. Think of the marble as balanced on the top of an inverted bowl. If jostled, it no longer moves back to where it was, but rather takes off down the side of the bowl with increasing speed. And there's no way to predict just where the marble will end up, since the destination will depend sensitively on the precise details of the jostle.

Conventional economics has focused on negative feedbacks and equilibria, Arthur says, because until recently those have been the most important factors in understanding the economy. In any market based on resources—farming, mining, the production of bulk goods—diminishing returns hold sway because resources are limited. If beef, for example, were to experience a sudden surge in popularity, that popularity would not build on itself until beef began to crowd out chicken, pork, and lamb in the meat market. Instead, the greater demand would lead to greater production, which would push up the prices of cattle feed and ranch land, making beef more expensive and stabilizing the yen for beef, perhaps at a somewhat higher level than before.

In many modern markets, however, information is more important than physical resources, and, Arthur says, when information is involved, increasing returns are likely to take a hand. Consider the personal computer. It is an extremely complicated machine, and designing and building just one demands an investment of tens or hundreds of millions of dollars in research and development. Building the second one, however, is relatively cheap—the only cost is for the parts and labor to assemble it. Building the thousandth or the millionth one is even cheaper, thanks to learning effects in manufacturing and economies of scale, and the benefits to an individual customer of the millionth personal computer are likely to be much higher than those of the first or second. With a million computers on the market, there will be much more software available, with the costs of individual programs shared among many users, and people can share their experiences to help each other get more out of their computers. In general, Arthur says, when the major investment in a product is the knowledge needed to make it—whether it's a computer, a drug, or an advanced material such as a high-strength ceramic—then the price per unit drops and the profit increases with greater production and sales. Unlike beef, sales of such knowledge-based products feed on themselves, with each success making it more likely that the successes will keep coming.

Arthur points to five separate ways that technologies can become more attractive as they become more widely adopted. First, the more of a product that a manufacturer makes and sells, the more it can improve the product by learning from both its own experience and that of its customers. This learning-by-doing and learning-by-using is, as we saw in the last chapter, particularly important in complex technologies. Second, consumers often find that it pays to belong to a large community of users of a particular product. This was why VHS won the battle of the VCR formats. Third, if economies of scale exist in making a product—which is often the case when research and development are a major part of the cost of the product—the price will come down with increased production. Fourth, many people will decide which product to buy based on how widespread it already is, figuring that lesser-used technologies are less well tested and thus riskier. And fifth, many technologies demand an infrastructure, which grows and becomes more entrenched as the technology is adopted. The automobile, for instance, relies on a network of oil refineries and gas stations, parts manufacturers and stores, repair shops, and so on. A competing technology without such a well-established infrastructure would be at a huge disadvantage.

Modern, complex technologies like the automobile and the computer often fall into several of those categories, making them particularly likely to have increasing, rather than diminishing, returns. But modern technologies are by no means the only examples. One of the best-studied cases took place more than a century ago with a relatively low-tech product: the typewriter.

Today, almost all typewriters and computer keyboards have the same key placement: the so-called QWERTY layout, named for the first six letters on the top line of letters on the keyboard. QWERTY is not a particularly efficient layout—a typist's fingers must move from their "home" position to hit many of the most common letters—and periodically people have tried to replace it with something better. The best-known alternative is the Dvorak Simplified Keyboard system, developed in 1932, which allows typists to type faster than with QWERTY—somewhere between 5 percent and 40 percent faster, depending on whose claims you believe. But QWERTY has had a lock on the market for a hundred years, thanks to historical accident coupled with increasing returns.

Christopher Latham Sholes, a printer from Milwaukee, was the inventor of one early version of the typewriter. He filed for a patent in 1867, but the machine was plagued by a number of faults, the most serious of which was the tendency of its typebars (the bars that fly up to strike the ribbon and print a letter on the paper) to hit each other and jam together if the typist struck the keys too quickly in succession. Instead of improving the typewriter design to avoid this problem, Sholes experimented with various keyboard layouts, looking for one that would keep the typist from picking up too much speed and thus minimize jams. In 1873, when he had something that worked reasonably well, Sholes and a partner sold the manufacturing rights to the arms maker, E. Remington and Sons. Remington in turn made a few minor changes to produce the QWERTY keyboard in essentially the same layout that exists today. One of Remington's modifications, for example, was to move the "R" up to the second row, replacing the period, so that salesmen could impress customers by typing out TYPE WRITER (the brand name) without their fingers leaving the row.

At the same time that Remington was offering its QWERTY typewriter, other companies were offering their own machines, each with its own keyboard layout. Some of the models avoided the problem with the typebars that plagued the Sholes design, which allowed them to place the keys in more convenient locations. One arrangement, the "Ideal" keyboard, put all the most common letters in the home row (DHIATENSOR), so that a typist could form 70 percent of English words without the fingers leaving that row.

It took more than a decade before the typewriter began to catch on, but it boomed in the 1880s, and soon touch typing became a valuable office skill. Once this happened, the selection of a keyboard layout was subject to a powerful positive feedback. Workers, if they were going to put in the time to learn to type, wanted to train in the system that was most widespread. This way they could more easily find jobs and would be less likely to have to learn a second system. Employers, in turn, had an incentive to buy typewriters with the most popular keyboard layout—it would be easier to find typists who already knew how to use that layout and so avoid having to pay to train them. With both workers and employers looking to use the most popular layout, positive feedback guaranteed strongly increasing returns. Whichever choice gained an initial lead would be likely to expand it, eventually putting its competitors out of business.

QWERTY may not have had the most sensible keyboard arrangement, but it had the advantage that it had been out longer than most other designs. Throughout the 1880s, typewriters with a variety of keyboard layouts were sold, but in the 1890s the market moved more and more toward QWERTY, and by the first decade of the twentieth century it had taken over. The takeover was similar to what happened with office word-processing software in the late 1980s, as WordPerfect became a de facto standard in offices across the country.

For almost a century now, nothing has overturned QWERTY's predominance. In the 1940s the U.S. Navy experimented with the Dvorak system and found that it speeded up typing. Indeed, a U.S. Navy report claimed that it accelerated typing speeds so much that an office that retrained its typists on a Dvorak keyboard would recoup the cost of that training within ten days. Other tests have seen less of an improvement in typing speeds, and a conservative estimate is that typists using Dvorak will, on average, be about 5 percent faster. Either way, QWERTY remains the office standard, and although Dvorak proponents argue—quite correctly—that theirs is the superior system, the upstart can make no headway.

A similar challenge faced the gas-powered automobile in the mid-1920s, with similar results. After the collapse of the Stanley company and other makers of steam cars, the industrialist Abner Doble decided to give it one more try. He had become convinced that the steam car, if it incorporated all the advances that had been made in controlling its burner, would be superior to gasoline cars. So he set out to build and sell steamers using all the tricks that had worked for the big Detroit auto manufacturers. He assembled a network of a thousand dealers, applied all the latest manufacturing techniques, and even set up production in Detroit to have easy access to skilled workers and all the little companies that had sprung up to provide parts and supplies for the big car makers. By all accounts, the Doble steam cars were excellent products—as good as or better than anything powered by an internal-combustion engine—and the company started out with $20 million in advance orders. If anyone could have broken the monopoly of the gas-powered car, it should have been Doble. But he couldn't. The Doble-Detroit Company died during the Depression, and for sixty years nothing has come close to replacing the internal combustion engine.

Faced with this sort of intractability, it's tempting to think that something is interfering with the workings of the free market. Perhaps the makers of the gasoline-powered cars colluded to sink Doble's company. Or perhaps, as August Dvorak once suggested, the typewriter makers conspired against his invention because they feared that speeding up typing too much would lower the demand for their products. But no. This "technological lock-in," as Arthur named it, makes perfectly good sense if looked at from the perspective of increasing returns.

Once a technology has reached a certain level of adoption, anyone who chooses an alternative technology ends up paying a premium. In the case of the typewriter keyboard, an employer would have to retrain existing employees and would find it more difficult to hire new employees, who would prefer not to have to learn a new system. Employees who moved from QWERTY would not only have to take the time to learn a new system but would also hurt their chances of getting a job at another company. A few employers and employees might judge the investment to be worth it, but not most. Each individual decision to stay with QWERTY is a rational one, but all those rational decisions add up to an irrational, or at least a less-than-optimal, choice.

All in all, it's a rather unsatisfactory situation from a classical economist's point of view. When new technologies compete with each other in an increasing-returns market, the ultimate winner of the competition will often be determined by random events ("random" in the sense that they have nothing to do with the technical merits of the technology). Depending on the particular path the competition takes, an inferior technology may well come out on top. And once this happens, that second-rate choice may become locked in. Positive feedbacks may have their advantages—such as dramatically lowering the costs of many modern technologies—but guaranteeing the best outcome isn't one of them.

NUCLEAR POWER AND THE MULTI-ARMED BANDIT

Because inferior technologies can lock out better choices, the early work on a technology is particularly important. Ideally, one would like to keep things fluid until the most desirable option is clear. But in free-market competition, that's hardly feasible. The pressure on each business is to lock up as much of the market as possible and not to worry if its competitors' products have more long-term potential. So an obvious question is: Can some central authority—a government agency or other decision-making body—step in to prevent the lock-in of inferior technologies? Robin Cowan, a student of Arthur's who is now at the University of Western Ontario in London, has studied that question closely, and his conclusion is sobering. Even with complete central control, he says, both economic analysis and experience demonstrate that it is possible to lock in on an inferior technology. To show why, he turns to the history of nuclear power.

As we saw in earlier chapters, the Atomic Energy Commission set out in the 1950s to determine the best technology for making nuclear plants. The AEC was not trying to avoid technological lock-in—that was a concept still three decades away. Instead, the commission had been given responsibility for developing nuclear power, and its members took that duty seriously. They planned to do the job right and figure out what the best option was before the United States began building large numbers of nuclear plants.

The choices were many, but from the beginning some looked better than others. Out of the hundred or so possibilities that had been catalogued at the 1955 Peaceful Uses of the Atom Conference, a dozen or so serious candidates had emerged three years later at a follow-up conference. That initial whittling down was made mostly on the basis of physical calculations and engineering judgment, but further narrowing would come only with actual experience in building and operating plants of different types.

The candidates varied in their fuel, their moderators, and their coolants. The "low-tech" choice was anything that burned natural uranium and avoided the need for the complex and costly uranium-enrichment plants. These included, for instance, the graphite-moderated, water-cooled reactors such as those built at Hanford during the war to produce plutonium.

A close cousin to the Hanford reactors became one of the leading candidates for generating nuclear electricity. As at Hanford, the fuel was natural uranium and the moderator graphite, but the coolant was a high-pressure gas. This type was a British specialty, and England's Calder Hall—the first reactor to generate electricity for commercial distribution—was a gas-graphite machine.

In the earliest versions of the gas-graphite reactor, the coolant gas was air—a poor choice, given that overheated graphite will burn in the presence of oxygen. In later gas-cooled reactors, gases such as carbon dioxide and helium became the standard.

No matter what the coolant, any graphite-moderated reactor has a major disadvantage: neutrons take a relatively long distance to slow down in graphite, so the reactor cores must be very large. Consequently, graphite reactors are big and rather costly machines with little design flexibility. On the other hand, the larger core offers extra safety. If the coolant is lost in an accident, the residual heat from the fuel is diffused through the graphite, making a meltdown less likely.

Early on, however, the gas-graphite reactor's two pluses outweighed its minuses. Not only did it run on natural uranium, but it also could be used to produce plutonium for bombs. For this reason, both Britain and France started their nuclear programs with these reactors, and Calder Hall was a "dual-use" plant, making both plutonium and power.

If one wishes to use natural uranium as a reactor's fuel but doesn't like graphite as a moderator, the only alternative is heavy water. On the surface, heavy-water reactors don't seem like a good option. They demand the same large cores as graphite reactors, and they also pose a couple of extra problems. Since heavy water makes up only a tiny proportion of normal water, its production demands large, expensive separation plants. And heavy-water reactors tend to be run at somewhat lower temperatures than other types, which hurts their efficiency. Nonetheless, Canada—the only country to develop heavy-water reactors—has had good success with its CANDU design.

The CANDU reactors, which were developed starting in the early 1950s, use heavy water as both moderator and coolant. Each reactor consists of a single large tank filled with heavy water and run through by hundreds of pressure tubes. These tubes contain uranium fuel bundles and also serve as conduits for pressurized heavy water, which rushes past the fuel elements and carries heat away to steam generators. Dispersing the fuel and coolant through these tubes has two advantages. It avoids the need for the large, difficult-to-build pressure vessel standard in light-water reactors, and it allows the reactor to be refueled while in operation. Fuel elements can be removed and replaced one at a time instead of all at once, as in a light-water reactor. Although the CANDU reactors, which have generated electricity in Canada for more than thirty years, may have been "low-tech" in using natural uranium as fuel, the rest of the design was state-of-the-art.

In the United States, and later in other countries, the availability of enriched uranium opened up nearly endless options for reactor design. CANDU reactors, for instance, are now constructed to use a slightly enriched fuel—just over 1 percent U-235—which allows them to have more compact cores and better operating efficiencies. And the gas-cooled, graphite-moderated reactor becomes an entirely different machine when it is designed to use highly enriched uranium (more than 90 percent U-235). The few such reactors that have been built typically get high efficiencies by heating their gas coolants to temperatures much hotter than are practical for water coolants.

But the major significance of enriched uranium fuel is that it made a host of reactor types possible that were not feasible with natural uranium fuel. With a higher percentage of U-235 in the fuel, more of the neutrons produced by splitting U-235 atoms could be lost without killing the chain reactor, so the choice of moderator became less sensitive. Furthermore, the greater the enrichment, the smaller and more compact the core could be and the greater the flexibility in its design.

With only slightly enriched uranium, it was possible to use light-water—the familiar, plentiful H_20—in place of heavy water as both moderator and coolant. The light-water reactor's great advantage is its potential compactness. Because neutrons slow down more quickly in water than in, say, graphite, a relatively small volume of water is needed as moderator. With enriched fuel, a light-water reactor can be built with a very small core and a high power density. It was this factor more than anything else that led Rickover to focus on light-water reactors for his submarines. Light-water reactors do have their negatives, however. Water is not a particularly efficient material for transferring heat, for instance, and both superheated water and steam are highly corrosive, particularly at the high pressures necessary in a reactor. Corrosion would turn out to be a constant concern in light-water reactors of all types.

One obvious way to avoid the corrosion problem was to work with liquid metal coolants. Sodium, for instance, melts at a low enough temperature to be liquid at the operating temperature of a reactor and, as a metal, it is not corrosive to metal pipes. Furthermore, it and other liquid metals make excellent coolants since they absorb and carry heat much better than water. Coupled with an efficient moderator, a liquid-metal reactor can be as compact as a light-water machine, and, indeed, a sodium-cooled, beryllium-moderated reactor was Rickover's second choice for a naval reactor. But since sodium reacts explosively with water, which is used to create steam and turn a turbine, even small leaks in the heat-transfer system are big trouble. It was difficulties with this system that led Rickover to abandon the liquid-sodium reactor, but sodium's outstanding thermal properties kept it in contention for other reactors, both in the nuclear-aircraft project and the civilian power program.

Organic liquids—those based on carbon atoms—offered another way to sidestep the corrosion problem, and, unlike liquid sodium, they would not react with water. Unfortunately, the organic liquids did not have particularly good heat-transfer properties. Furthermore, a prototype organic-moderated and -cooled reactor that was opened at Piqua, Ohio, in 1964 had to be shut down after only three years because the organic liquid was deteriorating.

There were other mix-and-match combinations of coolant and moderator. The Canadians, for example, eventually tried both boiling water and organic liquids as coolants in their heavy-water-moderated reactors, and they appeared to have solved the breakdown problem of the organic coolant before the heavy-water, organic-cooled reactor was canceled for political reasons. And scientists at Oak Ridge suggested getting rid of the coolant altogether. In their homogeneous reactor, the fuel was kept in a solution that circulated through a reactor, where it chain reacted and heated up, and then through a heat exchanger, where it boiled water to run a turbine.

Finally, many believed that the breeder reactor was the best hope for the future. The breeder is designed to create more fuel than it consumes by turning some of the nonfissile U-238 into plutonium (or, in other designs, turning thorium-232 into the fissionable U-233). A certain amount of "breeding" takes place in any reactor—a few neutrons are absorbed by U-238 nuclei without causing them to split, and the resulting uranium-239 atoms spontaneously transmute into plutonium-239—but a breeder reactor is designed to maximize this process. Since natural uranium is 99.3 percent U-238 and only 0.7 percent fissionable U-235, only a tiny proportion of uranium is useful for producing energy in a normal reactor. Theoretically, a breeder makes it possible to squeeze out more than a hundred times as much power from a pound of uranium, but at a price: breeders produce electricity less economically than other reactors, and the recovery of plutonium from the used and highly radioactive fuel is itself a complex and expensive process. It would all be worth it if uranium supplies were scarce—which appeared to be the case in the first years of atomic energy—but as more and more uranium ore was discovered, the breeder became less attractive. Later, some began to worry that the production of large amounts of nearly pure plutonium would make it too easy for rogue nations or terrorists to steal the plutonium and use it to make bombs. These concerns led the United States to drop its breeder program, but they have not stopped other countries.

On paper, depending on what assumptions one made, it was possible to make a case for just about any of the reactor types. The only way to settle the issue was to build prototypes and test them so that the arguments would be based on fewer assumptions and more evidence.

And that is what the Atomic Energy Commission set out to do. All through the 1950s and into the 1960s, it funded research and development of a number of promising reactor designs, with the philosophy, "Let the best reactor win." Working with many reactor manufacturers and utilities, the AEC sponsored a variety of reactors: light-water, including both pressurized and boiling-water designs; heavy water; graphite-moderated and sodium-cooled; graphite-moderated and gas-cooled; organic-moderated and -cooled; and fast-breeder. At least one power plant of each of these types was built somewhere in the United States.

Other countries, without access to enriched uranium, didn't have the luxury of choosing among so many different types of reactors, and their nuclear programs generally focused on one or two designs. Canada settled early on heavy-water reactors with natural uranium fuel. The United Kingdom chose to develop gas-cooled reactors, first using natural uranium and later enriched uranium, and later experimented with a heavy-water-moderated, light-water-cooled reactor similar to the CANDU. The French, like the British, began their program with graphite-moderated, gas-cooled reactors that burned natural uranium, and they later built a unique gas-cooled, heavy-water-moderated reactor. The Soviet Union came closest to the United States in its variety of designs. Having its own uranium-enrichment facilities, the country could try different reactor types, and it did: graphite-moderated, boiling-water-cooled (the Chernobyl reactor was of this type); pressurized-water reactors; and liquid-metal-cooled fast-breeder reactors.

Each of these countries was taking the chance of backing the wrong horse—putting its time and development dollars into technologies that might ultimately prove inferior to one they ignored—but there was little choice. Not only did most have no enriched uranium, but they were all playing catch-up with the United States, which had built reactors during the war and taken a giant lead in nuclear physics and engineering. If any of them wished to become a nuclear power, developing a homegrown industry and exporting nuclear know-how to other countries, the best bet was to pick one promising technology and bring it along as quickly as possible. Not the most desirable option, perhaps, but the best under the circumstances.

The United States, on the other hand, had the opportunity to "do it right." It had the experience, the infrastructure, and the resources to look into a number of options, taking the time to figure out which one would be the best. The problem, then, was how to go about it.

The situation was almost a textbook example of what Robin Cowan refers to as the "multi-armed bandit." Think of a slot machine with several arms. Each time you drop a quarter into the bandit, you pull down an arm and hope for a jackpot. The chances of winning are different for each arm, and you don't know what they are, but with each play you accumulate a little more information on the jackpot probabilities. The object is to figure out which arms to play and in which order so as to have the best chance of maximizing your winnings.

That challenge, which has been studied by probability theorists, is not a trivial one. One complication, for instance, is that the more you play one arm, the more you learn about that arm's payoff probabilities—which, if the payoffs are acceptable, gives it an advantage over the other arms, whose probabilities are less well known, Suppose you've played one arm enough to know that you're likely to get a small but positive payoff if you keep playing it. Do you switch to another arm whose probabilities you don't know as well? It might have a higher payoff, but it's more of a gamble since you haven't learned as much about it. The best strategy, mathematicians have proved, depends upon calculating payoff probabilities that take into account both what is known about the likely payoffs for each arm and how certain that knowledge is.

Choosing between competing new technologies is a similar problem, but with an added complication. In many technologies, increasing returns will change the payoffs as the game proceeds: the more you play one arm, the greater its payoff becomes. Build one light-water plant, and the lessons you learn will lower costs and increase efficiency for the next one. Therefore positive feedback plays two mutually reinforcing roles in the technological multi-armed bandit. First, the more experience is gained with one technology, the less uncertainty there is about its payoffs and the less of a gamble choosing it becomes. This puts other technologies, even those that are potentially better, at a disadvantage, because not as much is known about them. And second, experience increases the real payoffs of that first-tried technology, putting its competitors at even more of a disadvantage.

Not surprisingly, mathematical analysis shows that if one pursues a strictly rational strategy when playing the multi-armed bandit, eventually one will choose a single technology and stick with it. After enough testing of the different arms, one of them will inevitably appear to offer better chances for profits than the others. Once this happens, it makes no sense to pull on the other arms—or to invest in other technologies—unless you just like to throw money away. The result: technological lock-in.

Like his mentor Brian Arthur, Robin Cowan finds that the locked-in technology may not be the best choice, and this is true even when the players have, at every point, made rational decisions based on expected payoffs. The reason can be found in the power of increasing returns. Suppose that an inferior technology is the first one to be tried out in this multi-armed bandit game, and suppose further that it doesn't perform too badly. Then, whatever the reason it was initially tried, that reason is even stronger the second time around, since learning effects will have improved the technology's performance. Eventually, if the pattern is repeated, the learning effects will have accumulated enough so that even if the technology doesn't turn out to be as good as first hoped, it won't make sense to drop it and try another technology. Not only would the new technology be uncertain in its payoff, but investing in it would mean starting from scratch on the learning curve, paying all over again to learn the technology's strengths and weaknesses.

If the unfettered free market can lock in to an inferior technology, Cowan wondered, what happens if a central authority gets involved? Such was the case with the U.S. nuclear program, with the AEC subsidizing its technological development. Theoretically, Cowan says, the situation should improve with the involvement of a central authority. In a pure free-market setting, if one technology is performing acceptably well, economic competition will discourage companies from trying others. After all, a company has little to gain and much to lose by investing in what seems, at the time, to be a less-desirable technology. If after spending money on the technology the company finds that it was indeed a poorer choice, it has lost that investment. If the technology turns out to be a better choice, then the company has learned a valuable lesson about the relative payoffs of the technologies and can adjust its strategy accordingly—but the company's competitors will also profit from that knowledge, and they will not have paid anything to gain it. On the other hand, a government or other central authority (such as a research consortium funded by a number of companies) is more likely to spend the money to learn about other arms of the bandit and so is less likely to miss a more promising technology by getting locked in too early.

Even so, Cowan says, a central authority cannot guarantee that an inferior technology will not get a stranglehold on the market. "[I]f increasing returns are very strong, lock-in will occur very rapidly. If learning takes place rapidly enough, the first technology used will be the only one ever used." This is most likely to happen with very complex technologies, he notes, since learning is so much more important for them. If the early choices are the wrong choices, the right choice may never have a chance.

That's the theory, anyway. To see how closely experience matched theory, Cowan analyzed the history of commercial nuclear power in terms of the multi-armed bandit model. It turned out to be a pretty good fit.

When the AEC set out to learn about the different options for nuclear reactors, it was following exactly the sort of strategy suggested by mathematical analysis of the multi-armed bandit. By sliding quarter after quarter into the machine, pulling on this arm and that, and keeping track of which fruits appeared in the windows, it would eventually accumulate enough information to judge the potential payoffs of the various types of reactors. It might still make a mistake and pick a poor technology, but the chances of such an error would be minimized, and the ultimate choice would probably not be hugely inferior to another choice.

Unfortunately, the AEC hadn't been playing the game too long before others started inserting quarters and jerking arms—and their choices were based on considerations much different from the AEC's long-range strategy of finding the best technology for generating electricity. The most influential player was Rickover, who started out working the pressurized light-water and liquid-sodium arms equally, but who quickly decided that the light-water reactor was where his payoffs would lie. Then Eisenhower joined in with his request for a civilian reactor to serve as the flagship for his Atoms for Peace program. He wasn't concerned about the long-term outlook; he wanted the easiest, fastest, and least risky choice at the time. So, with the construction of Shippingport, more quarters were poured into the bandit, and the light-water arm was pulled some more. As a result, light-water reactors took a significant lead in the U.S. nuclear sweepstakes. Their pluses and minuses were better known, they were less likely to spring any surprises, and the experience in building and operating them was already paying off in better designs for the next generation.

In Europe, meanwhile, the French and British had been betting most of their quarters on the development of gas-graphite reactors. Other European countries, particularly West Germany and Italy, were eager for nuclear power but had no development programs of their own. They would have to get the technology from somewhere else, and both France and Great Britain hoped to be that source. In 1957, six European countries, including France, Italy, and West Germany, signed a treaty creating Euratom, an agency intended to promote the development of nuclear power in Europe. The French expected that Euratom would help them sell their gas-graphite technology to their neighbors, but the United States had other ideas.

At the time, things were moving more slowly in the United States than nuclear advocates wished. Industry was not rushing to embrace nuclear power, and Republicans had blocked Democratic calls for government-built power-generating reactors. One option was to open up a European market for U.S. nuclear technology. The European governments were, after all, not nearly so hesitant to engage in industrial policy. And exporting U.S. nuclear know-how to European allies was exactly what Eisenhower had envisioned for the Atoms for Peace program. Not only would it demonstrate the peaceful side of the atom, but it would also strengthen Europe economically by providing it with cheap power and prove to the world the superiority of American technology. Thus, encouraging European countries to invest in American nuclear reactors became a priority for both Congress and the executive branch.

In 1956, the six governments that would make up Euratom appointed a three-man commission to study nuclear power and to report on its future in Europe. The "Three Wise Men," as they were called, were Louis Armand, president of the French National Railroad Company; Franz Estel of Germany, the vice president of the European Coal and Steel Community; and Francesco Giordani, former president of the Italian Atomic Energy

Commission. When they visited the United States, the AEC treated them like royalty and offered them technical data that seemingly backed up the commission's claims about light-water reactors. The technology, according to the AEC, was on the verge of being economically competitive with coal—particularly in Europe, where costs for generating electricity from coal were much higher than in the United States. In 1957, the wise men's report, *A Target for Euratom*, concluded that Europe needed nuclear power and urged cooperation with the United States. The light-water reactor, the wise men found, was at least as good a bet as the French gas-graphite machine.

At about the same time, the Italian government decided that it would build a 150-megawatt nuclear power plant and asked for bids from companies in the United States, Canada, France, and Great Britain. The winner, announced in 1958, was an American light-water reactor.

Then in November 1958, the United States scored a major coup with its Atoms for Peace program. Euratom signed an agreement with the United States to collaborate on building 1,000 megawatts of nuclear generating capacity. In return for Euratom choosing American technology—light-water reactor designs from GE and Westinghouse, as it turned out—the United States would pay for about half the research and construction costs of the reactors as well as provide fuel at bargain prices. Eventually, three light-water plants were built in Europe under this agreement. The first was the Italian plant that had been announced before the U.S.-Euratom treaty; the second was a collaborative French-Belgian project set in France near the Belgian border; and the third was a German plant in Bavaria.

The U.S. strategy had worked: now American technology had a foothold in Europe. Even France, which had devoted itself to gas-graphite, was taking a look at light-water, although the reasons lay as much in internal politics as in U.S. blandishments. (Electricité de France, the French national electric utility, saw building a light-water reactor as a way to sidestep the French atomic energy commission, which had kept control of atomic matters to itself and which had focused on gas-graphite reactors.) Light-water did not hold the dominant position in Europe that it had in the United States, but it had caught up with gas-graphite reactors.

By this time, then, the AEC's cautious strategy of playing a number of arms to see which offered the greatest long-term payoff had been undermined by several factors: Rickover's choice of light-water, the building of Shippingport, and the desire to export nuclear technology to Europe. The light-water selection was getting many more quarters than the AEC's strategy would have apportioned. Nonetheless, several arms were still in play besides light-water—gas-graphite reactors in France and Great Britain, the breeder, and a number of other alternatives being explored in the United States.

It was, as we saw in chapter 3, the turnkey contracts offered by GE and

Westinghouse that finally made light-water the overwhelming technological choice. Because of the contracts' rosy economic implications, companies began pouring money into the light-water arm of the bandit and virtually ignoring the others. With each announcement of a new light-water nuclear plant, the reasons for the next player to choose light-water became stronger. Even the enormous cost overruns of the late 1960s and early 1970s didn't make light-water's competitors look more attractive. If it cost so much to get the kinks out of light-water technology, it seemed likely that the same would be true for any other nuclear technology. Choosing another reactor type would mean climbing up the learning curve all over again.

Gradually, the increasing experience with light-water technology led to its almost-complete domination of the world market. In 1969, France turned away from two decades of experimentation with gas-graphite reactors to build light-water reactors based on a design licensed from Westinghouse. Great Britain held out longer, first trying heavy-water reactors when its gas-cooled reactors did not generate electricity as cheaply as expected, but then switching to pressurized light-water reactors in the late 1970s. Today, of the world's 500 or so nuclear power plants, just over 100 are not light-water. Most of those are either gas-cooled reactors in Great Britain or heavy-water reactors of the CANDU type, the only technology still putting up even feeble resistance to light-water today. As Cowan's multi-armed bandit model predicted, the competition among nuclear technologies eventually settled on a single selection.

Was it the best selection? No one knows for sure, even now, but many nuclear engineers believe it was not. In the 1950s and 1960s, light-water reactors were widely seen as the "easy" option—they could be put to work more quickly and with less research and development than other types—but they were also thought to be less economically desirable in the long run than some of the more difficult-to-perfect alternatives. The AEC itself tacitly acknowledged this by its focus on a "second generation" of nuclear power, which was envisioned not as an improved version of the existing light-water reactors but as some totally different sort of machine that would eventually replace the stopgap light-water technology. And talk to nuclear engineers today—at least those not working on light-water reactors—and you'll generally hear one of two types of reactors mentioned as the best technical option: either liquid-metal reactors or high-temperature gas-cooled reactors.

Today's liquid-metal reactor is a far cry from the machine that Rickover spurned because of its many problems. It still uses liquid sodium as coolant, but little else is the same. The leaks that bothered Rickover, for instance, no longer appear to be a problem. The steam generators on one experimental liquid-metal reactor, the Experimental Breeder Reactor-II, have operated for twenty-five years without leaks. The liquid-metal design that has been studied most carefully is the Integral Fast Reactor, whose development was sponsored by the Department of Energy. The IFR, as it is known, should theoretically be more economical to operate than a light-water reactor because it is more efficient: where commercial light-water reactors burn up only about 3 percent of the U-235 atoms in their fuel before they must be replaced, the IFR is designed to use up 20 percent or more. The IFR's supporters also tout it as being much safer than current light-water reactors—even if the control rods are pulled completely out of the reactor core, the chain reaction will stay under control. The IFR should create less nuclear waste because the highly radioactive elements created during the reaction can be burned up in the reactor itself. And the IFR is a breeder, getting theoretically a hundred times as much energy from a pound of uranium as do light-water reactors. But to people concerned with nuclear proliferation, the IFR's breeding capability makes it dangerous, and those concerns led to its downfall. Congress ended the IFR program in 1994.

Many nuclear engineers see the high-temperature, gas-cooled reactor as probably the best alternative to light-water if both safety and economics are taken into account. The gas-cooled reactor is potentially more efficient than light-water reactors because gas can be heated to much higher temperatures than is feasible for pressurized water or steam, and the efficiency of converting heat into power increases with increasing temperature. Furthermore, the gas-cooled reactor has a number of safety advantages over light-water reactors, such as the core not being so sensitive to a loss-of-coolant accident. Indeed, gas-cooled reactors have been designed that are invulnerable to core meltdown, even in the worst possible accident. To accomplish this, the reactor must be made less powerful than usual—100 to 150 megawatts instead of 1,000 or 1,200—but a number of the smaller reactors could be strung together to create an electricity-generating plant of the same capacity as a standard light-water plant.

Other reactor types get an occasional mention as well. The PIUS reactor, devised by Swedish engineer Kåre Hannerz, is a light-water reactor that works in a completely different way from today's pressurized-water and boiling-water reactors. Its selling point is that it, like the small gas-cooled reactor, is inherently safe: it cannot suffer a meltdown. And Alvin Weinberg at Oak Ridge National Laboratory still believes that a design tested there in the late 1960s could be the ultimate reactor. The molten-salt reactor put its uranium fuel in a liquid solution of molten salts and circulated it through a system of pipes, passing first through the core where it would chain-react and heat up, and then on to a heat exchanger, where the molten salt would transfer some of its heat to a second system that could be used to run a turbine. This reactor—which was a breeder, creating uranium-233 from thorium-232—operated successfully at Oak Ridge from 1966 to 1969. But despite the success and despite great enthusiasm at Oak Ridge for the reactor, the AEC chose to cancel funding for it and concentrate its efforts on a fast-breeder reactor. The reason, Weinberg says, was that the molten-salt reactor was too different. By the late 1960s, nuclear technology had gone far enough in one direction that it was nearly impossible to push it onto a radically different course. He concludes:

> Perhaps the moral to be drawn is that a technology that differs too much from an existing technology has not one hurdle to overcome—to demonstrate its feasibility—but another even greater one—to convince influential individuals and organizations who are intellectually and emotionally attached to a different technology that they should adopt the new path. This, the molten salt system could not do. It was a successful technology that was dropped because it was too different from the main lines of reactor development.

It is unlikely that we will know anytime soon whether other technologies would have been superior to light-water. The lock-in is so complete that, even with the major problems facing nuclear power today, there are no serious attempts now being made to develop an alternative to light-water reactors. For more than a decade, the U.S. nuclear industry has been trying to figure out how to make nuclear power acceptable in the United States once again. No orders for new nuclear plants have been placed since 1974 (except for a few that were later canceled), and there are none on the horizon. An industry task force has put together a strategic plan calling for major changes in a dozen areas, including plant operations, licensing, waste disposal, and public acceptance, but one thing won't change. Light-water will remain the technology of choice. The reactor designs will be improved, simplified, and standardized, but they won't stray far from what the industry knows best. Thirty years after the start of the Great Bandwagon Market, light-water is still the only game in town.

Cowan's and Arthur's work goes a long way toward explaining how technological choices derive from a series of individual choices made by various players. Furthermore, the two researchers demonstrate something that would have seemed unlikely, if not impossible, to classical economists: even if every player makes completely rational decisions, the end result may be the choice of an inferior technology. Of course, real-life economic players do not make completely rational decisions, so there is even less chance than Cowan's and Arthur's models indicate of actually finding an optimal technology, or something close to it.

In the case of nuclear power, the deviations from rationality were, as we've seen, often quite dramatic. When GE and Westinghouse offered their turnkey contracts, business rivalries and technological optimism played as large a role as did objective economic analysis. But even more interesting is how far the AEC and the other government players strayed from rational decision making. If a central authority is to improve the chances of finding the best technology, then it must do so by providing what the free market can't: a commitment to maximizing the long-term rewards and a relative immunity to the lure of short-term results. Yet, at least in the case of nuclear power, the government had as difficult a time doing this as did the commercial nuclear industry.

The AEC did start off well. It approached the multi-armed bandit in a more rational way than the free market would have or could have. By subsidizing different technologies, the AEC allowed the nuclear industry to get information about likely payoffs from various types of reactors that the industry would not have gained otherwise. But while the AEC was trying to optimize commercial nuclear power over the long run, other government interests were hijacking the nuclear program.

The nuclear navy was the first. Rickover's decision to go with light-water tilted the commercial decision toward the same technology. Still, the edge that light-water got from the Navy was probably not an insurmountable edge. After all, everyone recognized that while light-water reactors were a good choice for sea duty because of their size, their strengths and weaknesses didn't match up well with commercial power needs.

Instead, it was the government's own desire to develop nuclear power quickly that forced the issue and gave light-water a lead it would never relinquish. When Eisenhower and the Joint Committee wanted a demonstration plant because other countries seemed to be getting ahead in the nuclear power race, light-water was the only option. The result: Shippingport and its long-lasting effects on the industry. A few years later, the president and Congress worried that the U.S. nuclear industry might get shut out of the European market. The result: the United States-Euratom treaty, AEC subsidies, and three light-water plants built in Europe. Meanwhile, the same people worried that the domestic nuclear industry wasn't moving fast enough. The result: subsidized nuclear research, subsidized fuel, subsidized insurance, and a threat (in the 1956 Gore-Holifield bill) to have the government build reactors if the private sector didn't.

With the exception of a few nuclear enthusiasts, no one in the utility industry was in a hurry to build atomic power plants. Without the push from the AEC and the Joint Committee, light-water would not have been locked in so quickly, and there might have been a real competition among it, the CANDU heavy-water technology, and the gas-cooled reactors developed by the British and the French. It's even conceivable that some other technology—organic-cooled, say, or molten-salt—would have gotten a chance to show what it could do, or that, without such a big push from the U.S. government, nuclear power would not have grown up so quickly and prematurely. In short, the government—which in Cowan's model was making rational decisions with an eye on the long-term payoffs—acted to lock in a particular choice even sooner than would have happened if things had been left up to private industry.

That shouldn't be surprising. In the ideal world of economics, the central authority may be an objective body, making its decisions on the basis of calculations and expert opinions in an effort to settle on the best technology. But in reality, government bodies are subject to all sorts of pressures, both internal and external, that urge them to push a technology in one direction or another. And the more important and powerful the technology, the greater these pressures will be.

It starts with the prejudices and motives of the people involved. Just as individual economic players in a market are not totally rational, neither are individuals in government bodies. The members of the Joint Committee, for instance, were driven by an enthusiastic desire to put the atom to work. They had no real interest in finding which reactor type would be the best choice in a utility-maximizing sense. They just wanted some reactor—any reactor—to be adopted by the electric utility industry. The commissioners and staff members of the AEC were more thoughtful advocates of atomic energy, but they were still advocates. As engineers and businessmen, they realized the importance of developing nuclear power with as much forethought as possible, but they also had a mandate to promote civilian nuclear power—and sooner rather than later.

Institutional inertia also often influences decision-making bodies. Once people have resolved to go in a certain direction or have spent a great deal of time and effort on one option, they tend to resist the alternatives. Changes threaten their ways of thinking, perhaps even their jobs. As Weinberg discovered in trying to convince the AEC to continue funding the molten-salt breeder, it's difficult to redirect a bureaucracy once it has built up momentum.

The most important pressures are the conflicting demands that arise from other parts of the government and from pressure groups outside government. An agency like the AEC may wish to determine and develop the "best" technology, but it probably won't be left alone to pursue that course. Once such power is vested in an agency, others want to use it for their own purposes. In the case of nuclear power, the AEC found itself called upon to help out with a number of tasks—making the country more economically competitive, boosting its prestige, and fighting the Cold War against the communists.

There is no reason, of course, that technology policy shouldn't take such things into consideration. It's a political question how much the government should seek to shape technology in pursuit of its own ends. But if the case of nuclear power is any indication, the U.S. government has a difficult time restricting itself to the purely economic goal of helping to determine the technological path that offers the greatest economic returns. And once political goals come into play, the government finds it difficult to guide the bets on the multi-armed bandit effectively.

The one remaining question is: So what? Granted that the market will sometimes lock in technologies that are not the best possible ones—QWERTY keyboards, VHS videotapes, perhaps the light-water reactor—but is it important? No one knows how common such inferior choices are, and no one knows just how much better some technologies could be, since it's normally impossible to compare an existing technology with one that was never fully developed. There are some examples of truly bad technological choices—the Russian RBMK reactor, for example, whose design flaws led to the Chernobyl disaster—and the tolls from such choices may be quite high, but they are relatively rare. Most technological decisions, particularly in open societies with free markets, seem to be at least defensible, if not close to optimal, and that may often be good enough.

It is the nature of economists to worry about absolutes: What is the best technology? or What is the best allocation of resources? But people in general are more likely to think in terms of whether something is acceptable and whether it's worth the effort to change it. Yes, Dvorak is a much more efficient system than QWERTY, but does the difference really matter? Most people would probably say no. If a locked-in technology is "good enough," people don't care whether it's the absolute best. If it's not good enough, it probably won't stay locked in for long.

Still, the existence of lock-in is of more than academic interest. Any time a complex, expensive technological innovation is under development, the chances are good that the best choice will lose out. Cowan calculates that when the choice is between two competing technologies, one of which has a potential payoff 10 percent higher than the other, the chances are about three in ten than the inferior selection will capture the market—even with a relatively small degree of increasing returns. Although Cowan cautions that these numbers come from the particular set of equations he uses to model economic choice and don't necessarily reflect a particular real-world market, they nonetheless offer something to think about. Three chances out of ten that a technology with a 10 percent lower payoff will prevail is probably a good ballpark figure for at least some technologies. And in the case of a major technology like nuclear power, which represents an investment of hundreds of billions of dollars, a 10 percent difference in return adds up to some real money. It may be impossible to guarantee an optimal choice on the multi-armed bandit, but knowing what's at stake may make everyone a little less likely to rush to a decision.

six

RISK

In June 1995, speaking to an audience of 250 fellow doctors and medical researchers, Steven Deeks described what he hoped would be a breakthrough treatment for AIDS. The human immunodeficiency virus, which causes AIDS, attacks key components of a person's immune system and gradually destroys the body's ability to fight off infection. Consequently, an AIDS patient generally succumbs to what doctors call "opportunistic infections"—invasions by viruses, bacteria, and other microorganisms that take advantage of the body's weakened defenses. If some way could be found to rebuild a patient's devastated immune system, Deeks said, it could be lifesaving news for the 100,000 or so Americans in the advanced stages of AIDS. It might even make it possible for people with AIDS to live relatively normal lives.

The treatment Deeks advocated was dramatic. He proposed extracting bone marrow from a baboon, separating out a special portion of it, then injecting that bit into a patient with an advanced case of AIDS. Because bone marrow contains the special cells that produce the immune system, Deeks hoped that the bone-marrow extract would create a baboonlike immune system in the patient. And because baboons are immune to AIDS, Deeks surmised that the patient's new immune system could survive and do the job that the old, AIDS-wracked system no longer could perform—fight off disease-causing invaders. The AIDS virus would still be present, lurking in the remnants of the patient's own immune system, but its main threat to the patient would have been deflected. For Deeks, a San Francisco physician who treats many AIDS patients, it was a gamble that had to be taken.

Deeks spoke at a conference on xenograft transplantation, the medical term for the transplant of organs or tissue from one species into another—

particularly, from animals into humans. The audience, most of whom were xenotransplant researchers, generally approved of Deeks's proposal, but there were dissenters. The most vocal was Jonathan Allan, a virologist at the Southwest Foundation for Biomedical Research in San Antonio. Pointing out that AIDS probably originated as a monkey virus that somehow jumped the species barrier into humans, Allan warned against transplanting tissue or cells from baboons into humans. Baboons could harbor an unknown virus as dangerous as AIDS, he said, and a treatment like Deeks proposed could transfer the virus directly into a human. "Are you going to take a bet that AIDS is only going to happen once?" he asked.

Thus provoked, the audience began to debate the merits of Deeks's idea. His proposed trial should not pose any real risk to the population, one doctor argued, since the subject of the trial will already have a deadly virus and will be taking precautions not to spread it. True, Allan responded, as long as Deeks treats only one carefully selected patient. "But if it works, you'll probably see hundreds or more [getting the baboon bone-marrow treatment] very quickly, and then you'll have problems." It is possible to screen baboons for known viruses, another pointed out, but there are undoubtedly many viruses infecting baboons that researchers have never discovered. Would it be possible to raise baboons in special disease-free colonies? Yes, but it would be expensive, and even then there would be no guarantee that the baboons didn't carry some virus that had no effect on them but that was dangerous to humans. If there were such a virus, wouldn't the people handling the baboons have contracted it? Not necessarily, since some viruses, such as the one that causes AIDS, are passed only through bodily fluids, particularly blood. What about monitoring the recipients of baboon cells or tissue to make sure they didn't come down with some new disease? It would be a good idea, but it wouldn't stop a new AIDS-type epidemic, since it takes many years for symptoms to appear in a person who has been infected with the AIDS virus.

Back and forth it went, with no clear resolution. What did become clear, however, is that there is a great deal of uncertainty about what might happen were doctors to start putting bits of baboons into humans. Is there another virus as dangerous as AIDS hiding in baboon blood? No one knows. And even if researchers did have a list of every virus that infects baboons, they still couldn't guarantee that the transplants would be safe. It's possible, one researcher pointed out, that a patient whose immune system is weakened from AIDS would offer an ideal place for viruses to reproduce and trade bits of genetic material. From this cauldron, new hybrid viruses might emerge, and some of them might be dangerous to humans.

The doctors in the audience seemed to have little patience with such hypothetical dangers. People are already dying, they said. How can we refuse to help them merely because someone dreams up a worst-case scenario that is better suited to the movies than to medicine? The public-health experts, on the other hand, were unwilling to ignore the possibility of unleashing a new epidemic. By the end of the meeting, a consensus of sorts had emerged: Experiments like Deeks's should go ahead, but with great caution. And before baboon-to-human transplants move from clinical trials to widespread use, researchers will need to learn much more about the potential dangers.

Some day, Deeks's experimental treatment may become a full-fledged medical technology, although as this was written, in late-1996, its benefits are still theoretical. In late 1995, Deeks performed the procedure on one volunteer, a San Francisco man named Jeff Getty who was in the advanced stages of AIDS. Getty was apparently not harmed by the treatment, but neither was he helped. Tests indicated that the baboon bone marrow never grew and functioned in its new body, perhaps because Deeks did not inject enough of the cells or did not weaken Getty's immune system enough before the bone marrow was put in to keep the immune system from rejecting the transplant. It will likely take many more trials to determine if this is a feasible treatment for AIDS.

Successful or not, the baboon-to-human transplant points to the most vexing problem facing technology: how to deal with risk. Much of modern technology is potentially dangerous. Not just big, scary things like nuclear power plants and oil tankers, but comfortable, everyday items like the cars we drive and the food we eat. Part of the difficulty in dealing with risk is knowing just what the dangers are. As the case of the baboon bone marrow AIDS treatment illustrates, it can be difficult to figure out ahead of time what can go wrong. Could a baboon virus kill the patient? Could a deadly baboon virus spread from the patient into the general population? Could a baboon virus swap DNA with a human virus in the patient—perhaps the AIDS virus—and create a new supervirus? How likely is it that any of these things might happen? And even if these questions can be answered, that's only half the battle. How does one balance the risk against the potential benefit? Suppose that the bone-marrow transfer poses a small, but not insignificant, risk to the public health. Do we give thousands of AIDS patients a chance to live, despite the risk of creating a new epidemic? And who decides?

Traditionally, scientists and engineers have seen risk as a purely technical issue, one that can be boxed off from the rest of a technology and handled separately. But for modern complex technologies this is often impossible. When risk is involved, technical and nontechnical issues can get tangled to the point that they are impossible to separate. Consider, for instance, the strange case of the high-tech cow.

BOVINE BROUHAHA

Dairy farming in the United States is a funny business. For decades, dairy farmers have produced much more milk than they can sell. In any other industry, either the companies would cut production or else the less efficient producers would go out of business. But not here. In the late 1940s, the federal government began a program to guarantee an adequate supply of milk, and by the 1980s the government was paying farmers more than $1 billion annually for their surplus butter, milk, and cheese. As a result, dairy farmers produced as much milk as they could, secure in the knowledge that they could always sell the extra to the government. Things got so bad that in 1986 and 1987, in order to cut down on the surpluses, the government paid 14,000 farmers some $1.8 billion to slaughter one and a half million of their cows and calves. That did cut the oversupply somewhat, but the government continued to support milk prices, ensuring that the nation's farmers would continue to produce more milk than the country needed.

Against this backdrop, four separate drug companies announced in the mid-1980s that they would soon offer a revolutionary new product that would—of all things—increase milk production. Recombinant bovine growth hormone, or rBGH, would boost the amount of milk each cow put out by anywhere from 10 to 25 percent. The drug companies, expecting hundreds of millions of dollars a year in sales, were each pumping in tens of millions of dollars in research and development costs, making rBGH the most expensive animal drug ever created.

The drug would, in effect, allow farmers to turn up the milk production dial on their cows. In nature, the amount of milk a cow produces is determined in part by how much bovine somatotropin, a natural protein, is manufactured by the cow's pituitary gland. Long ago, researchers showed that they could artificially increase a cow's production by injecting it with bovine somatotropin collected from slaughtered cows, but at the time this had no practical application because collecting the protein cost much more than the value of the extra milk produced.

This changed in the late 1970s, when Genentech scientists isolated the gene in cows that directed the production of bovine somatotropin. Now, with genetic engineering techniques, they could create bacteria that would turn out bovine somatotropin nearly identical to that from a cow, but at a much lower cost than by "harvesting" it. This recombinant bovine somatotropin—or, as it was more commonly called, recombinant bovine growth hormone—would have the same effect on cows as the natural protein. It would shift their udders into overdrive.

Each of the four companies that would offer rBGH—Monsanto, Eli Lilly, Upjohn, and American Cyanamid—had a slightly different version. Of them, only Upjohn's contained exactly the same sequence of 191 amino acids—the molecular building blocks of proteins—as the natural form. The others differed from the natural hormone by up to nine amino acids, although the companies said that the slightly different structure would have no effect on the hormone's function.

From the standpoint of the drug companies, rBGH seemed an ideal product. It was natural—no different from what cows themselves produced—and it would allow dairy farmers to increase their productivity without making major changes in how they ran their farms. Cows given the hormone would have to eat more, but the extra milk they produced would more than pay for both the extra grain and the drug. Granted, the United States didn't need any more milk, but that wasn't the drug companies' problem. "We saw this as a continuation of all the technologies that have helped the dairy industry," recalled Robert Schenkel, who headed up American Cyanamid's research on rBGH. So in the early 1980s, the four companies began the long process that they hoped would lead to approval from the Food and Drug Administration (FDA) to market the growth hormone.

Things went well at first. In 1984 the FDA concluded that the milk produced by rBGH-treated cows was safe for human consumption. Within a few years, all four companies were conducting field trials in which herds of dairy cattle were given the hormone. The results allowed the companies to tell farmers how much they could expect to boost their yields with rBGH and gave the FDA the data it needed to assess the drug's effects on the cattle. The milk produced from the test herds was sold alongside milk from untreated cows.

But meanwhile, in 1986, use of the drug ran into its first organized opposition. At about the time that dairy farmers began slaughtering cows as part of the program to reduce the milk surplus, a collection of animal-rights organizations, environmentalists, and farm groups asked the FDA to prepare an environmental impact statement considering the effects of rBGH. The petition claimed that making rBGH available would "damage the environment, cause unnecessary suffering to cows, and wreak havoc on the dairy community." It was the last claim that got the most attention. The coalition had relied on a study done by an agricultural economist at Cornell University with no ties to the anti-rBGH group. The study had concluded that introducing rBGH could cause a 25 to 30 percent drop in the number of the nation's dairy cows and farms. And the farms hardest hit would be the small farms, which were already having trouble competing with the large factory farms. Although the risk to small farms was not something the FDA by law could take into account, it struck a chord elsewhere. Editorial writers decried rBGH for the damage it would do to small dairy farming communities. In 1989, the ice cream company Ben & Jerry's started a campaign against rBGH, putting labels on its products that read, "Save family farms—no BGH." And in Europe the European Commission discussed an amendment to its drug approval process that would reject any drugs for use on farm animals that would have adverse socioeconomic effects. Although the proposed amendment was eventually dropped, the European Parliament did impose a two-year moratorium on the use of the drug in order to consider the possible consequences of rBGH for farmers.

The instigator and organizer of much of the opposition to rBGH was Jeremy Rifkin, a well-known crusader against biotechnology products of all sorts. In 1985, Rifkin had made headlines by trying to stop testing of a genetically engineered bacterium intended to prevent frost damage to crops. In 1986, he had challenged the use of a genetically engineered vaccine in pigs. And Rifkin saw rBGH as his most important cause yet. "This is for the big marbles," he told the *New York Times* in a 1988 interview. "If this product moves onto the market, it opens the floodgates for biotechnology on farms." It's hard to escape the conclusion that for Rifkin, if not for many other opponents of rBGH, the issue wasn't the demise of small dairy farms, harm to the cows, or possible damage to the environment. It was biotechnology itself, and Rifkin would use any argument he could find to try to stop the genetically engineered hormone. It didn't take Rifkin and his allies long to figure out that the most effective tactic against rBGH was to question its safety.

Although the FDA had already decided that humans were at no risk from drinking milk from rBGH-treated cows, opponents of the drug began to bring up issues they said the FDA had missed or not considered adequately. The most vocal scientist to join Rifkin's crusade was Samuel Epstein, a physician and professor of occupational and environmental medicine at the University of Illinois College of Medicine. Epstein, an environmentalist, published a paper that challenged the FDA's conclusion that rBGH was safe. As *Science* magazine reported, Epstein had done no experimental studies on the hormone, but he said that looking at studies by others had convinced him "that there are unresolved questions, including whether rBGH stimulates premature growth in infants and breast cancer in women." In particular, Epstein was concerned because cows treated with rBGH had higher-than-normal levels of insulin-like growth factor 1, or IGF-1, in their milk. It is IGF-1 that instructs a cow's lactating cells to produce milk; in humans, it triggers the production of human growth hormone. Epstein claimed that IGF-1 from cow's milk might make its way from the stomach into the bloodstream and, by increasing the amount of growth hormone, cause unnatural growth.

To the public, Epstein's charges seemed serious—they came, after all, from a physician and scientist—and they seemed scary. Could milk from hormone-treated cows cause children to grow abnormally and women to develop breast cancer? On the surface, it looked possible. Yet when other scientists examined Epstein's scenario, they pooh-poohed it. Yes, the FDA acknowledged, milk from rBGH-treated cows did have 25 percent or so more IGF-1 than milk from untreated cows, but even if IGF-1 had an effect on people, it was no more serious than the consequences of drinking five glasses of milk a day instead of four. Furthermore, FDA scientists pointed to studies in rats showing that IGF-1 is broken down in the digestive tract before it gets into the bloodstream. The extra IGF-1, it seemed, would have no effect on humans.

Epstein and others also questioned what rBGH might do to the health of the cows. Citing documents leaked from inside the FDA, the opponents reported that cows treated with rBGH had a higher level of infections. This, they charged, would lead farmers to use more antibiotics on the cows and lead to high levels of antibiotics in the milk. And any increase in the amount of antibiotics that humans are exposed to could ultimately harm the antibiotics' effectiveness in fighting infection by leading to strains of microorganisms resistant to the antibiotics. Once again, however, the charges were less serious than they sounded. Cows were indeed more likely to contract mastitis, an inflammation of the udder, when treated with rBGH, but at least some of the studies demonstrating this were done at doses five times higher than would normally be administered. And the issue of antibiotics in milk was a red herring. Because farmers normally treat cows with antibiotics to fight infection, government regulations already existed to make sure that no milk containing antibiotics got to the public. Farmers must discard the milk from cows being treated with antibiotics until treatment has stopped and the antibiotics have had time to leave their systems. Furthermore, all milk is monitored for the presence of antibiotics, and no contaminated milk is allowed to be sold.

To address the safety concerns, Congress asked the National Institutes of Health to put together an independent panel to study the possible health effects of rBGH on humans and cows. Its conclusion, released in November 1990, was that both milk and meat from rBGH cows were safe for human consumption and that the drug had no major effects on the health of the cows. To the FDA and the drug companies, it seemed that this should put the debate to rest. The scientific community had spoken. But the FDA and drug companies were assuming that the risk controversy was purely a technical issue, one that could be answered by reaching a scientific consensus. It wasn't, and the controversy did not go away.

In 1989, pressure from Rifkin had led five large supermarket chains to swear off using milk from rBGH-treated cows in any dairy products with their stores' brand names. It wasn't much of a sacrifice, since rBGH was still undergoing field tests and only a few farms were actually using the hormone. But five years later, after the FDA had approved rBGH for commercial sale, grocery stores remained shy of the hormone. The FDA decision came on November 5, 1993, and the drug went on sale February 5, 1994, following the expiration of a three-month moratorium imposed by Congress. By that time, Rifkin had assembled the Pure Food Campaign, a coalition of farm groups, animal rights organizations, and consumer groups that had protested the FDA decision with milk-dumping ceremonies in a number of cities. The protest campaign, Rifkin told the *New York Times*, had eighteen full-time employees and was spending about $100,000 a month. And it was working. Both the Kroger Company, which ran the largest supermarket chain in the country, and the Southland Corporation, parent of 7-Eleven convenience stores, announced they would tell their suppliers they didn't want any milk from rBGH cows. Several smaller grocery chains followed suit.

The companies professed not to be worried about the safety of the milk. Instead, they said, they were responding to the worries of their customers. Enough doubts had been raised about the use of the hormone that people didn't want to take chances, so the groceries took a cautious approach. Citing similar reasoning, a number of school boards opted to keep milk from rBGH-treated cows out of school lunches.

In April, two months after the drug went on sale, both Maine and Vermont passed labeling laws aimed at rBGH. Vermont demanded that any products made from rBGH-treated cows be identified as such. And Maine established an official label to distinguish products free of milk from hormone-treated cows. At the same time, companies in other states were developing their own labels to assure consumers that their milk was not tainted by any hormones. Monsanto, the first company to put its rBGH on the market, responded by suing some of the labelers. "We contend that these companies falsely implied to consumers that their products are more wholesome and in other ways superior to products that come from milk from BST-treated cows," a company spokesman said.

What had gone wrong? Why hadn't the controversy faded away after the scientific community gave its seal of approval to rBGH? To the technically minded supporters of rBGH, the reasons seemed clear. Rifkin and a few others were consciously misrepresenting the science for their own ends, and much of the public wasn't savvy enough to see through it. Jerry Caulder, president of the Industrial Biotechnology Association, spoke for many when he blamed the continuing controversy on the public's lack of understanding. It was, he said, "the result of living in a society that's relatively scientifically illiterate." If only the public could be educated enough to comprehend the scientific data, there would be no disagreement.

Certainly much of the public didn't understand the science behind the debate. But would a more scientifically literate public have behaved any differently? There is no reason to think so. The opposition to the drug had its roots in factors that were independent of scientific knowledge—things such as opinions of how the world works and beliefs about the best way to deal with uncertainty.

Much of the doubting public, for instance, was suspicious of the government and big business. When the *New York Times* interviewed Julie Korenstein, a member of the Los Angeles board of education who had played a major role in keeping milk from hormone-injected cows out of city schools, she explained that she didn't trust the FDA to tell the whole story. "I think a lot of it was swept under the rug. I don't think we have all the information that is out there." Kathy Steele, a dietitian, told the paper that she only bought milk that she knew was not produced with the hormone. "I know the government says things are safe that turn out not to be safe. You can't always trust what they say."

The organizations opposing rBGH fanned these suspicions by painting the FDA as being too cozy with the drug companies it was regulating. The groups revealed papers leaked by FDA insiders, which seemed to imply that the FDA was trying to hide damaging evidence. And the conspiracy charges got a big boost in November 1989 when the FDA fired the veterinarian who had been in charge of the rBGH review from 1985 to 1988. Richard Burroughs said that he had been kicked out not because of incompetence, as the FDA stated, but because he had pointed out problems in the way the drug companies had done their safety studies. The FDA, he said, was too eager to help the companies get their drugs to market.

More fundamentally, rBGH opponents claimed that the scientific evidence about rBGH was not trustworthy because most of the studies had been paid for by the drug companies. "This creates an inherent conflict of interest, and a conscious and subconscious effect on those receiving the funds," charged Epstein, the scientist who had raised many of the early questions about risk. His implication was that researchers whose salaries were paid by drug companies were unlikely to discover anything that would not be pleasing to those companies.

Such arguments, anathema to scientists, were convincing to many outside the scientific community. Arthur Kaplan, director of the University of Minnesota's Center for Bioethics, spent a lot of time at public hearings where scientists tried to convince their audiences that milk from rBGH-treated cows was no different from regular milk. Often, he said, the public didn't buy it. "People just look at the scientists and say, 'I don't believe you.'"

This was not something that could be fixed by making sure everyone in the general public had a working knowledge of biochemistry. As David Garry, a molecular cardiologist at the University of Texas Southwestern Medical Center, told one reporter, "I distrust scientists too, and I am a scientist." The problem, he said, is that scientists are motivated by their own agendas, not the public's. Scientists are eager to push ahead in their own areas, and they take it for granted that the loose ends, such as questions about safety, can be tied up later.

In short, the disagreement had its roots in a clash of world views. The drug companies and the FDA saw the analysis of rBGH's risk as a purely scientific process, rational and value free. Others perceived it differently. They took it for granted that science—particularly science done for drug companies and the like—is skewed by the beliefs and motivations of its participants. From this standpoint, science needed to be constantly questioned and challenged and its claims taken with grains of salt.

As this is written, it seems as if the anti-rBGH movement has faded away and rBGH will become an accepted part of the dairy industry. Many farms, particularly large farms, are using it because it improves their profits, and most consumers don't seem to care. A market has grown up for dairy products that are certified to come from rBGH-free cows, but it is a small market. Jeremy Rifkin has moved on to other battles. For better or worse, most cows in the future will be getting a boost from rBGH and producing 20 percent more milk than they would normally.

The rBGH debate was never just about risk, of course. As risks go, the potential dangers of using rBGH seem relatively minor and unlikely, and if the threat to small farms had not existed, it's difficult to believe that the controversy would have been a major one. Nonetheless, the clash over rBGH is indicative of the sorts of battles that will become increasingly common in the future, particularly over biological technologies.

Traditionally, engineers have treated risk as one more technical issue to be solved. It has been their responsibility to guarantee that their products are safe, or, if there is some unavoidable risk, to weigh the risks and the benefits of the product and determine whether to proceed. Risk decisions have been made inside the system. But two characteristics of modern technology have thrown open the process. The first is the way that many technologies are a potential threat to innocent bystanders—that is, to those who had no say in the choice to use the technology. Transplanting baboon bone marrow into a human might be dangerous to the patient, but it might conceivably threaten the general public health as well. Thus the public may well demand a say as to whether that procedure becomes an accepted treatment for AIDS. The second characteristic of modern technology that threatens to take risk decisions out of the hands of engineers is its complexity. Because of that complexity, assessing and minimizing risk becomes a very uncertain process.

This is nowhere more evident than in technologies that manipulate living things. Whether it is injecting extra hormones into cows, putting genetically modified viruses on crops, or transplanting baboon bone marrow into humans, scientists can never say with certainty what will or won't happen. Living organisms are simply too complex. Furthermore, it's impossible to consider organisms in isolation. They are unavoidably part of an intricate web of life, constantly acting on other organisms and being acted upon in turn. And living organisms have the unfortunate characteristic that they tend to multiply whether you want them to or not—and, in the cases of bacteria and viruses, multiply very quickly. Build a nuclear power plant and you will have exactly one nuclear power plant now and in the future, unless you build more. But create one new bacterium or virus, and under the right circumstances it will shortly be billions of bacteria or viruses. And, since mutations are likely, those later generations will probably be different from the original version in unpredictable ways.

Michael Hansen, a scientist with Consumers Union (the publisher of *Consumer Reports*) and an opponent of rBGH, offered one example of how tampering with the web of living organisms might prove dangerous. In a March 1994 letter to the *New York Times*, he repeated the assertion that rBGH would cause additional infections in cows, leading farmers to dose them more often with antibiotics. But instead of worrying about the effects of the antibiotics on people, Hansen focused on what they might do to the bacteria that infect cows. The increased use of antibiotics might lead the bovine bacteria to develop extra resistance to the antibiotics, and somehow these bacteria might transfer this resistance over to bacteria that infect humans, creating new strains of human pathogens that could not be easily treated with drugs. It was scientifically plausible—no one could say it wouldn't happen—but what were the chances? Most scientists would write the scenario off as no more likely than a cow jumping over the moon, but they couldn't *prove* it. And even if someone could prove this chain of events was impossible, Hansen or Epstein or someone else could come up with another, and another, and another. The system of cows, humans, and bacteria was simply too complex to analyze in any but the crudest detail. The calculation of risk could only be an approximation.

In such cases, where there is no clear right or wrong answer, people tend to rely on their instincts, biases, and gut feelings about how the world works. Some scientists, call them "builders," see the world as a place to be explored and manipulated. They trust in the power of human reason and believe that acquiring knowledge and putting it to use are acts of virtue. Others, call them "conservers," are less sanguine about human reasoning ability and suspect that, unchecked, human manipulation of the environment will lead to disaster. They see nature as something to be protected and to be respectful of, and they consider unrestricted growth and development not just dangerous but often morally wrong.

When presented with the same evidence about potential risks, a builder and a conserver may come up with opposite conclusions. The builder, whose bias is toward going forward whenever possible, is satisfied with what he considers proof beyond a reasonable doubt. The conserver has the nagging suspicion that one day the world will spring a nasty surprise on the human race. The builder acknowledges the possibility of surprises but is sure they can be dealt with. The conserver isn't so sure, and counsels caution. The mistakes of the builder are generally sins of commission—going one step too far, overlooking warning signs that will look obvious in retrospect. The conserver specializes in sins of omission—lost opportunities and missed chances from being overcautious.

When these two sorts of scientists speak with one another about risk, they use the language of science—hypothesis, experimentation, cause-and-effect—and so it may seem that their disagreements are rational ones that can be settled by some further discussion or by some additional experiments. But in effect they often talk past one another because they have different sets of assumptions about proof and about what sorts of questions are worth pursuing. Eventually, debates like the one over rBGH come down to a matter of where to draw the line. When has enough evidence been gathered and enough scenarios considered that one can conclude that the risk is worth taking? Such line-drawing decisions are not, and can never be, purely technical.

And once the debate leaves the realm of the purely technical, it also stops being something that scientists and engineers can claim a monopoly on. Science has no special ability to decide what to do in the face of uncertainty. Margo Stark of the Minnesota Food Association, an activist group opposing the use of rBGH, made precisely this argument in a 1991 interview with *Technology Review*. "What they [the FDA and drug companies] fail to address in the assertion that all policy should be science-based is the fact that we have to make public policy choices based on a certain amount of ignorance on everybody's part, including the scientists. How we want to proceed in the face of unknown risk, such as in the case of rBGH, really becomes a matter for democratic policy-making."

In technology after technology—nuclear weapons, biotechnology, medicine, and others—members of society have used this argument to step in and make their voices heard in discussions over risk. How should decisions be made concerning risk, and who should make them? We will look at these questions in the next chapter.

For now, however, we will transfer our attention to nuclear power. No technology has been so wracked with questions about risk. Part of the reason is the nature of the beast. Nuclear power is inherently risky, since it generates radioactive materials that are very dangerous, and this risk extends to people who live hundreds of miles away from a nuclear plant. But just as important, nuclear power was a path breaker. Much of today's thinking on risk did not exist when nuclear power was being developed because the issues simply had not come up. So the nuclear pioneers made a number of mistakes and miscalculations that would be less likely made today. Indeed, much of today's thinking on risk was catalyzed by the mistakes and miscalculations that dogged the development of nuclear power.

SEARCHING FOR SAFETY

Looking back from the vantage point of a post–Three Mile Island, post-Chernobyl world, people sometimes wonder why we ever went ahead with nuclear power. Didn't anyone realize how dangerous it was? Didn't anybody think about the risks to people living close to nuclear plants? Didn't anyone consider the implications of generating so much nuclear waste? These things seem so obvious today.

But it was a different world in the years after World War II. For starters, the environmental consciousness so prevalent now did not exist, nor was there nearly as much concern about small risks to the public health. Indeed, people were cavalier about radioactivity to an extent that seems shocking today. As late as 1958, for example, the Atomic Energy Commission was exploding bombs above ground at its Nevada test site, and many tests in the 1950s created measurable amounts of radioactive fallout that landed outside the test area. After one test, people were warned to stay inside because the wind had shifted and carried a cloud of radioactive dust over a populated area. The exposures were probably too small to be dangerous, but it's impossible to imagine such tests being carried out in the 1990s.

The people in charge of the country's nuclear programs were not unconcerned with risk, however. They simply had a different perspective from that of today. No one worried much, for instance, about very small doses of radiation. They knew that humans are exposed to low levels of radiation from a number of natural sources, such as cosmic rays, and figured that a little extra didn't matter much. More important, forty and fifty years ago people had a tremendous faith in technology and in their ability to handle technology. They recognized that nuclear reactors were potentially dangerous and generated vast amounts of radioactive waste, but they assumed that nuclear scientists and engineers would find solutions. It wasn't even something they thought to debate—Can we make reactors safe? Can we figure out a way to dispose of nuclear wastes?—but instead was a silent premise that underlay all thinking about risk. Disputes concerning risk were always about "how," never about "can."

It is strange, even discombobulating, to read the descriptions and discussions of nuclear power from that era, so full of enthusiasm and so lacking in self-doubt. Could this really have been us only two generations ago?

But what is perhaps even stranger is how little input the public had—or wanted to have—in any of it. Risk was considered a technical issue, one best left to the experts. And for the most part, the engineers agreed. Risk was a question whose answers could be found in facts and figures, in engineering calculations and materials testing. And so for twenty-odd years, the engineers and other experts debated among themselves how best to keep the nuclear genie safely in its bottle.

In the beginning everything was new. No one had built, much less operated, nuclear plants before, and the people worrying about safety were flying by the seats of their pants. What types of things could go wrong? How likely were they? What were the best ways to prevent them? There was precious little information with which to answer these questions, and so, not surprisingly, the answers that people came up with depended on their biases and points of view. In particular, there were three main groups that influenced the debate on nuclear safety—the AEC, the Advisory Committee on Reactor Safeguards, and the national laboratories—and each had its own personality and predilections.

The AEC wished to promote nuclear power. That was the assignment given to it by the Atomic Energy Act, and if the progress toward a nuclear future seemed too slow, members of the Joint Committee on Atomic Energy were sure to complain. At the same time, however, it was the AEC's responsibility to make sure that nuclear power plants were safe. There was a tension between the two goals; paying too much attention to safety would slow development, while paying too much attention to development might compromise safety—and many thought that safety was the loser. The AEC denied this, and indeed it did impose many regulations that the nuclear industry and the Joint Committee thought were needlessly conservative. On the other hand, the AEC regularly disregarded the advice of its own experts who would have liked much more stringent requirements.

Among those experts, the most important were the members of a group of scientists originally assembled in 1947 to counsel the AEC on reactor safety. Known first as the Reactor Safeguards Committee and later as the Advisory Committee on Reactor Safeguards (ACRS), the panel offered the AEC an independent source of expertise on nuclear safety. Its influence grew after 1957, when Congress mandated that it review all applications to build and operate nuclear power plants. While the ACRS's recommendations were not binding on the AEC, they generally carried a lot of weight, particularly in the early years. The members of the ACRS, being scientists, were most interested in the general question of how to ensure reactor safety, and they helped establish the philosophy underlying the AEC's safety regulations.

Where the ACRS was concerned with the big picture, the national laboratories were involved with the nitty-gritty of reactor engineering. What would fuel elements do if the core heated up past its normal operating temperature? What pressure could a reactor containment vessel be expected to withstand? Would an emergency core-cooling system behave as designed? The engineers at the national labs epitomized the technical approach to reactor safety. They asked questions and set out to answer them. By understanding all the technical details of a reactor, they believed that they could determine whether the reactor was safe. More than anything else, clear, unequivocal data were important.

All three groups were dominated by technological enthusiasts. They all wanted to see nuclear power plants built. But the enthusiasts were of two types, what the sociologist James Jasper calls professional and nonprofessional technological enthusiasts. At the ACRS and the national labs, the people were professional scientists and engineers. Their enthusiasm for nuclear power was tempered by an intimate familiarity with the technical details and a resulting appreciation of how difficult it was to assure safety. "The practice of engineering," Jasper says, "leads to a kind of artisanal pride as well as to technological enthusiasm: engineers are confident that they can build anything, but they want to take the time to build it properly." The AEC, on the other hand, was manned more by nonprofessionals, and the AEC commissioners were more likely to be lawyers or businessmen than scientists. Even the few scientists who served as commissioners generally came from areas outside nuclear physics. Thus the technological enthusiasm at the AEC tended to be a blind faith, unmoderated by the engineering realities.

In the early days of nuclear reactors, safety was assured by a simple scheme: put the reactors far away from where people lived. The Hanford site, where plutonium was produced for the atomic bomb during World War II, was a large government reservation in a desolate area. If a major accident struck, the only people at risk would be the workers at the plant. In 1950, the Reactor Safeguards Committee formalized this approach with a rule of thumb linking the power of the reactor with an "exclusion radius"—the size of the area around the plant in which people could not live. The radius was calculated so that even in the case of a major accident causing the reactor to spit out all its fuel into the atmosphere, people living outside the radius would not get a fatal dose of radioactivity. The committee's rule of thumb was a blunt instrument. It made no attempt to look at how likely it was that such an accident would occur, or even if such an accident was physically possible, but merely assumed the worst and planned for that.

Quite naturally, the larger a plant was, the larger its exclusion radius had to be. For small, experimental reactors, a typical exclusion radius was a mile or two—a workable size. But for the larger plants that would generate electricity for a city, the exclusion radius would be ten times that. And that was a problem. If a reactor had to be surrounded by empty land for ten or twenty miles in all directions, it would be impossible to put one anywhere near a city. A utility would have to pay to transmit its nuclear electricity long distances to its customers. Furthermore, the cost of buying the land for an exclusion zone would add even more to the price. If nuclear power were to be economically practical, a different way to assure safety would have to be found.

Within a year or so, reactor designers at General Electric had come up with a solution: they would put their reactor inside a large steel shell—a "containment" building—which would keep the reactor's radioactive materials from escaping in case of an accident. The safeguards committee accepted this idea, and containment became a standard feature of nuclear plants, beginning with a test reactor that GE built in upstate New York. With containment, the AEC was willing for reactors to be built much more closely to the surrounding population, although the commission did insist on keeping a certain distance between them.

To build a containment vessel, the engineers had to figure out the worst possible accident that might occur—the so-called maximum credible accident—and then design the vessel to withstand it. And so began a pattern that would characterize thinking about safety for decades to come. Someone—the reactor designers, the ACRS, the researchers at the national labs—would come up with a scenario for what might go wrong in a reactor, and then the engineers would be asked to design safety features to make sure such an accident didn't jeopardize public safety. In general, little thought was given to how likely such an accident was to occur. If it was physically possible, then it had to be taken into account. And in this approach lay the seeds of a tension between the AEC and its technical advisers that would grow with the size of the nuclear program. The AEC, whose job it was to see that nuclear power plants got built, had to draw the line somewhere. It did not wish to put off building plants while every possible thing that could go wrong was addressed. The ACRS and the national labs did not feel the same pressures. Their natural inclinations were to answer each technical issue as completely as possible. This struck many in the nuclear industry as extreme. A 1966 issue of *Nucleonics* quoted one industry representative's opinion of the ACRS: "Like Rickover, they say, 'prove it.' So you do, and they say, 'prove that'—and you can go on doing arithmetic forever. . . . They should have to prove the justification of their question. We always have to prove—they can just think, opine. Maybe they should be asked 'where?' 'when?' 'why?'" When the AEC sided with industry, as it often did, the nuclear scientists and engineers in the ACRS and national labs became frustrated with what they saw as a too-careless attitude toward safety.

As the nuclear industry began to sell more and more reactors in the 1960s, the AEC and its advisers struggled to address a host of safety issues. What, for example, were the chances that a reactor's pressure vessel would burst and release large amounts of radioactivity into the containment, and would the containment hold if the pressure vessel did burst? The pressure vessel was a large steel unit that surrounded the reactor core of a light-water reactor. It held the high-pressure water that was fed through the core to carry heat away from the fissioning uranium atoms. If the pressure vessel ruptured, it would release a tremendous amount of highly radioactive and highly dangerous material, and it might even send pieces of steel flying that could break through the containment. The issue has remained a tricky one because, although engineers know the capabilities of normal steel quite well, they're less comfortable with steel that has been bombarded by neutrons for a number of years. Neutron bombardment can make steel brittle and liable to break under conditions in which it would normally be quite strong.

But the most difficult-to-resolve issues centered on the so-called loss-of-coolant accident, or LOCA. If the supply of cooling water to a reactor core stops for some reason—a break in a pipe, for example—then part or all of the reactor core can come uncovered, and without water surrounding the core there is nothing to carry off its intense heat. In case of a loss-of-coolant accident, reactors are designed to shut down automatically by inserting all the control rods into the reactor core. The control rods absorb neutrons and stop the chain reaction. Unfortunately, the core stays hot even after the chain reaction is killed since once a reactor has operated for a while, radioactive "fission products" build up in the fuel, and they decay spontaneously, generating heat. The control rods do nothing to prevent this spontaneous decay and its raising of the core's temperature. The result is that without cooling, the fuel can get so hot that it melts, even if the chain reaction has been shut down.

Before 1966, the AEC assumed that even if the flow of coolant stopped and the reactor core melted, releasing radioactive gases, the containment building should keep the gases from escaping. But in that year the AEC realized that in a large reactor—1,000 megawatts and more—the fuel could get so hot after a loss-of-coolant accident that it might burn through the concrete floor and into the earth. This was dubbed the "China syndrome," a humorous reference to the direction the fuel would be taking, but not, as some would later think, an allusion to where the fuel would end up. Actually, it seemed unlikely that the fuel would descend much more than 100 feet into the earth before solidifying—and perhaps much less. But whether the fuel made it to China was beside the point. Such an accident could breach the containment and release radioactivity into the surrounding area. Even more worrisome was the possibility that a loss-of-coolant accident might cause a break in the containment somewhere besides the floor. If the hot fuel melted through the pressure vessel but not through the floor of the containment building, it might somehow generate enough gas pressure to blow a hole in the containment above the ground. This would release much more radioactivity into the atmosphere than would the China syndrome.

The ACRS and AEC agreed that a way had to be found to keep a loss-of-coolant accident from causing a breach in the containment, but they disagreed on the tactics. Several members of the ACRS wanted to explore ways to keep the containment intact even if the reactor core melted. It might be possible, for instance, to build a large concrete structure—a "core-catcher"—underneath the containment building. If the core melted through the floor of the containment, it would be captured and held here. But the AEC, under pressure from the industry not to delay the building of nuclear plants by adding extra requirements, wasn't interested. Instead, it would rely on two other strategies, demanding improved design of cooling systems to make loss-of-cooling accidents less likely, and focusing on preventing a core meltdown if the flow of coolant stopped. In practice, this second strategy meant an increased emphasis on emergency core-cooling systems, devices that could quickly flood the reactor core with water in case of a loss-of-coolant accident.

This may have seemed like the most direct route to solving the problem of a core melt, but it would actually make life much more complicated for the AEC and the ACRS. Now the safety of a reactor depended explicitly on engineered safeguards performing correctly in an emergency. Instead of relying on distance or containment, the engineers had to guarantee that, no matter what else happened, the core wouldn't melt. To this end, they designed more and more sophisticated systems and added backups to the backups. But would they work as planned? No one knew for sure. The forces generated in a reactor by a loss of coolant, sudden heating, and then a flood of cold water from the emergency core-cooling system would be violent and unpredictable. What would happen, for example, to the fuel rods when they were heated and cooled so brutally? It would take a great deal of testing and studies to find out.

At the time, however, the AEC was cutting back on research into such basic safety questions. Milton Shaw, the head of the AEC's Division of Reactor Development and Technology, was convinced that such safety research was reaching the point of diminishing returns. An old Rickover protege, Shaw saw light-water reactors as a mature technology. The key to the safety of commercial power plants, he thought, was the same thing that had worked so well for the Navy reactor program: thick books of regulations specifying every detail of the reactors, coupled with careful oversight to make sure the regulations were followed to the letter. Shaw rejected the doomsday scenarios that the ACRS had been working with as academic fantasies. The worst-case loss-of-coolant accidents, for example, envisioned a major cooling pipe breaking in two. It was a scientist's approach to safety: figure the maximum credible accident, prepare for it, and everything else will be automatically taken care of. But Shaw contended that nuclear accidents were more likely to be the result of little breakdowns that snowballed. Take care of the little things, and the big things would take care of themselves; that was Shaw's safety philosophy, and he was in charge of all the AEC's safety research.

Believing that the light-water reactor needed little more basic research, Shaw had turned most of his attention to what he thought would be the next-generation reactor: the breeder. He pushed the AEC to move its attention in that direction and leave the light-water reactor to the manufacturers and the utilities. And he applied much of the AEC's spending on safety to the breeder. In 1972, for example, with a hundred light-water reactors either built or on order in the United States and no commercial breeders, Shaw split his $53 million safety budget down the middle—half for light-water and half for the breeder.

Shaw's autocratic personality and attitudes toward safety quickly estranged many of the personnel working on safety at the national labs. Like the scientists on the ACRS, the engineers at the labs could see many things that might go wrong in a reactor, and they thought it important to track them down. It was in their nature to want answers, and they couldn't understand how Shaw could ignore all these potential problems with the reactors. Many came to believe that Shaw was consciously suppressing safety concerns to protect the interests of the reactor manufacturers. Eventually, this dissension would explode into a very public controversy.

It began in the spring of 1971 when several men associated with the Union of Concerned Scientists, originally founded to oppose various weapons such as antiballistic missiles, challenged the licensing of a nuclear plant in Plymouth, Massachusetts. In the argot of the AEC, they became "intervenors," which meant they were full parties in the licensing hearings. The three intervenors, Daniel Ford, James MacKenzie, and Henry Kendall, began looking for ammunition with which to oppose the reactor, and they soon settled on the emergency core-cooling system after discovering what seemed to be damning evidence against it. A few months earlier, the AEC had performed scale-model tests of an emergency core-cooling system and it had failed miserably. The emergency cooling water flowed out the same break in a pipe that allowed the original coolant to escape. Because the tests were not very realistic, no one expected a real emergency core-cooling system to perform the same way, but the failure underscored a striking fact: there was no hard evidence that such a system would work the way it was supposed to.

As Ford, MacKenzie, and Kendall delved deeper into the physics and engineering of emergency core cooling, they decided to go directly to the source of much of the information: Oak Ridge National Laboratory. There they were surprised to find many of the lab's researchers sympathetic to their arguments. The Oak Ridge scientists did not believe nuclear reactors were dangerous, but neither did they believe there was enough evidence to deem them safe. They were upset with the AEC's level of funding for safety research, and they were particularly incensed that funding for a series of experiments on fuel rod failure during a loss-of-cooling accident had been canceled in June 1971. The Oak Ridge researchers spoke at length with the three intervenors and sent them home with stacks of documents.

Meanwhile, concerns about the emergency core-cooling system were surfacing at hearings for several other reactors around the country, and, hoping to calm the criticism, the AEC decided to hold national hearings on the issue. The hearings began in January 1972 and ran on and off until July 1973, and they were an embarrassment for the commission. A number of engineers from the national laboratories testified that their concerns about safety had been ignored or papered over by the AEC, and Shaw himself came off as arrogant, overbearing, and unwilling to listen to criticism. By the end of the hearings it was clear that the AEC could not guarantee that emergency core-cooling systems would work correctly in the worst-case scenario, the complete break of a pipe supplying cooling water. Was that important? The AEC thought not, declining to make any major changes to its rules governing the emergency core-cooling system. But the revelations created an impression among many that the AEC's reactor development division was ignoring its own experts and catering to the wishes of the nuclear industry. In May 1973 the AEC took away responsibility for safety research on light-water reactors from Shaw's office and placed it in a newly formed Division of Reactor Safety Research. Shaw resigned a few weeks later. The following year the AEC itself was dismembered, its regulatory arm being reincarnated as the Nuclear Regulatory Commission and its research and development activities handed over to the Energy Research and Development Administration, later part of the Department of Energy.

The AEC was hoist with its own petard. It had based its safety thinking on maximum credible accidents because they were a convenient analytical tool. But they had come to take on a life of their own, as many of the safety experts came to see their job as making sure that a reactor could survive a worst-case accident without risk to the outside. Because safety had been presented as an absolute, there was no way to draw a line and say, "Past this we won't worry about any more what-ifs." And when Shaw did try to draw a line, he was an easy target for people opposing nuclear power plants since they could always point to questions that had not been answered. It made no difference how unlikely the maximum credible accident might be; the logic of the AEC's approach to safety, taken to its extreme, demanded that such an accident be accounted for and either prevented from happening or kept from doing any damage outside the nuclear plant.

If any lesson can be drawn from these first two decades of thinking on nuclear safety, it is the difficulty of trying to ensure safety on the run. Nuclear power was evolving rapidly. In the mid- and late-1960s, the size of the reactors being offered jumped yearly as reactor manufacturers competed to offer what they thought would be ever-increasing economies of scale. The part of the plant outside the reactor never became standardized, because each utility that built a nuclear plant chose its own designers, suppliers, and builders. Meanwhile, the AEC's ideas on safety were evolving too. Each year the commission demanded something different and something more than the year before. The result was a confused and confusing situation, with the nuclear industry complaining it was being harmed by overzealous regulation and the AEC's advisers at the ACRS worrying that too little was being done to make sure the plants were safe. Often it seemed the AEC was trying to split the difference. It would demand extra precautions from industry, but only if those precautions were not so expensive or time-consuming as to slow the growth of nuclear power. It was a crude sort of cost-benefit policy-making: by setting safety policy so that the yelps from industry and the ACRS were approximately equal in volume, the AEC could balance the benefits of extra safety measures against their cost.

In the late 1960s, with the appearance of the China syndrome, people in the nuclear community began to realize that they needed a new way to think about reactor safety. It no longer seemed possible to assure absolute safety—to guarantee that in case of a worst-case accident no radiation would escape to threaten the public. Safety now depended on the performance of active safety features such as the emergency core-cooling system. If they failed, there was a chance that the containment could be breached and radioactive materials get out. "We then had to change the basis of our claim that reactors were 'safe,'" recalls Alvin Weinberg. "Instead of claiming that because reactors were contained, no accident would cause off-site consequences, we had to argue that, yes, a severe accident was possible, but the *probability* of its happening was so small that reactors must still be regarded as 'safe.' Otherwise put, reactor safety became 'probabilistic,' not 'deterministic.'"

The approach that the nuclear community gradually adopted was called probabilistic risk assessment. It evaluated risk by taking into account both the probability of a certain accident occurring and the consequences of that accident. Thus an accident that was expected to occur only once in a million years of reactor operation and that might kill a thousand people could be treated as equivalent to an accident calculated to happen once in a thousand years with only one death expected. Each worked out to one expected fatality per thousand years of reactor operation. This was something that engineers could appreciate. Instead of endless arguments over whether some potential mishap was worth worrying about, they could establish a numerical target for reactor safety—say, one expected death from radiation release for every thousand years of reactor operation—and set out to reach that goal. To pronounce a reactor safe, engineers would not have to guarantee that certain accidents could never happen, but only to ensure that they were very unlikely.

The probabilistic approach had the advantage of putting the maximum credible accident to rest, but it had several weaknesses. One, it explicitly acknowledged that major accidents could happen. Although it was fiction to claim that even a maximum credible accident posed no threat to the public, it was a useful fiction. Nuclear plants could be considered "safe" in a way the public understood: engineers said that major radiation-releasing accidents were impossible. With probabilistic risk assessment, the engineers now said that major accidents were indeed possible. And although such an accident might be a one-in-a-million shot, much of the public didn't find this comforting.

But the major weakness of probabilistic risk assessment was a practical, technical one. Calculating the probabilities for various accidents was next to impossible. In theory, it was done by identifying the possible chains of events that might lead to an accident—a break in a pipe, the failure of a sensor to detect it, an operator throwing the wrong switch, a backup generator not starting up—and estimating how likely each of the events in the chain was. By multiplying the probabilities for the separate events, one calculated the probability that the whole chain of events would occur. Then by adding the probabilities for various chains of events, one arrived at the probability for a given accident. But because nuclear power was still a new technology, identifying the possible chains of events and estimating the probability of each event involved a good deal of guesswork. The reactor designers could list all the sequences they could think of that might lead to an accident, but inevitably they would overlook some, perhaps many. And while experience in other industries might help in estimating how likely it was that a pipe would break or a pump would fail, much of the equipment in a nuclear plant was unlike that found anywhere else, and it operated under unique conditions. Inevitably, until there was much more experience running nuclear power plants, much of the probabilistic risk assessment would be done by guess and by gosh.

This became brutally clear after the release of the so-called Rasmussen report, a study released in 1974 that attempted to calculate the likelihood of a major nuclear accident. The commission wished to fight rising concern over the safety of nuclear power with an objective report, which the AEC fully expected would validate its position. It hired MIT nuclear engineering professor Norman Rasmussen to put the study together, and asked him to do it quickly. Rasmussen and coworkers drew data from two reactor sites that were taken as representative of U.S. nuclear plants in general, and then analyzed the data extensively. Much of what the researchers did was extremely valuable. By envisioning various ways an accident might happen, for instance, they identified types of accidents that had been ignored in the past, such as mid-sized loss-of-coolant accidents, in which the coolant was lost from the reactor core gradually instead of suddenly, as would happen with a major break in a pipe or failure of the pressure vessel. The report focused attention, for the first time, on the types of accidents that could be caused by the coming together of several relatively minor mishaps, none of which would be a problem by itself. But the AEC was not interested in highlighting this part of the Rasmussen report. Instead, in its public discussions, it focused on a much weaker and more speculative part, the estimates of the probabilities of accidents in nuclear plants.

Indeed, even before the study was released, AEC officials were boasting that it showed a core meltdown could be expected only once in every million years of reactor operation, and that a major accident only once in one to ten billion years of reactor operation. When the report was officially made public, the AEC summarized its findings by saying that a person was about as likely to die from an accident at a nuclear power plant as from being hit by a meteor.

The Rasmussen report had its desired effect. The media reported on it widely, accepting its conclusions at face value. But a number of scientists took a closer look at it and didn't like what they saw. Some of them were committed skeptics or opponents of nuclear power, but a particularly convincing criticism came from a respected, nonpartisan source—the American Physical Society, the major professional organization for physicists. At the request of the NRC, which had taken over the safety duties of the AEC, the physics society assembled a study group to review Rasmussen's work, and it returned a mixed appraisal. Although the group found Rasmussen's use of probabilistic risk assessment valuable in pinpointing ways that accidents might happen, they had serious reservations about how he had estimated the likelihood of major accidents. The group concluded that the report contained a number of flaws that made its conclusions much less trustworthy than they should have been. Nuclear power might be safe, but few believed that the Rasmussen report proved it.

Nothing the critics said, however, could match the dramatic repudiation of the report delivered less than five years later in Middletown, Pennsylvania. On March 28, 1979, Unit 2 of the Three Mile Island nuclear plant had a major accident. The reactor core melted partially, and some radioactivity was released into the atmosphere, although most of it stayed inside the containment building. The probabilistic risk assessment in the Rasmussen study had concluded that a meltdown in U.S. light-water reactors could be expected only once in every 17,000 years of reactor operation (not once in every 1,000,000 reactor-years, as the AEC had advertised before the report was released). Since only a few dozen nuclear plants were in operation at the time, the United States might expect to go centuries without a meltdown. It was instead only years.

The TMI accident triggered an orgy of rethinking on nuclear safety. It shocked the public, but it shocked the nuclear establishment equally. Roger Mattson, who directed the Nuclear Regulatory Commission's division of systems safety, described it this way:

> There is a difference between believing that accidents will happen and not believing it. The fundamental policy decision made in the mid-1970s—that what we had done was good enough, and that the goal of regulation ought to be to control the stability of the licensing process—that the body of requirements was a stable and sufficient expression of "no undue risk" to public health and safety—and if new ideas come along we must think of them as refinements on an otherwise mature technology—that's been the policy undertone. That's the way I've been conducting myself. . . . It was a mistake. It was wrong.

More than any other event, the partial meltdown at Three Mile Island has shaped recent thinking on risk—and not just in the area of nuclear power. The lessons that emerged from analyzing what went wrong at TMI have application to almost any risky, complex technology.

Before Three Mile Island, most of the AEC's and NRC's safety efforts had been aimed at the equipment. The prevailing philosophy was: Make sure that everything is designed, built, and maintained properly, and safety will follow. And indeed, a piece of malfunctioning equipment did play a key role in the accident. But the Kemeny Commission, the presidential board that investigated TMI, concluded that problems with equipment played only a small part. More worrisome was the performance of the operators running the reactor. They had been poorly trained and poorly prepared for an emergency of the type that caused the accident. Not only did they not take the correct steps to solve the problem, but their actions made it worse.

The shortcomings of the operators were, however, just part of a more general failing. The Kemeny Commission argued that running a nuclear power plant demands different sorts of management and organizational capabilities from those needed to operate a fossil-fuel plant. Utilities tend to run coal- and oil-fired plants at full power until something breaks, then fix it and start up again. There is little concern about safety, little concern about preventive maintenance, little concern about catching problems before they happen. This works fine for these plants, which are relatively simple and endanger no one when they break down. But many utilities—not just the one running TMI—had carried over these attitudes into the management of their nuclear plants, and that didn't work. Successfully operating a nuclear plant demanded an entirely different institutional culture (a culture we'll examine more closely in chapter 8).

But perhaps the most important lesson from TMI was the realization that big accidents can be triggered by little things. Until then, much of the thinking on nuclear safety had focused on responding to major failures, such as a break in a large pipe. By piecing together the chain of events that led to the TMI accident, investigators showed how a number of seemingly minor foul-ups can create a major disaster.

It began at 4 A.M. when the pumps sending water into the reactor's steam generators shut down as a result of a series of human errors and equipment malfunctions. The TMI reactor was a pressurized-water type, which meant there were two main systems of water pipes. One, the reactor-cooling system, carried water through the reactor to a steam generator and back to the reactor. This cooling water carried heat away from the reactor core, and that heat was used to boil water in the steam generator. The second system of pipes sent water into the steam generator, where it was turned into steam, which powered a turbine. After passing through the turbine, the steam was condensed into water, which then was routed back to the steam generator. It was this second system that stopped working when the pumps shut down.

With the secondary system shut down, the primary cooling system had no way to pass on the heat from the reactor core, and the water in that primary system began to heat up and expand, increasing the pressure in that system. As planned, the reactor immediately shut down, its control rods slamming down into the reactor core to absorb neutrons and kill the chain reaction. At about the same time, an automatic valve opened to relieve the pressure in the primary cooling system. So far the plant had responded exactly as it was supposed to. But once the relief valve had released enough pressure from the coolant system, it should have closed again. And indeed, an indicator in the control room reported that a signal had been sent to close the valve. But, for some reason, the valve did not close, and there was no way for anyone in the control room to know that. So for more than two hours, first steam and then a mixture of water and steam escaped through the open valve. This caused the pressure to continue dropping in the primary cooling system. After a few minutes, the pressure had dropped enough to trigger the startup of high-pressure injection pumps, which sprayed water into the system. But because the operators misread what was happening in the reactor, they shut down one pump and cut back on the other to the point where it could not make up for the water being lost as steam through the open relief valve. They thought they were doing what they had been trained to do, but they were making matters worse. Gradually, the coolant in the primary system became a turbulent mixture of water and steam. Then, an hour and a half into the accident, the operators decided to turn off the pumps circulating the coolant through the reactor to the steam generators and back. Again, because the operators didn't understand exactly what was going on inside the reactor, they believed they were following standard procedure, but cutting off the reactor pumps removed the last bit of cooling action in the core. Soon half the core was uncovered, and its temperature shot up, melting some of the fuel and releasing highly radioactive materials. Some of the radioactive gases escaped from the containment building, but fortunately not enough to threaten anyone in the surrounding areas. Finally, nearly two and a half hours after the accident started, someone figured out that the pressure relief valve had never shut and closed it off. It took another twelve hours to reestablish cooling in the core and start to bring the system back to normal temperatures.

It was a chain of events that no one performing a probabilistic risk assessment would likely have imagined ahead of time. Upon hearing the chain of events described, it's easy to conclude that the operators were mostly to blame—they did, after all, overlook the open pressure relief valve, turn down the injection pumps, and shut off the reactor coolant pumps. But there was plenty of blame to go around. The designers had failed to include anything in the control room to show whether the pressure relief valve had closed; the one indicator told only that a signal had been sent to close it. The NRC had known about a similar accident eighteen months earlier at another reactor, where a relief valve had stuck open, but it had not notified other nuclear plants that the valve might be a problem. The plant management had operated the reactor in such a way that little problems seldom got corrected, and these little things created a sloppiness that contributed to the accident. Prior to the accident, for instance, there had been a steady leak of reactor coolant from a faulty valve; during the accident, when the operators saw abnormal readings that should have told them the pressure valve was open, they thought the readings were caused by the leak.

But more than anything else, the culprit was the complexity of the system. It created a situation where a number of seemingly minor events could interact to produce a major accident, and it made it next to impossible for the operators to understand what was really going on until it was too late.

The political scientist Aaron Wildavsky has suggested that, because of this complexity and the way that different components in a nuclear plant interact, adding safety devices and procedures will at some point actually decrease safety. The TMI accident offers a number of examples of how this works. The control room, for instance, had more than 600 alarm lights. Each one, considered by itself, added to safety because it reported when something was going wrong. But the overall effect in a serious accident was total confusion, as so many alarms went off that the mind could not easily grasp what was happening.

Charles Perrow, a sociologist at Yale, has taken this sort of reasoning one step further and argued that such complex, tightly interconnected technologies as nuclear power are, by their nature, unsafe. With large numbers of components interacting, there are so many different ways an accident can happen that accidents are an inevitable feature of the technology—what Perrow calls "normal accidents." The technology cannot be made safe by adding extra safety systems, for that only increases its complexity and creates more ways for something to go wrong.

It's impossible to do justice to Perrow's complex, tightly argued thesis in just a few sentences, but in essence he argues that the requirements for successful control of certain technologies contain inherent contradictions. Because what happens in one section of a nuclear plant can dramatically affect events in others, some central control is needed to make sure that actions in one place don't cause unanticipated consequences in another. This control might be in the form of a central management that approves all actions, or in the form of a rigid set of rules governing actions throughout the plant. On the other hand, because the technology is so complex and unpredictable, operators need the freedom to respond quickly and imaginatively to special circumstances as they arise. Both rigid central authority and local discretion are needed, and, Perrow says, it is impossible to have both. Thus a nuclear plant will always be vulnerable to one type of accident or another—either one caused by a failure to adapt quickly to an unanticipated problem, or one created by not coordinating actions throughout the plant.

Perrow argues that a number of technologies besides nuclear power face the same inherently contradictory demands: chemical plants, space missions, genetic engineering, aircraft, nuclear weapons, and the military early warning system. For each, he says, accidents should be considered not as anomalies but as a normal part of the process. Their frequency can be reduced by improved design, better training of personnel, and more efficient maintenance, but they will be with us always. Perrow goes on to suggest that society should weigh the costs of these normal accidents against the benefits of the technology. For chemical plants, the costs of accidents are relatively low and are usually borne by the chemical companies and its workers, while the cost of shutting down chemical plants would be rather high since there is nothing to replace them. But nuclear power is different, Perrow says. The costs of a major accident would be catastrophically high, while the cost of giving up nuclear power would be bearable: other ways of generating electricity could take its place.

So far, of course, that hasn't happened. But what has happened is that the way in which people think about safety has changed, particularly in the nuclear industry, but in others as well. No one believes any longer that it is possible to engineer for complete safety, to determine the maximum credible accident and then assure that it won't threaten anyone. The best that can be done is to try to make dangerous accidents very unlikely.

People have also come to appreciate how complexity changes the risk equation, how it makes risk harder to calculate by making it difficult to understand all the ways that things can go awry. But equally important, complexity can amplify risk. The more complex a technology, the more ways something can go wrong, and in a tightly coupled system the number of ways that something can go wrong increases exponentially with the number of components in the system. The complexity also makes a system more vulnerable to error. Even a tiny mistake may push the system to behave in strange ways, making it difficult for the operators to understand what is happening and making it likely they'll make further mistakes.

Since Three Mile Island, the Nuclear Regulatory Commission has tried to assure safety by looking to the one truly successful model in the history of nuclear power: Rickover's nuclear navy. This has led to ever more rules and specifications covering ever more minor matters, along with mountains of paperwork intended to assure the quality and design of the different components of a nuclear power plant. By all accounts, this has rid the industry of much of the sloppiness that plagued it before TMI.

But there has been a growing realization in the industry that, ultimately, the safety of such a complex, risky technology lies in human hands. It may be useful to improve the technical aspects of a nuclear plant—the machinery and the regulations that specify how to operate it—but its complexity guarantees that there will always be surprises. And in the case of surprises, the best defense is human competence, expertise, and imagination. As we'll see in chapter 8, such an emphasis on the human element may hold the best hope for assuring safety in complex, risky technologies.

No matter how good our technological managers get at controlling risk, however, the issues of assessing risk and deciding how much is acceptable will remain a difficult issue. On these matters there will always be disagreement, because people's beliefs about risk depend greatly upon the lens through which they view the world, and people use very different lenses.

THROUGH THE LENS OF CULTURE

In his book *The Nuclear Energy Option*, physicist Bernard Cohen lays out the scientist's case for nuclear power. In chapter after chapter, he argues with statistics and logic to convince the reader that nuclear power is needed, is safe, and is superior to the alternatives. His chapter 5 is particularly memorable. Here he shows that in various highly publicized nuclear accidents, the public was typically exposed to about 1 millirem of radiation or less. How much, he asks, is 1 millirem? It is the amount of radiation one gets from a dental x-ray. It is the amount of radiation that one gets from living in a brick house for a month or so, since bricks and stone are slightly radioactive. It is the amount of radiation one gets from one's own body over the course of a couple of weeks, since living organisms have trace amounts of radioactive materials. It is the amount of radiation one receives by taking a coast-to-coast flight, since flying increases one's exposure to cosmic rays. And, Cohen asks, how dangerous is 1 millirem of radiation? Using estimates of its risk that are almost certainly exaggerated, he concludes that 1 millirem of radiation reduces one's life expectancy by about two minutes—the same risk posed by crossing the street five times, taking a few puffs on a cigarette, or driving five miles in an automobile. Yet, he grumbles, each time there is a small nuclear accident, the newspapers and television jump on it as if it were a major catastrophe instead of the equivalent of a few people taking a drive to the grocery store or sitting next to a smoker in a restaurant.

If Cohen's complaint sounds familiar, it should. For decades, supporters of nuclear power have contended that if the public would just take the time to learn the facts, then they would realize that nuclear energy is a good thing. In 1964, some twenty-six years before Cohen's book came out, Westinghouse vice president John Simpson expressed similar sentiments:

> It would seem to me that an industry-wide program of education and rebuttal is called for. Our whole industry is being attacked; the whole industry should answer. The nation's utilities and reactor manufacturers should join forces to spearhead this industry effort. The only weapon needed is *truth*—well presented and widely disseminated.

But despite decades of public education efforts and more than a few books like Cohen's, a large percentage of the population has remained dubious of nuclear power. Many fear it. Others judge it to be a technology of last resort, to be used only to avert global warming or if nothing else is available to produce the growing amount of electricity that society consumes. One study after another finds that the public rates nuclear power as one of the most dangerous threats to the public health, comparable with smoking, handguns, and motor vehicle accidents. It does little good for Cohen to argue that burning coal to generate electricity is responsible for tens of thousands of deaths per year, while nuclear power is responsible for at most tens of deaths per year. People prefer the risks of coal to the risks of the atom.

Puzzled by the discrepancy, social scientists have probed and prodded groups of laypeople to discover why their assessments of nuclear power and other potential risks differ from those of the scientists and engineers who study such things. The answer, they've found, is that people who haven't been trained in a technical area—and even many who have—use different equations to calculate risk than do specialists. The experts tend to use things that can be easily measured—deaths, injuries, monetary damages—

and to judge risk mathematically. By assigning a number to everything, they can make comparisons in a simple, direct manner. Choice A is riskier than Choice B if the expected number of deaths from A is higher than that from B. But most people don't think this way.

In one study, for example, a team of risk specialists asked randomly selected people from three groups—college students, businessmen, and members of the League of Women Voters—to rank the risk of various activities and technologies: nuclear power, X-rays, pesticides, smoking, automobile accidents, motorcycles, handguns, and quite a few others. They also asked a dozen experts to do the same thing. When they compared the rankings, they found that the experts rated nuclear power as only the twentieth most dangerous of the thirty choices. The lay groups saw it differently: Two of the three groups thought nuclear power was the most dangerous choice, while the third (the businessmen) placed it eighth. To find out why, the researchers asked the people in the three groups to estimate how many people died in an average year from each of the thirty activities and choices. Strangely, on this measure, the members of the public believed nuclear power to be safer than the experts. The experts reckoned that the nuclear industry was responsible for about 100 deaths in an average year from accidents at nuclear plants, exposure of uranium miners to radioactivity, cancers caused among the public by radiation from nuclear plants, and so on. The group members, on the other hand, figured that nuclear power caused only a couple of dozen deaths in an average year. This was fewer estimated deaths than for any other item on the list, including food coloring and spray cans. In short, according to the expected number of fatalities in an average year, the groups rated nuclear power as the least risky item on the list.

Clearly, then, it was not the number of deaths in an average year that made nuclear power seem so risky. So what was it? The answer appeared when the researchers asked how many deaths might be expected from each item on the list in a particularly disastrous year. For almost every other item, the respondents figured that in a really bad year the death rate might double or triple. But for nuclear power they guessed that a calamitous year might see thousands, tens of thousands, even hundreds of thousands of deaths. This was much higher than the official figure from the NRC, which estimated that the worst possible accident in a U.S. nuclear plant would kill 3,300 people immediately, with others dying years later. The NRC calculated the chances of such an accident in any given year as one in three million. The researchers did not ask the participants in the survey how likely they thought such an accident might be, but it seems probable they would have offered odds that weren't quite so long.

Nuclear power's disaster potential seemed to explain much of its perceived risk, the researchers found. Those respondents who estimated large numbers of fatalities in a disastrous year were likely to rate nuclear power as riskier than the other choices. Besides its potential for disaster, people were also taking into account nuclear power's scariness. Those who said that members of the public have great dread of nuclear power were more likely to see it as very risky than those who thought the public had learned to live with it.

In short, the members of the public viewed the risks of nuclear power differently from the experts because they were using a different set of factors to judge it. To the experts, an item's risk was essentially synonymous with the number of deaths expected in an average year and was figured in the same way an actuary or accountant might calculate it. But the public didn't care that, on average, nuclear power was quite safe. What it cared about was that if things went really wrong, a lot of people could die all at once, and it hardly mattered that the worst-case accident was highly unlikely. The possibility of thousands of people dying at once made nuclear power, in the eyes of the people answering the survey, as dangerous to the public health as cigarettes or automobiles.

Over the past fifteen or twenty years, researchers have spent a good deal of time trying to figure out just how the public weighs risk. It's been obvious for some time that laypeople don't see eye to eye with the experts, and the original assumption was that it was due to ignorance on the public's part. It's clear now that this was wrong. Yes, there's plenty of ignorance to be found, but it doesn't always explain the difference of opinions. Instead, even if members of the public were working from the same set of numbers as the experts, they'd likely come up with different conclusions because they attach a different importance to the numbers. Consider, for instance, a worst-case accident that would kill 30,000 people and happen once every 30,000 years. A risk expert would divide the total deaths by the number of years and pronounce this accident as having an average fatality rate of one death per year. To most people, however, the possibility of an accident killing 30,000 people, even if the odds of it happening this year are one in 30,000, carries much more weight than a single death. The public doesn't simply divide one number into the other. It takes the sizes of the numbers into account.

But the difference between the experts and the public goes deeper than this. To the experts, risk is a simple, well-defined concept. It includes death, injury and illness, and, sometimes, damage to property. The regulator, the actuary, and the engineer work with these concrete, measurable costs instead of with more abstract, difficult-to-measure items because they are convenient and because historically they have been the most important harms from technology. The mathematically minded seldom look past these when assessing risk. But the general public, not having been indoctrinated into this way of thinking, sees risk as a broader thing. It is not merely something that can be entered into a calculator.

The historian Spencer Weart, in his book *Nuclear Fear*, describes how in the 1950s radioactivity came to be seen as "unnatural," as an insult to or an assault on nature. The images from movies of the time are particularly telling—many of them feature monsters that are created by radioactivity and then attack humans. Today, much of public attitude toward nuclear power is influenced by the sense that it is a threat to the natural world. This is particularly true in discussions of nuclear waste storage. The idea of stuffing tons of highly radioactive material into an underground vault and leaving it there for millennia seems like a slap at Mother Earth, and all the assurances by scientists that it will be safe can't calm the uneasy feelings.

Besides the risk to the environment, people also worry about the social costs of technology. Will it affect various classes or groups of people differently? Will it create economic inequities? Will its costs be distributed equally and fairly? Will our children have to pay for our actions? Such concerns are particularly important to groups that are skeptical about the modern world's emphasis on material goods. The neo-Luddite movement, for instance, appeared a few years ago arguing for a thoughtful, conscious decision on whether to accept new technologies rather than simply grabbing every one that came along. As one sympathetic writer described them,

> Neo-Luddites judge the acceptability of a technology not merely by its impact on human health and the environment but also by its effects on human dignity and traditions of society. When they oppose nuclear power, for instance, they note not only the threat of radioactive contamination but also the threats to democratic institutions that could result from a power supply directly linked to the most awful tools of destruction ever created. The lethal potential of radioactive materials as a weapon means that nuclear plants must be constantly kept under armed guard.

But it is not merely such fringe groups as the neo-Luddites that take social factors into account when evaluating risk. According to Steve Rayner and Robin Cantor, two analysts at Oak Ridge National Laboratory, the public at large places great significance on the social contexts in which decisions about risky technologies are made. In particular, the researchers say that the public focuses on three factors in deciding whether a risk is acceptable:

1. Is the process by which decisions are made about the technology acceptable to those who stand to be hurt by an accident?
2. In case of an accident, does everyone agree ahead of time who is responsible for what?
3. Do people trust the institutions that manage and regulate the technology?

If the answer to all three questions is yes, Rayner and Cantor say, then the public will likely not worry about the sorts of low-probability, high-consequence accidents that have plagued nuclear power and genetic engineering. Where the experts emphasize risks that can be quantified and like to argue about "How safe is safe enough?" the public is not interested in quibbles about the exact value of a risk and instead asks, "How fair is safe enough?"

These differing attitudes toward risk reflect different cultures. The engineer's culture is an aggressively rational one. Problems are solved with a logical application of scientific principles. This is how engineers attack designing a new computer chip or a jet aircraft engine, and they put the same rules to work in assessing risk. They see their approach as so reasonable, so obviously *right,* that everyone else must—or at least should—think the same way. But not everyone does, and it's not necessarily because they are irrational. They have a different sort of rationality from engineers, one that starts from a different set of assumptions about how the world works and how it should work.

Perhaps the most striking examples are the people who belong to what the anthropologist Mary Douglas and the political scientist Aaron Wildavsky called "sects." These are organizations whose members see themselves as set off from the main part of society, and who believe that society is headed for a disaster that only they see clearly. Traditionally, sects have been religious groups such as the Amish who hold themselves separate from the world and who have a nonhierarchical organization in which each member has an equal say. In the absence of any leaders or hierarchy, what holds a sect together is the separation it feels from the rest of the world and the members' sense of having a higher spiritual value. Today, Douglas and Wildavsky say, certain environmental organizations such as Friends of the Earth have a sectarian culture very similar to that of the Amish and other religious sects. And this results in beliefs about the world that have a number of parallels with those religious sects:

> The sectarian cosmology expects life in the future to undergo a radical change for the worse. It is not confident that the disaster can be averted. There may be no time left. But it knows how the disaster has been caused: corrupt worldliness; that is, ambition for big organization has endangered mankind and new technology represents all that is most reprehensible—social distinctions, the division of labor, materialist values, unfeelingness for individual suffering. . . . The sect espouses the widest causes, all mankind rather than a section. They refer to God or nature as arbiters to justify their rightness, and every tragedy that can be attributed to worldliness as here defined is chalked up as one more divine or natural warning.

In short, members of a sect see risk very differently than do engineers. All the calculations of accident probabilities and possible loss of life are meaningless to them. Their opinions have a different foundation altogether. They see a greater truth—danger to the entire world—and the particular details don't really hold much interest for them. They may exploit this or that engineering detail when it helps their cause, but their arguments stem not from the rationalism of science and engineering but from appeals to a higher authority.

In analyzing the antinuclear movement, Douglas and Wildavsky show how the beliefs and goals of the various antinuclear groups are correlated with the groups' cultures. The more traditional, hierarchical organizations work within the system and are generally willing to compromise. The Audubon Society, for instance, has indicated that it is not irrevocably opposed to nuclear power—if it could be made safe and economical, and if it were shown to be superior to other energy choices, the group might support it. But such sects as the Clamshell Alliance have no interest in compromise. One does not compromise with the devil. "The Clamshell Alliance," Douglas and Wildavsky say, "views all economic and social ills as stemming from the distribution of energy in favor of large corporate and governmental interests. Their aim in blocking nuclear power is not merely to safeguard themselves from the possibilities of exposure to hazardous radiation but to break the stranglehold which they consider such interests have on society." Such beliefs are so far from the rationalistic, materialistic world view of the engineer that when members of the two cultures try to communicate, they find they have little common ground.

But it's not necessary to venture so far out into the fringes to see how culture shapes attitudes toward technology and technological risk. Throughout society, people's opinions on nuclear power, genetic engineering, and other technologies are closely related to political ideology and various other beliefs that, on the surface, seem to have nothing to do with technology.

In 1990, two political scientists, Richard Barke and Hank Jenkins-Smith, surveyed more than a thousand scientists and engineers about their perceptions of risk, particularly the risks associated with nuclear wastes. They also polled the general public and members of the Sierra Club. As a group, the scientists found nuclear waste to be less risky than did the public. On an aggregate risk scale from 1 to 17, which lumped together the answers to several questions, the scientists rated the risk to be a 9, on average, while the public put it at 12. Sierra Club members gave it a 13. This isn't surprising. You would expect scientists and engineers to have a greater faith in technology than people from other walks of life. What was surprising was the large difference among the scientists' opinions, which varied according to their fields and also according to the types of organization for which they worked.

Physicists were the most optimistic about the hazards of nuclear waste, rating the risks to be about 7.3 on the same 17-point scale. Next came engineers, chemists, and earth scientists (geologists, atmospheric scientists, and the like), ranging from 8.3 to 8.5. Situated farther along on the spectrum, biomedical researchers and biologists averaged 10.1, and doctors gave it a 9.3. In other words, researchers who dealt with living organisms found nuclear wastes to be riskier than did those who dealt with inanimate objects.

When Barke and Jenkins-Smith analyzed the scientists' responses according to their institutional affiliations, they found a pattern of a different sort. Researchers employed by federal agencies and by government-owned research labs gave the lowest risk ratings, along with scientists working as business consultants. The scientists who rated nuclear wastes as most risky came from universities and state and local agencies. The survey was restricted to scientists in Colorado and New Mexico, and in those two states, researchers who work for a federal agency or lab are likely to be employed at an institution, such as Los Alamos National Laboratory, with a stake in nuclear research. On the other hand, scientists working for state and local agencies are more likely to be involved in protecting public safety. It's hard to know whether the different attitudes toward nuclear power among these scientists reflect attitudes they learn at their institutions or merely self-selection in where they work. Perhaps the scientists who are more skeptical of nuclear waste and technology in general are less likely to go to work for a federal agency with an involvement in nuclear research and gravitate instead to universities and state and local agencies.

What is clear, however, is that the differences of opinion on nuclear waste were part of a consistent pattern of varying opinions on technology, the environment, and the role of the federal government. The life scientists, who saw environmental problems as much more serious than did the physical scientists, were more likely to believe that the government should regulate nuclear energy and nuclear wastes. (By contrast, the biologists and biomedical researchers saw less need for the government to regulate genetic engineering.) Furthermore, the biological researchers were less willing for society to impose risks on individuals without their consent—as would be the case, for instance, with the building of a nuclear waste repository in an area where many of the residents were opposed.

The results imply that scientists are influenced by their overall belief systems—their cultures—when they form their opinions on risk, and that scientists in various fields of research have different belief systems. Even among scientists, the assessment of risk is not a purely rational pursuit but depends strongly on the lens through which the researcher views the world.

This is even more true among nonscientists, according to studies done by political scientists Stanley Rothman and Robert Lichter. They polled more than a thousand people in positions of authority or influence, including journalists, military leaders, congressional staff and top-level government bureaucrats, lawyers, and leaders of public-interest groups. The survey included one question about the safety of nuclear power plants and a series of questions designed to tease out the respondents' political beliefs. When Rothman and Lichter analyzed the answers, they found that more than 40 percent of the differences of opinion among the respondents could be attributed to their ideologies. The two important factors were a person's socioeconomic liberalism and his feelings of alienation from government. The more liberal a person was or the more alienated from government, the more likely he was to believe that nuclear power was risky.

Much like Barke and Jenkins-Smith, Rothman and Lichter found that scientists rated nuclear power as less risky than did the nonscientists (except for military leaders, who were mostly unworried about the safety of nuclear plants). Furthermore, the scientists who were more familiar with nuclear power were more optimistic about its safety. Sixty percent of all the scientists sampled thought nuclear power was safe, 76 percent of scientists specializing in energy research did, and nearly 99 percent of the nuclear-energy experts rated nuclear plants as safe. Furthermore, ideology played less a role in shaping scientists' attitude toward nuclear power than it did for nonscientists.

The picture that emerges from the Rothman-Lichter and Barke-Jenkins-Smith studies is of a technical consensus that nuclear plants are safe coupled with the presence of nontechnical doubts. Among nuclear-energy specialists, there is almost complete unanimity about the safety of the plants. The worries about risk appear mostly among people who are not steeped in the technical lore and who are not intimately familiar with the workings of nuclear reactors. Without technical knowledge to guide their opinions, these nonspecialists are more likely to fall back on ideological prejudices when deciding what they think about the safety of nuclear power.

It is this situation that leads Bernard Cohen and other nuclear proponents to attempt to educate the public. Clearly, it seems to them, knowledge about nuclear energy leads one to conclude that nuclear plants are safe. After all, those closest to and most familiar with the technology believe in its safety. If the public knew as much as the experts, they would agree with the experts. But this line of reasoning makes a crucial—and flawed—assumption. It supposes that the opinions of the experts are the product of a completely rational process, that the experts' trust in nuclear power exists solely because they have examined all the evidence thoroughly and dispassionately and then reached their conclusions using logic and nothing more. If so, however, it would be hard to explain the accident at Three Mile Island. Nuclear experts in the 1970s were just as convinced of nuclear power's safety as those interviewed a decade later by Rothman and Lichter, and they reached their conclusions in essentially the same way. It was not—could not be—a completely rational process. The technology held and still holds a certain amount of uncertainty, and at some point the experts must leave the realm of evidence and logic and offer their judgment calls. It is here that the biases, predispositions, and hidden assumptions of the nuclear engineering culture make their appearance. In hindsight, it is easy to see how the culture of the nuclear community caused it to ignore the flaws that allowed TMI. It is not so easy to know what the nuclear community is overlooking or misjudging now because of its particular way of seeing the world.

There is no such thing as a perfectly rational approach to risk. Engineers hate to admit that, but it's true. Every judgment about risk, even those made by scientists, reflects the culture in which it was formed. In *The Atom and the Fault,* Richard Meehan tells how engineers and geologists, two very different scientific cultures, tended to come up with very different scientific judgments when asked about the safety of nuclear plants sited near fault lines in California. They had access to the same information, but viewed it through different lenses.

It is tempting to conclude, therefore, that the rationalistic approach of the scientist and engineer is no better than any other approach to risk assessment. Each is influenced by the underlying assumptions of its culture, assumptions that can be neither proven nor disproven. And, indeed, many opponents of technology—and of rationalism in general—argue in exactly this way. The judgment of the engineer, they say, is ultimately no more credible—or "true"—than that of, say, a member of the Clamshell Alliance. Both are culturally produced, and it should be up to the members of the larger society to decide which it accepts.

But this ignores a crucial difference between the rationalistic culture and every other one. An important, explicit ingredient in the culture of science and engineering is learning from mistakes and modifying one's views in light of experience. One tests theoretical predictions against experimental evidence and revamps the theory accordingly. The scientific ideal, if not always the reality, is that a person must be willing to change his mind as soon as the data point in a different direction. There is little or none of this in other ways of looking at the world. A sect like the Clamshell Alliance, for instance, has few or no beliefs that can be tested against empirical evidence. The rationalistic approach offers the best chance of making a better choice today than one could have made yesterday.

An ideal approach to risk assessment would thus be to lean heavily on science and engineering while keeping their shortcomings in mind. They cannot offer absolute answers, and sometimes their practitioners may not even realize what they don't know. Indeed, it seems part of the engineers' culture to underestimate the challenge of complex systems and to be overly optimistic about their ability to predict and control the behavior of such systems. It is up to the larger society to weigh the engineers' opinions about risk and decide how much faith to put in them. To this end it would be helpful if everyone—engineers and public alike—had a better grasp of the strengths and the weaknesses of the scientific approach.

Beyond that, a democracy must allow the public to decide which risks to consider and how much emphasis to place on each. It is in the nature of engineers to consider only the tangible ones, since that is what they are trained to do, but this is not a country of engineers. If the general public wishes to take into account such intangibles as the threat of nuclear power to democratic institutions or the hazard that rBGH holds for the family dairy farm, that is their right. Engineers can argue otherwise, but when it comes to such matters of subjective opinion, they have no more expertise than anyone else.

seven

CONTROL

Texas Utilities is a big company. Through its subsidiary, TU Electric, it provides electric service to a large chunk of Texas, including the Dallas-Fort Worth metropolitan area. It employs some 10,000 people. Its sales are around $5 billion a year. It has assets near $20 billion.

Yet this corporate Goliath was brought to its knees by a single determined woman, a former church secretary named Juanita Ellis. For nearly a decade, Ellis fought Texas Utilities to a standstill in its battle to build the Comanche Peak nuclear power plant. During that time the cost of the plant zoomed from an original estimate of $779 million to nearly $11 billion, with much of the increase attributable, at least indirectly, to Ellis. Company executives, who had at first laughed at the thought of a housewife married to a lawn-mower repairman standing up to their covey of high-priced lawyers and consultants, eventually realized they could go neither around her nor through her. In the end, it took a negotiated one-on-one settlement between Ellis and a TU Electric executive vice president to remove the roadblocks to Comanche Peak and allow it to begin operation.

No one was really happy with the outcome. Antinuclear groups denounced the settlement as a sellout and Ellis as a traitor. Texas Utilities bemoaned the years of discord as time wasted on regulatory nit-picking with no real improvement in safety. And the utility's customers were the most unhappy of all, for they had to pay for the $11 billion plant with large increases in their electric bills. So it was natural to look for someone to blame. The antinuclear groups pointed to the utility. TU Electric, they said, had ignored basic safety precautions and had built a plant that was a threat to public health, and it had misled the public and the Nuclear Regulatory Commission. The utility, in turn, blamed the antinuclear groups that had intervened in the approval process and a judge who seemed determined to make TU Electric jump through every hoop he could imagine. The ratepayers didn't know what to believe. Was the plant safe? Dangerous? And why did it cost so much money to find out?

The answer lies not so much with the utility or its opponents but with the system that had been set up to assure the safety of nuclear plants. If a society feels threatened by a technology, be it pesticides, pharmaceuticals, or nuclear power, people will create some way to oversee it and keep the threat under control. The easiest way to control a technology is to kill it; then it can't possibly do any harm. But if the public wants the benefits from the technology, it must live with some risk and so must decide on a tradeoff between the technology's benefits and its dangers. Each society does this in a different way, depending upon its attitude toward technology and its political culture.

When a society exerts control over a technology, the development of that technology becomes a collaboration. And the form of that collaboration—the particular ways in which the public interacts with the designers and engineers doing the technical work—can greatly influence the technology's ultimate outcome. In the case of nuclear power, the awkward system of control that the United States adopted was in large part responsible for the technology's bust in the late 1970s and 1980s.

THE HOUSEWIFE AND THE NUCLEAR PLANT

It seemed like a good idea at the time. In 1971, when TU Electric decided to build a nuclear power plant, it offered several convincing reasons. First, the company was generating nearly all of its electricity from natural gas. In this it was no different from many other Texas utilities, since natural gas had been plentiful and cheap around the state, but natural gas was about to become less plentiful and much more expensive. The known gas reserves in the state were declining, and, as a result, suppliers were getting nervous about offering long-term contracts. And, indeed, events would prove that TU Electric was right to worry. In 1972, a couple of suppliers would default on contracts with electric utilities, in one case leaving Dallas Power & Light without fuel for its gas-fired electrical-generating plants through much of December. In 1973, worried about having enough natural gas for other uses, the Texas Railroad Commission (which regulates energy production in Texas) would begin pushing utilities to use less gas in generating electricity. By 1975 the commission would outlaw the construction of new gas-fired electric plants. Meanwhile, the OPEC oil embargo would be driving the cost of both oil and gas up, up, up.

So, with customer demand steadily increasing, Texas Utilities had begun to look at other fuels for generating its electricity. One obvious choice was lignite, a low-grade coal that is abundant in Texas, and the company built a number of lignite-fired plants, with the first one opening in late 1971. But burning lignite is a dirty business, and the federal government was starting to crack down on air and water pollution. Eventually the company would be forced to install expensive scrubbers on its lignite plants to clean up their exhausts.

The other obvious alternative was nuclear power. In 1971, the second phase of the nuclear bandwagon market was just getting under way. The nuclear industry, chastened by cost overruns in the first wave of plants, was sure that it knew what it was doing now, and nuclear power was looking better and better when compared to gas, oil, and coal. So TU Electric, like many other utilities around the country, decided to go nuclear. The public announcement came in August 1972: Near the town of Glen Rose, thirty-five miles southwest of Fort Worth, the company would build a two-reactor plant that would be called Comanche Peak.

Construction began in late 1974, and for the next several years everything seemed to be going smoothly. The one recurring headache for TU's construction team was the steadily evolving set of requirements from the NRC. Engineering practices and designs that were acceptable one year were taboo the next as the NRC set out its specifications in more and more detail. Even worse, from the utility's point of view, the NRC required an increasing amount of proof that everything was done properly. Companies building nuclear plants had to document everything in great detail, test much of the equipment after it was installed and record the results of those tests, and justify every design change that was made to the original plans. The last was particularly burdensome. Any project as complex as a nuclear plant inevitably has a great deal of modification in the field, as the builders discover things that don't work exactly as they were drawn out. For any other type of industrial plant, the necessary changes are made with little fuss. But in a nuclear plant, work must stop while the engineers rework their plans to accommodate the problem.

Because NRC requirements were evolving as the plant was being built, TU found itself redoing work it had thought was finished. Sometimes this meant performing a new set of engineering calculations to show that work conformed to the revised regulations. And if the work did not conform, it had to be ripped out and redone. After the 1979 accident at Three Mile Island, the NRC requirements changed even more rapidly and became even more stringent.

As would become apparent later, Texas Utilities and its contractors had trouble taking the NRC regulations—and indeed the whole issue of nuclear safety—seriously enough. Like many other companies from the fossil-fuel culture, they saw little reason for all the picky rules, and sometimes, if the requirements were too inconvenient and didn't seem important, they might simply bypass or ignore them. In this, TU Electric was little different from some of the other utilities building nuclear plants at the time, sure that a reactor was just another way to heat water. But other utilities didn't have a Juanita Ellis to make them pay for their hubris.

To all appearances, Ellis does not seem to be the sort who could stop a major corporation in its tracks. Her college education consisted of two quarters at a junior college. At various times she has worked as a secretary in a Presbyterian church, for her husband at his nursery, and at an insurance company. She and Jerry Lee Ellis live in a small brick house in the Oak Cliff subdivision of Dallas, one that, from the outside at least, seems no different from thousands of other small brick houses around Dallas. But over the years, spurred by worries about the safety of Comanche Peak, this quiet woman transformed herself into what one observer called "an anti-licensing expert, a master at throwing wrenches into the vast and complicated machinery of nuclear regulation."

It began in December 1973, when Ellis read an article about nuclear power in a local gardening magazine. Until then she had known little about nuclear plants and had not been too concerned by TU Electric's plans to build Comanche Peak. But the article disturbed her enough that she contacted its author, Bob Pomeroy, to learn more. Convinced that TU was not giving the public the whole story, in January 1974 she joined with Pomeroy and four friends to form the Citizens Association for Sound Energy, or CASE. At the beginning their goal was to push TU Electric into providing more information about its planned nuclear plant.

Gradually their campaign against the utility became broader—and more personal. Pomeroy learned that the Texas Department of Public Safety had been keeping a file on him. Because he was seen as an antinuclear activist, the department had labeled him a "subversive." And the more Ellis dealt with TU Electric, the more she sensed that it was indeed trying to hide something. "The utility seemed reluctant for people to look at what they were doing," she told a reporter for *Texas Monthly*. And so, she said, "I felt they needed looking at closer." After construction started on Comanche Peak in October 1974, Ellis and the other members of CASE began to hear of problems at the plant. Quality-control inspectors were raising concerns that the utility's management was ignoring. Ellis managed to contact some of them and took affidavits which she sent to the NRC.

In February 1978, as the two units of Comanche Peak were beginning to take shape, TU Electric applied to the NRC for its operating license. At the time it was predicting that Unit 1 would go into operation in 1981 and Unit 2 in 1983. That timetable got extended by the NRC's one-year moratorium on licensing in the wake of the March 1979 accident at Three Mile Island, as well as by all the work the utility was doing to comply with the new regulations, and in October 1981, TU was predicting the two units would begin operations in 1984 and 1985.

Public hearings on Comanche Peak's operating license got under way in December 1981. Three outside groups had filed petitions to take part in the hearings: CASE, which by this time was run almost completely by Ellis; the Citizens for Fair Utility Regulation (CFUR); and the Texas Association of Community Organizations for Reform Now (ACORN). Such hearings are run much like a trial. Acting as judges, a three-member Atomic Safety and Licensing Board listens to the evidence and the arguments and makes a decision on the operating license. The board hears testimony from the NRC staff, the utility, and from the "intervenors"—the members of the public who have petitioned to take part. As in a trial, the parties can call witnesses and cross-examine witnesses called by others. Without intervenors, the matter of whether to grant an operating license is between the utility and the NRC, with no involvement by a licensing board and no public participation.

The hearings can be exhausting, physically, psychologically, and financially. They entail weeks of excruciatingly detailed engineering testimony, combing through piles of documents, and arguing over minutiae and the subtleties of technical interpretations. In this setting, the intervenors are at a tremendous disadvantage. The utility has millions of dollars to spend on technical experts and lawyers, while the intervenors are generally public-interest organizations dependent on donations and volunteers. So it wasn't surprising that by the end of the first year of the Comanche Peak hearings, ACORN had dropped out. It had no money to continue. Later, CFUR gave up. The surprise was that Ellis and CASE hung on. As the sole intervenor left, Ellis was the only thing keeping the public involved in the licensing hearing.

Utility officials made no attempt to hide their opinions of Ellis. "They slurred Juanita in the elevators as a little old housewife," one observer told *Texas Monthly*. "They would sit in the audience and make rude remarks. They thought she was an idiot." But Ellis, who had become obsessed with the case, was stronger than anyone realized. And although she may not have understood all the technical details, she had one weapon that evened the odds between her and TU Electric.

It arrived on her doorstep one day in summer 1982 in the form of Mark Walsh, an engineer who had quit his job at Comanche Peak a few weeks earlier. Walsh told Ellis what he had told his bosses—that some of the pipe supports at the plant were defective. His bosses hadn't listened, so he'd quit. But Juanita Ellis listened. The pipe supports are a mundane but critical item in a nuclear plant. They do just what their name suggests: hold pipes in place. A typical nuclear plant has thousands or even tens of thousands of pipe supports, and some of them are critical. If the supports are not engineered and installed correctly, a sudden stress might cause a pipe to break and cut off the flow of cooling water to the reactor. Ultimately, problems in the pipe supports can conceivably lead to a major accident—a core meltdown. When Ellis heard what Walsh had to say, she had no way of knowing whether he was right or his bosses were right, but she knew she had been handed an explosive issue to throw into the hearings. It got better when Ellis heard from Jack Doyle. He too had worked at Comanche Peak as an engineer and had worries about the pipe supports. From Walsh's and Doyle's allegations, plus charges from other whistleblowers who claimed the pipe supports had not been installed correctly, CASE assembled a 445-page document that laid out the case against the pipe supports.

TU Electric was caught flat-footed. When the licensing board heard the allegations about the pipe supports, it suspended the hearings until the NRC staff could look into the matter. In May 1983, the hearings reconvened. The NRC staff had sent a couple of different inspection teams to Comanche Peak to investigate the pipe support problem as well as the overall quality of construction at the plant. Although the teams had found problems, the NRC judged that they were not serious enough to deny an operating license. The utility could be trusted to correct them.

And there the matter might have rested, had the Atomic Safety and Licensing Board been willing to take the advice of the NRC staff. But by the time the board had reconvened, two of its original members had resigned, so the hearings took place with two new judges. And one of those judges was, to put it mildly, not what the NRC and TU Electric were used to dealing with. Peter Bloch took over as chairman of the licensing board in April. His resume described him as a Harvard-trained lawyer and an administrative law judge, but it was his outside interests that caught the attention of the staid, technically minded members of the NRC and utility staffs. While the licensing hearings were going on, Bloch was taking various courses in "personal growth," including yoga and est, the trendy California self-exploration course created by Werner Erhard. By 1988, he had founded his own organization, Foundation for the Future, to hold personal development workshops including meditation and exercises designed to help people act out their feelings. One of the exercises was the Nuclear War Process, in which participants contemplated their own deaths at the hands of atomic weapons.

The reconstituted licensing board, under the direction of Judge Bloch, sifted through the inspection reports, mulled over the recommendations of the NRC staff, TU Electric, and CASE, and on December 28, 1983, awarded a great—and unexpected—victory to Juanita Ellis. The board rejected the assurances of the NRC staff and the utility that the problems with the pipe supports were as good as solved. And, furthermore, it found that TU Electric had not followed proper procedure in handling design changes in the field, even though the NRC staff had previously okayed the utility's approach. To prove to the licensing board that the plant was safe to operate, the utility would have to reinspect almost the entire plant, as well as review much of its design and construction.

It took TU's management a year to realize just how much work would be needed to satisfy the licensing board. In January 1985, the utility requested that the hearings be suspended while it attacked the problems. Over the next several years it redesigned and reengineered every part of Comanche Peak that had anything to do with the safety of the plant, with the sole exception of the reactor and steam generator provided by Westinghouse. Everything was done to meet the strict requirements laid down by the licensing board, which meant justifying every step in great detail—indeed, in far greater detail that had been demanded of any other nuclear plant. A summary by Cygna Energy Services, a consulting firm hired by TU Electric to help obtain the operating license, describes the "unprecedented design revalidation program on Comanche Peak":

> The [Licensing] Board's procedures and the expectation of hearings on Cygna's work required a far more extensive examination and discussion than is industry practice for such a review. The extremely detailed reviews that were necessary required the investigation of the technical adequacies of design and engineering practices that have long been considered a standard in the industry. To the experienced engineer, the soundness of the design of a system or component is intuitively obvious in many cases. However, when the proof of the soundness of that design practice is questioned and required to be proved with scientific-like accuracy, the associated cost and effort involved are enormous.

It was, by a large margin, the largest reworking of a nuclear plant—or of any industrial plant—in history.

By the middle of 1988, it was apparent that TU Electric would probably be able to satisfy the board and obtain an operating license. Even the whistleblowers Walsh and Doyle had agreed that the revised plans for the pipe supports would be acceptable. The one remaining obstacle was CASE and Juanita Ellis. They probably could not prevent Comanche Peak from starting up, but they could still delay it considerably—and each year that the plant was held up was costing TU Electric about $1 billion.

Enter Bill Counsil. In May 1985 he had been hired away from Northeast Utilities to head a new management team at Comanche Peak. He brought with him a different philosophy for dealing with the public. Instead of TU Electric's traditional us-against-them approach, he believed in communication and cooperation. He began courting Ellis, meeting her clandestinely to talk and sometimes deliver boxes of internal TU Electric documents.

Eventually he broached the idea of a settlement, and they discussed what it would take for CASE to drop its opposition to an operating license for Comanche Peak. Ellis was willing to listen. Unlike many of her allies, she was not antinuclear. She merely wanted to make sure that Comanche Peak was safe. The breakthrough came in April 1988. After having Ellis sign a confidentiality agreement, Counsil showed her a letter he had written admitting that TU Electric had made mistakes at Comanche Peak. To Ellis it seemed that the Comanche Peak management had truly come around.

On July 1, Counsil and Ellis announced an agreement. CASE would drop its opposition to an operating license for Comanche Peak, which would lead to the dissolution of the public hearings. In return, TU Electric would pay CASE $10 million, $4.5 million for its expenses in the hearings and $5.5 million to compensate a group of fifty whistleblowers who had been fired from the plant. Counsil read his letter in public acknowledging Comanche Peak's problems. And CASE became an insider at the plant—it was awarded a seat on the utility's independent safety review committee, given access to the plant on forty-eight hours notice, and allowed to participate in inspections and attend meetings between TU Electric and the NRC.

That was that. Comanche Peak got its operating license without much more trouble. Unit 1 began commercial operation in August 1990, and Unit 2 three years later. Juanita Ellis, for years the darling of the antinuclear movement, became an instant pariah. She had few regrets and no apologies, however. She had forced TU Electric to do everything right, and when she could accomplish no more as an outside adversary, she had reserved herself a place as an inside watchdog. Not bad for a former church secretary who got her first lesson on nuclear power from a gardening magazine.

A SYSTEM UNLIKE ANY OTHER

Like the stories of Horatio Alger or Huck Finn, the saga of Juanita Ellis versus Comanche Peak is truly an American tale—it captures something essential about the country, its institutions, and its people. Every modern society regulates hazardous technologies, and, in particular, every country that has embraced the nuclear dragon has set up some system to keep it in check, but the details of that control vary from place to place. How a society chooses to regulate nuclear power or any other technology depends on its legal and political systems, its culture, and its past experiences with risky technologies. Americans, with their exceptional ideas about how a country should be run, have created a regulatory system unlike any other.

The most striking feature of that system is its openness. The access that Juanita Ellis exploited is unparalleled in the rest of the world. It has its roots in the Atomic Energy Act of 1954, which set out the ground rules for the commercial development of nuclear power in the United States. In that act

Congress required that the AEC, when considering a construction or operating permit for a nuclear plant, should hold a public hearing "upon the request of any person whose interest may be affected by the proceeding." By 1969, the process had been amended to its current form, in which, before granting a construction permit for any nuclear plant, the AEC must hold a public hearing before an Atomic Safety and Licensing Board. No one need request the hearing. And before granting an operating permit, the AEC must hold a second hearing in front of an Atomic Safety and Licensing Board, but only if there are intervenors contesting the awarding of the license. Otherwise, it becomes a matter between the utility and the NRC.

Legally, these intervenors are on an equal footing with the "official" parties: the utility and the NRC staff. And, as Juanita Ellis's success in front of the licensing board showed, this equality is not simply theoretical. The arguments and evidence offered by the intervenors are taken seriously.

This is not a common attitude in other countries. In France, for instance, the public has essentially no access to or influence over decisions concerning nuclear power. Regulatory policy is set in discussions among the French Atomic Energy Commission (CEA), the government-owned utility, Electricité de France, and the government-owned nuclear vendor, Framatome. The CEA does not have to make public its safety studies of individual reactors, as the NRC must, and there are no public hearings in which citizens can make their objections to a planned nuclear plant.

The difference in the two approaches reflects differing attitudes toward government and the role of the citizen. Americans have a hands-on approach to government. They believe that official meetings should be as open as possible. They believe that politicians and bureaucrats should be accessible to the public. When they visit Washington, DC, they want to be able to walk into their congressional representative's office and discuss his or her vote on farm subsidies or foreign aid. And they believe that people should have a direct voice in decisions that may threaten their health or well-being. They would not likely accept the French system, in which the decision to build a nuclear plant can be made without the public's voice being heard.

A second unusual feature of the U.S. nuclear regulatory system, less obvious but more significant, is its adversarial nature. This can be seen in the trial-like conduct of the licensing hearings. The participants—the utility and the intervenors—are adversaries. Each attempts to make its own points and rebut those of the other party. Neither has any interest in finding common ground. In general, the proceedings end with a winner and a loser.

More important, the relationship between the U.S. nuclear industry and the government regulators is adversarial, whereas in other countries the regulators and the regulated generally see themselves as partners. They may not always agree, and their relationships may sometimes be strained, but they assume that they are working toward the same goals. And indeed, there is often a coziness between the regulators and the industry they regulate that would bring cries of outrage from an American public.

In the early years of the nuclear era, however, things were not so adversarial in the United States, either. With the Atomic Energy Commission pushing the commercialization of nuclear power, the reactor manufacturers and utilities understood the government to be an ally. But in the late 1960s and early 1970s, the partnership split. With growing commercial acceptance of nuclear power, the AEC (and later the NRC) began devoting more of its attention to safety issues, and soon its relationship with industry was transformed from cooperation to confrontation. Utilities chafed under an ever-expanding set of requirements, many of which seemed to add little to the safety of a plant, while regulators sensed that many utility managers did not understand the difficulty of keeping nuclear power under control. By the late 1980s, a group of nuclear experts at MIT described the situation this way: "Many industry people think NRC staff members are technically inept or promoting a hidden agenda. Conversely, regulators often believe that utilities managers are less than forthright in responding to requests and orders, and that some do not make safety a priority." TU Electric wasn't the only utility to be cavalier about following certain regulations. Because the utilities often viewed the regulators as misguided, if not malicious, and because they viewed many of the regulatory requirements as excessive and not important to safety, they saw no reason to follow the rules to the letter. If neglecting to do everything by the book would save money for the utility—and its shareholders and customers—without compromising safety, why not? The regulators, on the other hand, seldom tried too hard to convince the nuclear industry that their demands were reasonable and justified. The result was a great deal of head butting and talking past one another, with the industry blaming overzealous regulation for its problems, and regulators believing they had to get tough because industry couldn't be trusted to take care of safety by itself.

The nuclear industry's experience is not unique, of course. Strained relationships between the regulators and the regulated seem to be the rule, rather than the exception, in American industry. Pharmaceutical companies regularly accuse the Food and Drug Administration of costing lives with an overly cautious approach to approving new drugs. Chemical companies complain about the burden imposed by the Environmental Protection Agency. And nobody likes the workplace regulations set by the Occupational Safety and Health Administration.

Why should things be so contentious in the United States? In *Nuclear Politics*, James Jasper suggests that, in the case of nuclear power, the American regulators' habit of regulating "from the outside" prevented the government and the industry from working together to find solutions to the problems that became prominent in the 1970s. In other countries, the line between government and industry can get quite blurred, with government representatives becoming insiders in the industries they are overseeing. (Indeed, in France the nuclear industry is part of the government.) But in the United States, businesses and individuals alike prefer to keep the government at arm's length.

And, more than most other peoples, Americans tend to distrust government and dislike government intervention in their lives. In their guts they know that "We're from the government, and we're here to help you" is an oxymoron. Even when everyone recognizes the need for government involvement—as in regulating nuclear power—nobody likes to hear that knock on the door. And so there is a certain unavoidable tension to American efforts at controlling technology. Over time, the regulators and the regulated may be able to reach a mutual understanding, but the process can be painful.

Besides its openness to the public and its adversarial nature, the U.S. nuclear regulatory system had a third feature characteristic of the broader society: a fondness for explicit, written rules. Americans are a legalistic people. They go to court to haggle over the precise terms of a contract. They draw up by-laws for even the smallest club or organization. They read the fine print on 25-cent grocery coupons. They sign prenuptial agreements. But this rule-mania is strongest in relations between the government and the governed. Thanks in part to the wariness with which they view authority, Americans like to have their rights and responsibilities written down. In recent years, it has become a national joke how government regulations, codes, standards, guidelines, and the like have piled up in a mountain of red tape. But the nuclear industry fell under that mountain's shadow more than two decades ago, and it was no joke.

Beginning in the 1960s, when Rickover protege Milton Shaw was in charge of the AEC's reactor safety programs, the government crafted an increasingly large and detailed set of requirements for nuclear power plants. Such an approach had worked well with Rickover's nuclear navy, which had an excellent safety record, but it was not well suited for the commercial nuclear industry. Rickover's crew had settled on a few standardized designs early on, and the experience with them had allowed Navy engineers to write an effective set of requirements. But commercial reactors were evolving too quickly in the 1960s and 1970s for any set of regulations to keep up with them. The AEC was forced to change its safety rules constantly, and the utilities felt as if they were standing in quicksand. A design that was acceptable one year was judged not safe enough the next, and even after a plant was built, the AEC might demand changes to meet the latest thinking on safety. The utilities, many of which had never built a nuclear plant before, had a hard time believing that all the nit-picky details were important. They were accustomed to fossil-fuel plants, where all the details weren't important, and now they were being forced to dot every *i* and cross every *t*.

Not surprisingly, the utilities responded by doing what they were forced to do—usually—but nothing more. When the Kemeny Commission studied the nuclear industry in the wake of the accident at Three Mile Island, it found that safety had been reduced to simply following the regulations. Joseph Rees, a political scientist at the University of North Carolina at Chapel Hill, summed it up like this:

> [N]uclear utility officials were so consumed by the enormous task of complying with a "voluminous and complex" maze of NRC rules that satisfying regulatory requirements—going by the book—was equated with safety. As Thomas Pigford, a member of the Kemeny Commission and a professor of nuclear engineering, tried to explain it: "The massive effort required to comply with the vast body of [NRC] requirements and to demonstrate compliance therewith . . . foster[ed] . . . [the] complacent feeling that all of that work in meeting regulations must somehow insure safety."

The AEC and NRC, by regulating from the outside and promulgating specific requirements, had taught the nuclear utilities to worry about following the rules instead of making sure their plants were safe.

It was different elsewhere. In Canada, in Britain, in France, in Germany, and indeed in nearly every other country that built nuclear power plants, regulation focused more on ends than on means. In 1982, for instance, British reactor operators had forty-five pages of safety guidelines they were obliged to follow, compared with the 3,300 pages of detailed regulations confronting operators in the United States. "In Britain," said Walter Marshall, chairman of that country's Central Electricity Generating Board, "we concentrate on the question, 'Is the reactor safe?' rather than on the question, 'Have the regulations been satisfied?'" Jon Jennekens, who headed up the organization overseeing nuclear safety in Canada, made a similar point in a speech the year after Three Mile Island. "The Canadian approach to nuclear safety has been to establish a set of fundamental principles and basic criteria. . . . Primary responsibility is then placed upon the proponent to develop the competence required to show that the proposed plant will not pose unacceptable or public health and safety risks." Canadian regulators, he said, were careful not to write such detailed rules that they were, in effect, taking over much of the design job from the utilities. "The primary responsibility for safe design and operation rests with the licensee, and every effort must be made to guard against destroying his initiative and ingenuity."

It's difficult to know how much of this difference between the U.S. and other nuclear regulatory systems was due to differences in political culture and how much to differences in the nuclear industries. In other countries, nuclear power was developed by a few large concerns that accumulated a great deal of experience in the nuclear arena. In France, for example, just two partners ran the show for the entire country: Framatome, which designed and built the reactors and steam systems for the plants, and Electricité de France, the monopolistic, government-owned electric utility, which did the rest of the construction and operated the plants. In West Germany, one large company built almost all the nuclear power plants, which were owned and operated by nine large utilities. In Japan, nine utilities owned all the nuclear plants, and three companies did most of their construction. By contrast, in the United States several dozen utilities ran nuclear plants, with many of them owning only one, and the design and construction of the plants was spread out among several reactor manufacturers and a number of architectural/engineering firms and construction companies. As a result, U.S. regulators never knew what to expect from a nuclear plant. Some were designed and built to high standards by companies with considerable experience in the field, while others were problems from start to finish. Faced with such inconsistency, the AEC and NRC could either work closely with the utilities—regulate from the inside—or else try to specify everything so carefully that even a utility building its first nuclear plant could make it safe just by following the rules. Given the arm's-length relationship between American business and government, regulating from the inside was never really an option.

As it developed, the American system of control proved spectacularly unsuited to the industry it was regulating. For starters, the openness of the system offered antinuclear activists a chance to drag out licensing procedures, run up construction costs, and sometimes even kill nuclear projects. At least two nuclear plants were canceled after battles similar to those that TU Electric faced against Juanita Ellis. In Midland, Michigan, environmental activist Mary Sinclair waged a successful fifteen-year fight against Dow Chemical's planned Midland plant, and in Moscow, Ohio, Margaret Erbe, a mother of ten and head of the local PTA, faced down the Zimmer plant. And although Ellis seems to have been motivated by a true concern about the safety of the plant, many intervenors were ideologically opposed to nuclear power. This can be seen in the way activists denounced Juanita Ellis for dropping opposition to Comanche Peak when it seemed as if TU Electric had addressed her concerns. When *Business Week* interviewed Jane Doughty, director of a group resisting the Seabrook nuclear plant in New Hampshire, after Ellis's rapprochement with the utility, Doughty disparaged Ellis's willingness to compromise: "Our object is to stop Seabrook. I can't see what is to be gained by talking to the utility about that." In such cases the hearings become not so much a chance to air out safety concerns as a way to discourage the utility from continuing with its plans. Since nuclear plants are so complex and NRC regulations so extensive, it's almost inevitable that there will be a few issues upon which determined activists can seize.

This is not necessarily bad. The hearings can call attention to real problems in a plant's construction, and they can also point out weaknesses in the NRC's regulations. But because the intervenors are often either uninterested in the difference between true safety concerns and trivialities or unable to tell them apart, the two tend to get mixed in together, and this makes nuclear power much more expensive than it needs to be. At hearings, a utility can spend a great deal of time and money arguing over minor problems.

People who have studied the U.S. nuclear industry agree that the licensing delays caused by intervenors and others have wounded it noticeably. Without Juanita Ellis's dogged opposition, Comanche Peak would likely have been licensed years and billions of dollars sooner. And when the state of Massachusetts objected to evacuation plans at the Seabrook nuclear plant, it cost the utility a six-month, $300-million delay. But such delays were not fatal to the industry. Instead, it was the constantly changing regulations of the 1970s and 1980s that killed the nuclear option in the United States. They drove up the cost of nuclear power in the United States to the point that utilities no longer considered ordering nuclear plants.

The situation got particularly bad after the incident at Three Mile Island because the NRC, shaken to realize that a serious accident really could happen, unleashed a barrage of new regulations. Even the best, most careful utilities were not immune. Duke Power, for example, was recognized throughout the industry for its expertise in nuclear construction. But Bob Dick, Duke's vice president for construction, told *Forbes* that the utility was constantly frustrated by changing NRC demands. "At the McGuire plant, we were really putting in with one hand and taking out with the other." Utilities with less experience than Duke were hit even harder. The results were predictable—and devastating. Plants originally expected to cost $600 million or $800 million ended up at $2 billion, $3 billion, or more. And according to a careful analysis by Charles Komanoff, the shifting requirements of the AEC and NRC were primarily responsible for these cost overruns.

With costs skyrocketing and with the growth in electrical demand slackening as the United States got serious about energy conservation in the 1970s, utilities took a hard look at the nuclear option and decided they didn't need it. The nuclear price tag was too high, and, equally important, the uncertainty was too great. Would a plant cost $1 billion, or $2 billion, or maybe even $5 billion? Would it take five years to finish, or eight, or fifteen? The unstable regulatory environment and the always-present threat of intervenors dragging out the licensing process made it increasingly difficult for utilities to justify investing in nuclear power or to find financing for building the plants. In the late 1970s and into the 1980s, utilities canceled many of the nuclear plants they had ordered in the early and mid-1970s. Eventually, every plant ordered after 1974 in the United States was canceled, and of the 200 or so U.S. plants that were either in operation, under construction, or on order in the mid-1970s, just over 100 are now generating electricity.

The changing regulations that killed the industry were the product both of the times and of the system of control that the United States had chosen. In the 1970s and 1980s, nuclear power was an evolving technology, one that was still climbing the steep early incline of the learning curve. The truly large plants—those capable of generating more than 1,000 megawatts—only began coming on line in the mid-1970s, and for the next decade or more, the utilities and the regulators were still learning just what to expect from them. Much of the evolution in safety regulation was due simply to the process of gaining experience with the technology.

But the regulatory changes in the United States were much more drastic than can be explained by changes in the technology. Other countries were faced with the same technical environment—rapidly evolving knowledge about nuclear reactors—but did not saddle their nuclear industries with the same torrent of shifting requirements. It was the particular type of regulatory system that the United States had established that led inevitably to an ever-growing and ever-changing set of rules. Intervenors, for example, pushed the AEC and NRC to create a variety of new regulations that made no economic sense—the additional margin of safety was so small that the money would have been better spent in some other way. The intervenors did not limit their complaints to areas in which changes could truly improve the safety of the plant, and the regulators made no effort to determine which changes would be worth the extra costs they would impose. After all, the NRC was charged with making nuclear reactors safe, not with making nuclear utilities profitable.

But more important than the role of the intervenors was the choice to regulate from the outside with explicit, detailed rules. The United States had locked itself into a system in which the government told the nuclear industry exactly how to build plants that were safe enough to get a license. The utilities had to do very little safety thinking for themselves. But then, as the thinking on safety evolved, most of the design changes necessarily came from outside the utilities. If the utilities and regulators had been collaborators instead of adversaries, then the utilities would have had input into the changes and they would not have been caught by surprise. Furthermore, without the rule-bound, one-size-fits-all approach of the AEC and NRC, adapting to the changing thinking on safety would not have been so painful for the utilities. If everything had not been spelled out in great detail, utility engineers and NRC staff could have worked together to find solutions that would answer both the NRC's safety concerns and the special needs of a particular plant. As it was, however, because explicit rules were imposed by the regulatory agencies on the utilities, there was no room for compromise. Either a utility complied or it didn't. And if it complied this year, it might not be in compliance next year.

Could it have been done differently? Could the United States have avoided its problems by adapting, say, a regulatory system more like that of France? It sounds appealing, particularly to nuclear power proponents frustrated that the technology has not been as successful in this country as in France, where nuclear power now accounts for about 80 percent of all electricity. But such speculations overlook an important feature of the regulatory system. Its purpose is not merely to make the technology safe, but also to make it acceptable—to "legitimize" it, in the language of the political scientist. Every four years when Americans elect a president, a large percentage of the population disagrees with the choice. But everyone accepts it because the method of selection—popular vote in each of the fifty states leading to a vote in the Electoral College—is seen as legitimate. Similarly, unpopular laws passed by the Congress and signed by the president are accepted, even while people may work to change them. The features of the U.S. nuclear regulatory system—its openness to intervenors, its adversarial nature, its collections of rules—all serve to make decisions about nuclear power legitimate in the public eye. They signal to the public that a nuclear plant has jumped through the proper hoops and so can be accepted, if not liked.

In France, different things are needed to legitimize decisions about nuclear power. Less public involvement is necessary. Citizens there accept a stronger government and less emphasis on individual rights, so they don't expect the decision process to be as open. And there is no need to keep the government regulators and the nuclear industry at arm's length with adversarial relationships and books of rules.

It is an open question how far the U.S. system could move toward that of France and still maintain its legitimacy. In the early 1960s, when trust in both the government and the nuclear industry was still relatively high, legitimization was not quite as hard to attain as it is now. But until that trust is renewed, it seems unlikely that Americans will accept much less distance between the regulators and the regulated or less access to the system for members of the public.

A deeper question is whether the troubles of the U.S. nuclear industry flowed inevitably from the characteristics of the American legal, economic, and social systems. Some in the nuclear industry have suggested that there is a fundamental mismatch between American culture and the demands of a hazardous, complex technology like nuclear power. To master the technology, all the players in the nuclear game—utilities, reactor manufacturers, construction companies, and regulators—need to work together, sharing information and building on one another's experience. In France, this is exactly what has happened. The regulators and the government-owned companies of the nuclear industry have seen themselves as partners. And indeed, a study of nuclear industries around the world found that the key factor explaining which were successful was the level of industrywide cooperation. The more cooperation, the more success. But cooperation is not always easy for Americans, who pride themselves on their individualism.

Looking back, it's easy to convince oneself that the fundamental features of American society fated it to mishandle nuclear power: given our culture and our political system, the nuclear power shipwreck was inevitable. But the question of inevitability is, at heart, an empty one. It's impossible to repeat history with a few changes to see if things might have worked out otherwise. A more important question is whether the American system can change enough to handle nuclear power effectively in the future. And the answer to that question is, as we'll see in more detail over the next two chapters, a tentative yes.

Since the 1979 accident at Three Mile Island, regulators and industry have gradually come to recognize the need for cooperation and collaboration. Shortly after TMI, the utilities created their own private regulatory organization, the Institute of Nuclear Power Operators (INPO), and over time it has brought the industry together, fostering a sense that "We're all in this together" and "We're only as strong as our weakest link." Observers have given INPO credit for much of the improvement in U.S. nuclear plants over the past decade or so. The NRC has continued to regulate from the outside, and its requirements have become even more numerous and detailed, but the industry has gradually come to accept them and to learn to live with them, if not always like them or agree with them. And, in preparing for a possible next generation of commercial reactors in the United States, the NRC and the nuclear industry have worked together to settle on safety requirements ahead of time instead of dealing with them piecemeal as they come up. All it took was a near-disaster and the realization that, if things continued as before, the U.S. nuclear industry was doomed to die a slow death as its nuclear plants shut down and were not replaced.

TECHNOLOGY IN THE COURTS

In the early 1980s, when the activist Jeremy Rifkin was looking to strangle the newborn recombinant DNA technology in its cradle, he had few of the weapons available to Juanita Ellis. There were no licensing hearings in which citizens could intervene. There were no reams of regulations whose violation could shut down an offender. Indeed, there was nothing analogous to the Nuclear Regulatory Commission watching over the safety of this new technology. So Rifkin's battle against genetic engineering was waged in the courts. Like Ellis's battle against Comanche Peak, superficially it was a fight against a hazardous technology, but on a deeper level it was a struggle about who should control this hazardous technology and how decisions concerning it should be made.

Recombinant DNA, or rDNA, technology is a method of manipulating the genetic material of living organisms. The characteristics of an animal, plant, or microbe are determined by its DNA, the long, twisted molecule that contains the organism's genes. Modify that DNA—recombine it, in the scientific jargon—and you have created a new organism. Scientists, when they are attempting to reassure the public, often argue that rDNA technology does nothing that hasn't been done for years with traditional plant and animal breeding. When breeders select for a particular trait—a new color of tulip, or a cat with no hair—they are creating a line of plants or animals with a new set of DNA. And this is true, as far as it goes. But to liken genetic engineering with traditional breeding by saying both modify an organism's genes is like equating a modern computer with an abacus because both help people with their arithmetic. Breeders must work with what nature gives them. If a mutant cat is born with no hair, a breeder can use that cat to produce a line of hairless felines, but he cannot decide ahead of time to create that trait in cats. With recombinant DNA technology, however, scientists have a whole new level of control. They can, in principle, insert any gene they want into any organism. Genetic engineers can take a gene from a sheep and insert it into a rat, transfer a lettuce gene into a bacterium, or put a bit of human into a radish. It is even possible to modify the genes themselves, creating entirely new genes unlike anything that nature ever imagined. By manipulating the basic stuff of life, scientists open the door to countless applications: improving crops and farm animals in ways impossible with normal breeding, creating bacteria that produce chemicals less expensively than can be done in traditional chemical plants, or curing genetic diseases in humans. The potential power of genetic engineering is almost impossible to overstate.

But then so are the potential risks, and twenty five years ago, this power made rDNA researchers stop and think about what they were doing. It was 1971, and the Stanford scientist Paul Berg was planning to insert genetic material from an animal tumor virus into the common bacterium *Escherichia coli*. But other researchers were troubled. Since *E. coli*'s usual home is the human gastrointestinal tract, was it possible that the genetically engineered bacteria might escape from Berg's laboratory and take up residence in an unsuspecting human host? If so, then they might multiply in their human host and create a colony of bacteria containing Berg's tumor-virus gene, which in turn might spread to other humans and somehow trigger a cancer epidemic. No one knew how likely this was, but that was really the point: the risk was unknown. Some of Berg's colleagues argued that until they had a good idea of what the risks were and how to handle them, it would be better to hold back.

Berg agreed and put his work on hold, while scientists in the field debated what to do. For the next several years, other researchers working on recombinant DNA experiments also held off, waiting for a consensus to develop. Finally, in 1976, the Recombinant DNA Advisory Committee announced a set of guidelines for rDNA research. The committee had been set up by the National Institutes of Health (NIH), the federal agency that funds much of biomedical research. Its guidelines called for recombinant DNA research to be performed with two types of precautions. Scientists should use special strains of *E. coli* or other organisms that were unable to survive outside the laboratory, and the laboratories themselves should have physical safeguards to prevent the escape of any genetically modified organisms. Laboratories with the riskiest research should have the tightest controls: protective clothing for workers, air locks, and negative air pressure so that in case of leaks the air would rush into the laboratory, not out. The scientists in the field were justifiably proud of their effort at self-regulation, and to this day it remains a model for how researchers can anticipate possible risks of their work and come up with ways to deal with them.

The original 1976 guidelines had called for a complete prohibition of "deliberate releases"—taking genetically engineered organisms from the lab into the outside world—but the researchers soon realized this was too stringent if their techniques were ever to have practical application. So a revision in 1978 allowed deliberate releases, if the director of the National Institutes of Health approved. The revision did not, however, offer explicit guidance about when the director should make such an exception to the general rule, and this would become an issue when the NIH began approving rDNA experiments that included releases into the environment.

In 1983 the NIH gave the go-ahead to Steven Lindow and Nickolas Panopoulos, two University of California scientists, to spray genetically engineered bacteria on a potato patch. The bacteria, called "Ice-Minus," were designed to prevent frost formation on crops. But before the trial could go ahead, Rifkin and his organization, the Foundation for Economic Trends, filed suit in federal district court to stop it. Rifkin argued that NIH had not prepared an environmental impact statement and that, since the proposed trial posed a potential threat to the environment, such a statement was required by federal law. He also asked the court to force NIH to assess the environmental effects of deliberate releases in general before it went ahead with any such field trials. The judge, John Sirica, agreed with Rifkin and ordered the NIH to prepare such assessments before proceeding with the frost-prevention experiment or any other release of genetically modified organisms. In particular, Sirica was unhappy that the NIH had not set forth explicit standards for when it would allow the release of genetically modified organisms. The NIH appealed and got a partial victory: an environmental impact statement would be required for the experiment at hand, but not for the entire future rDNA program.

The scientific community was not pleased. The Recombinant DNA Advisory Committee had considered the potential risks of the field test and had agreed unanimously, 19-0, that they were negligible. This, the NIH argued, was in essence the environmental assessment that Rifkin had asked for. To rule as the courts did seemed to ignore scientific consensus in favor of legal technicalities. In an article in the *Yale Law and Policy Review*, the biologist Maxine Singer slammed the decision:

> The plaintiff's arguments that were accepted by Judge Sirica may make sense from a narrow legal perspective, but they make little sense in the context of available scientific knowledge. I cannot help but wonder whether the case might not have been found frivolous to begin with, if the Judge had understood the science.

Eventually, after jumping other legal hurdles erected by Rifkin, the frost-prevention experiment did go ahead, with no discernible effect on the environment or on public health. To scientists who saw the delays as a waste of time and money, it was difficult to understand how a single person—and a single person with no scientific expertise—could shut down the work of an entire community of researchers.

As the Rifkin story illustrates, although scientists and engineers may chafe under the constraints of regulatory agencies like the NRC and FDA, they grow absolutely indignant when courts start throwing their weight around. Leave science to the scientists, they mutter. But it is not that easy. When science and technology harm—or threaten to harm—the public, the public often insists on doing something about it. If a regulatory agency is already on guard, that may be enough. But if there is no such agency, or if that agency does not seem to be doing its job, the public looks elsewhere for help. And for Americans in particular, that generally means going to court.

Americans see the courts as protectors of their rights, as guarantors of proper procedure, and as neutral arenas in which individuals can confront the largest corporations or government agencies on an equal footing. If they have been harmed, they look to the courts for compensation. If they are about to be harmed, they look to the courts for defense. All of this makes the courtroom a natural place for the public to confront worrisome science and technology and work out a modus vivendi. In recent decades, as people have become more concerned about technological threats and more skeptical about how well government or private industry will protect them from those threats, they have gone to court in increasing numbers to challenge technology—not just nuclear power and genetic engineering, but also food additives, automobile design, medical procedures, and much more.

Generally speaking, the courts' role in shaping technology is a reactive one, in contrast to the proactive role of regulatory agencies. The courts do not determine which direction a technology should be heading and then push it that way. They can act only when someone brings a problem or complaint before them. On the other hand, courts are "responsive"—unlike the legislative or executive branches, which can ignore issues that don't seem important, courts must offer judgments on the items brought before them. This means that courts are often the first social institution to weigh in on a technology. Before legislatures pass their laws or regulatory agencies create their codes, the courts can be providing a rough, case-by-case control. For that reason, the legal system may have a larger role in shaping a new technology than one might guess.

However, as tools for controlling technology, the courts are a relatively blunt instrument. They are not designed for the job—they're meant to settle legal questions, not technical ones—and they are strongly limited in the sorts of issues they can consider and actions they can take. When Rifkin challenged the rDNA experiment, for instance, he used the best available weapon: a law that demanded environmental impact statements for government activities. But that framed the battle as an issue of environmental protection, and many of the real issues were elsewhere: Could genetically engineered bacteria threaten public health, for example, and should scientists be monkeying around with genes like that in the first place?

But to the technically minded, the most distressing thing about the legal system is the way it handles technical evidence and arguments. Too often, it seems, cases are decided in ways that ignore or fly in the face of scientific reasoning. Consider, for example, the bankrupting of the Dow Corning Corporation by large numbers of women claiming to have been injured by breast implants.

In the early 1990s, a few researchers uncovered evidence that seemed to link silicone breast implants with certain autoimmune diseases such as rheumatoid arthritis and lupus erythematosus. In the United States, the Food and Drug Administration responded by suspending the use of the implants in January 1992. Very quickly tens of thousands of women—some of them attracted by attorneys advertising for anyone who thought she might have a case—filed suit against the makers of the implants. At the time, the link between the silicone implants and the various maladies claimed by the women was tenuous at best, but the manufacturers decided to settle out of court, agreeing to pay $4.25 billion to women allegedly injured by the implants. Although this was the largest payout for a class-action lawsuit in history, thousands of the women who had filed suit thought it wasn't enough and refused to join in the settlement. Threatened with even more billions in liabilities, Dow Corning, the largest maker of the implants, filed for bankruptcy in May 1995. By August, the settlement was falling apart, as some 440,000 women had registered to claim pieces of a pie that had been designed for a tenth as many. As this is written, the fate of the settlement is up in the air.

Ironically, even before Dow Corning had filed for bankruptcy, the scientific case against breast implants was falling apart. In June 1994 a study published in the *New England Journal of Medicine* found no increased risk for connective-tissue diseases in women with breast implants. Such diseases formed a major part of the legal case against the implants. One year later, and a month after Dow Corning's bankruptcy, an even larger study also exonerated breast implants from causing connective-tissue disease. After several years of study, the opinion of the scientific community was that there might be some harmful effects of breast implants—it would be impossible to prove that the risk is zero—but that the risks are so small as to be difficult to detect. Most, if not all, of the women filing suit would likely have had the same health problems without the implants.

Still, few observers believed that the breast implant makers would win in court, which is why they were willing to pay out billions of dollars for claims that had so little scientific merit. Historically, courts have not done a good job in handling scientific arguments and evidence when they are complex or else contain some uncertainty, and the evidence about the breast implants was both complex and uncertain. There are two reasons for this weakness in dealing with science.

First, judges and juries seldom have the expertise to evaluate technical arguments on their technical merits. Instead, when faced with contradictory scientific testimony, they often choose what to believe based on factors that have little to do with the science. This is why good expert witnesses are so valuable to attorneys—they come across as knowledgeable and trustworthy, which can make the difference between winning and losing a case.

In her book *Science at the Bar*, Sheila Jasanoff describes a product liability case in which the plaintiffs alleged that a spermicide produced by Ortho Pharmaceutical produced serious birth defects in a baby girl. The judge, apparently unable or unwilling to weigh the arguments of the plaintiff's and defendant's witnesses on their scientific merits, relied instead on character judgments. In his decision, the judge explained why he gave more weight to the arguments of a scientist testifying for the plaintiffs:

> His demeanor as a witness was excellent: he answered all questions fairly and openly in a balanced manner, translating technical terms and findings into common, understandable language, and he gave no hint of bias or prejudice.

By contrast, one of Ortho's witnesses lost credibility with the judge because he had a "less-than-certain tone" during cross-examination, while another's opinions were downgraded because of "the absolute terms in which he expressed his conclusions." The judge made no attempt to determine if a consensus existed in the scientific community about the likelihood of spermicide causing birth defects, and he did not try to weigh the scientific arguments offered by the two sides. Instead, he based on his assessment of the personal credibility of the witnesses and awarded the girl and her mother $5.1 million.

This lack of technical sophistication is compounded by the way in which trials are carried out. Courts are designed to settle legal issues, not technical ones, and the two demand quite different ways of thinking. Scientists determine their "truths" by a process of data gathering, hypothesis formation, and hypothesis testing. As a community, scientists will explore different possibilities and debate interpretations of the evidence, gradually homing in on a consensus. By contrast, legal proceedings in U.S. courts are adversarial. To determine a legal "truth," two sides take opposing positions, and each attempts to prove its case by bringing in witnesses to buttress its claims and to tear down those of its foes.

The adversarial process is poorly suited for settling technical issues in an objective way. Its most obvious weakness is that the two sides can bring in approximately equal numbers of expert witnesses, and there is no easy way for judge or jury to sort out their competing claims. The research community may be nearly unanimous in favoring one side, but in court it seems as if scientific opinion is evenly split. A deeper flaw is that the adversarial system cuts and shapes the technical arguments to fit its own needs. For instance, the only scientific evidence that the court hears is what one side or the other deems important to making its case. A court has no independent evidence-gathering ability. Thus a great deal of information can be left out—information that may be key to understanding the controversy—if it doesn't fit with the points that the two parties are trying to make.

Joel Yellin, a legal scholar at the Massachusetts Institute of Technology, offers an example of how badly the adversarial system can twist the arguments. In an analysis of *Duke Power v. Carolina Environmental Study Group*, he finds that each side tried so hard to make its case that it ended up contradicting itself. The background is this: In an attempt to stop Duke Power's construction of its McGuire and Catawba nuclear plants, the Carolina Environmental Study Group had filed suit alleging that the Price-Anderson Act was unconstitutional. The act, originally passed by Congress in 1957 to encourage development of nuclear power and later extended in 1975, capped the amount of damages that a utility had to pay in case of a nuclear accident. The study group argued that the AEC and NRC themselves had estimated that a worst-case accident could cause damages far higher than the Price-Anderson cap, so that the act violated the group members' due process rights under the Fifth Amendment.

Before anything else, the plaintiffs had to demonstrate that they had the legal standing to challenge the Price-Anderson Act. Otherwise, their case would be thrown out of court, no matter what its merits. To prove their standing, they argued that without the act's limitation on liability, Duke Power would never have decided to build the two plants and threaten the health of the area's citizens, including the plaintiffs. Thus the Price-Anderson Act affected them and gave them the right to challenge it in court. On the other hand, to show that the act violated their due-process rights, the plaintiffs had to argue that it had not been "rationally" enacted by Congress, and as part of their arguments, they claimed that Congress could have found other ways to encourage nuclear power. In short, when making their case on standing, the group argued that the Price-Anderson Act was essential to the construction of Duke's nuclear plants, but when arguing against the constitutionality of the act, they contended that it was not essential. Duke Power countered with its own contradiction: the plaintiffs had no standing because the Price-Anderson Act was not crucial to the utility's decision to build the plants, but Congress had behaved rationally in passing the act because it was essential to the development of nuclear power.

The central legal issue, Yellin notes, was whether the Price-Anderson Act was a prerequisite for the development of nuclear power. But neither side addressed that issue too closely because each needed to have it both ways. The adversarial nature of the proceedings made it almost impossible for the court to get a good idea of the act's significance. As it was, the district court that tried the case found for the plaintiffs, ruling the Price-Anderson Act unconstitutional. The Supreme Court disagreed, however, and Duke Power was allowed to go ahead with its plants.

Episodes like this or the breast-implants case provide plenty of fodder for critics who paint the legal system as incompetent to deal with technical matters. In response, a number of people have called for the development of "science courts" or some other such innovation that would improve the legal system's handling of scientific and technical issues. The science court, as it is usually envisioned, would be charged with separating issues of scientific fact from policy and then addressing those factual issues, leaving the legal and policy reasoning to the normal court system. Or in lieu of a science court, special masters or technically trained law clerks might assist judges in cases with complex technical issues.

Such suggestions are most appealing to scientists, engineers, and others with a rationalistic turn of mind. They are troubled by the way courts deal with science and technology, and they are convinced that the legal system would work better if it improved its technical literacy. But would it? That depends in part on how one defines "better."

The areas in which courts need help with technical matters are actually very limited. Whenever possible, the courts tend to leave technical judgments up to the relevant regulatory agencies and instead focus on matters of law. Over the entire U.S. history of nuclear power, for instance, only one court decision has had a significant effect on the nuclear industry. In 1971, in *Calvert Cliffs Coordinating Committee v. AEC*, a U.S. Appeals Court ruled that the National Environmental Policy Act applied to nuclear power plants and that the AEC had to provide environmental impact statements as part of its licensing process. Until then, the AEC had complied with the act in only the most desultory fashion, fearing that vigorous environmental reviews might discourage utilities from going nuclear. The court lambasted this attitude: "We believe that the Commission's crabbed interpretation of NEPA makes a mockery of the Act." The case had little effect on the nuclear industry, however. The amount of time it took to get a construction permit increased by about a year, but only temporarily—by 1976, the permit process was almost as fast as it had been in 1971. Past this, the courts did little to challenge the policies or the judgments of the AEC and the Joint Committee on Atomic Energy.

Only when the legislature has not devised some other means of control do the courts wade deeply into technology policy. As Americans' first line of defense against scientific and technological threats, courts often find themselves deciding cases on the cutting edge of science and technology. If a science court is to be useful, it must be able to perform its job there, on the frontiers of knowledge. But it is precisely there that a science court becomes problematic.

The assumption behind most of the proposals for rationalizing the court's approach to technical issues is that these issues can be boxed off away from everything else and handed over to the scientific community for consideration. The relevant scientists or engineers would provide independent, objective responses that could be plugged into the judicial proceedings. The court would be responsible for interpreting the law, for balancing competing interests, and for weighing the reliability of testimony from various parties, and so on; the science court would provide the relevant scientific facts for the case. When the scientific community could not offer a universally agreed-upon answer—which would be quite often, given the types of cases that courts are asked to decide—then the science court would admit to the uncertainty, give the community's best guess, and indicate how comfortable scientists were with that guess. The court could decide how it was going to handle the uncertainty, but the scientific community would be responsible for defining exactly what the uncertainty was.

The problem is that on the frontiers of science, there is no such thing as an independent, objective answer. Scientists are working toward such answers, but along the way they depend on a host of assumptions and biases, some of them conscious but many of them not. Consider the 1974 Rasmussen report on reactor safety. It was the sort of thing that a science court might turn out—an attempt to calculate the probabilities of various types of accidents in a reactor, using a rigorous mathematical approach (probabilistic risk assessment) combined with a state-of-the-art understanding of what goes on inside nuclear plants. In retrospect, the Rasmussen report was almost laughably optimistic, but it was not out of line with what the nuclear engineering community believed at the time. Were there a science court at the time, it might not have come up with a figure of one major accident in a billion years of operation, as Rasmussen did, but it would certainly have reported the chances of a major accident to be far lower than they really were.

And, more important, Rasmussen's mistake was not a random one. It was a product of unexamined assumptions held by most of the nuclear community at the time—assumptions that fit quite nicely with what they wanted to believe. This bias went unrecognized until 1979, when the accident at Three Mile Island forced nuclear engineers to realize how badly they had miscalculated.

Given this history, it is understandable that in 1983 a court would not simply bow to the opinion of a nineteen-member committee of biologists that it was safe to release genetically engineered microorganisms into the environment. Scientists have their own unconscious biases, which generally cannot be recognized until after the fact. The result of creating a system of science courts or something like them would be to give these biases a preferred position in the legal system.

In *Science at the Bar*, Jasanoff argues that there are really two different sorts of knowledge at work here—scientific knowledge and legal knowledge—and that both of them are legitimate in their own spheres. Scientists construct their knowledge in ways that reflect the goals and methods of the scientific community, while the legal system constructs its own knowledge in another way altogether.

Consider breast implants. To the scientific community there is a simple question, or set of questions: Can breast implants harm a woman's health? Researchers address the question with a variety of studies and provide answers in terms of statistics and probabilities. The legal system asks a different question altogether: Should women who received breast implants and who later had health problems be compensated, and if so, by how much? In the process of answering that question, a court will examine the scientific evidence, perhaps by listening to different experts and gauging their reliability, but it will bring in much more: testimony from the affected women and their doctors, evidence as to whether the implant makers did the proper tests before putting their products on the market, and so on. When the legal issue is decided, either by the court or by settlement, then the public is provided with an answer: Yes, women who received breast implants and who later had health problems should be compensated for between $10,000 and $1 million apiece, depending on the severity of the problem (or whatever the final resolution of this case turns out to be). This legally constructed answer to the breast-implant question has more meaning to the general public than the scientifically constructed answer, which is likely to be stated in mathematical terms with various caveats attached.

This, in one sense, is what happens when the courts weigh in on a particular technological issue: they help construct a social attitude toward that technology. If people feel threatened by a technology, they will not be content to let the scientists and engineers make their decisions for them. They will insist on having a say in its direction. The courts then become a place that society and technology accommodate themselves to each other. The courts take the technical knowledge provided by scientists and engineers and place it in a broader legal and social context to create a construction of that technology that is meaningful to the larger public. "Litigation becomes an avenue for working out, often at an early stage, the compromises necessary for securing social acceptance of a new technology," Jasanoff writes. "The courts serve in this way as indispensable (albeit uncomfortable to some) forums for accommodating technological change."

In other words, like regulatory agencies, courts help to legitimize technology. In the modern world, assurances from experts are not enough to make the public accept a potentially risky technology. The technology must also be vetted by the courts or some other institution that is a representative of the public. The scientists and engineers involved in a technology often fail to appreciate the value of legal hoops through which they must jump, since the technology is already legitimate in their eyes, and the extra work seems like a waste of time. But for the larger society to accept it, the people must believe that it is under control in a way they understand.

NEGOTIATION

From August 1990 to June 1993, David Leroy held the loneliest job in government. As the nation's Nuclear Waste Negotiator, he was charged with finding a place that would accept nuclear wastes from the rest of the country for temporary storage. But it's hard to negotiate if no one will talk to you. When Leroy first came into office, he had planned to go to the governors of the fifty states and ask each of them what it would take to allow the federal government to build a nuclear-waste storage facility in his or her state. He quickly found, however, that the people he wanted to talk to wouldn't be caught dead speaking with him. And he understood why. As he told *The New York Times,* if he were to meet with a governor to talk about a possible waste site, "I would instantly have created the principal issue in that governor's next re-election campaign." Leroy wasn't exaggerating, Utah's lieutenant governor Val Oveson told the paper: "His fear is accurate. If someone knows who he is, and what he's doing, and finds him on the schedule of any politician, it's going to be a major news story."

As Leroy discovered, the nuclear-waste issue is today's most explosive and intractable technological problem. Each of the nation's more than 100 nuclear plants creates a steady stream of used-up fuel rods as well as other waste materials that are radioactive. The spent fuel is the bulk of the waste. By the end of the century, some 40,000 tons of it will have accumulated, most of it in pools at nuclear plants, where the water keeps the fuel from overheating and absorbs the radiation. But those pools are rapidly filling up, and the utilities are looking to the federal government—which has promised for decades that it would do something about reactor wastes—to take the fuel rods off their hands.

The obstacles to solving the nuclear-waste problem are, for the most part, not technical. High-level nuclear waste, the highly radioactive waste such as spent fuel rods, must be stored away for thousands of years before it has lost enough of its radioactivity that it is no longer dangerous. The usual scheme is to put the wastes below ground in places that will be safe from disturbance. It's important to keep the wastes away from groundwater, for instance, since over time the water might conceivably eat through the casks holding the waste and carry it away through the surrounding rock, perhaps to contaminate drinking water somewhere. As difficult as this sounds, most experts in nuclear-waste disposal believe that it is not at all impossible. The Swedish nuclear industry, for example, has devised a method of storing nuclear waste in copper canisters that are entombed in granite bedrock. The canisters are expected to remain intact for at least a million years, and if waste escapes a canister, it will not go far in the surrounding granite. Scientists cannot guarantee that this or any other waste-disposal scheme is foolproof, but most believe that 10,000 years of undisturbed rest is something that can be accomplished.

The public, however, sees it differently. Survey after survey shows that people find a nuclear-waste site more frightening than any other creation of modern technology: chemical-waste landfills, plants manufacturing deadly chemicals, even nuclear power plants. One telephone study asked respondents to describe what came to mind when thinking about an underground nuclear-waste repository. The responses were almost uniformly negative: "dangerous," "toxic," "death," "environmental damage," "scary," and the like. The researchers summed up their findings this way: "[T]he results demonstrate an aversion so strong that to call it 'negative' or 'dislike' is clearly an understatement. What these images reveal are pervasive feelings of dread, revulsion, and anger—the raw materials of stigmatization and political opposition."

Such attitudes frustrate those in the nuclear industry. Sir John Hill, the chairman of Britain's Atomic Energy Authority, once complained, "I've never come across any industry where the public perception of the problems is so totally different from the problems as seen by those of us in the industry who are actually dealing with them. . . . The public are worried about [things] which, in many ways, are nonproblems." But there seems little that Hill or other nuclear experts can do to bring the public around. Their fears are too deeply ingrained to be erased by the reassurances of nuclear industry experts.

Compounding this dread of nuclear waste is a conundrum that faces many modern industries: how to site an undesirable facility. It is not just nuclear-waste sites and nuclear power plants that arouse opposition. It is incinerators, chemical plants, landfills, prisons, and anything else that people see as damaging their quality of life. When the Walt Disney Co. announced it was planning to build an American history theme park in northern Virginia, not too far from Washington, DC, many of the area's citizens protested vehemently. They didn't want the extra traffic on their roads that such an attraction would bring, and they didn't want all the commercialism they expected to spring up around the park: hotels, restaurants, gas stations, shops aimed at tourists, and the rest. Disney eventually announced it would go elsewhere.

The difficulties in siting a nuclear-waste dump, a prison, or a Disney theme park all have their roots in a simple and unavoidable dilemma: the facility will benefit a large group of people to the disadvantage of a few. Those close to the site will see their property values decline, their quality of life decrease, and their peace of mind diminish. In some cases their health or safety may be put at risk. Understandably, they don't like it. Why us? they ask. Why should we have to pay the price so that others can have nuclear electricity, keep their streets free of criminals, or learn about Pocahontas while munching popcorn and wearing Mickey Mouse ears? NIMBY, they say: Not in my backyard.

In recent years political scientists and policy experts have thought quite a lot about getting around the NIMBY problem. In the past, the so-called DAD approach—decide, announce, defend—worked fine. The relevant officials from business and government would decide among themselves where to put an airport or an industrial plant, and when everything was set, they would tell the public. If they were worried about opposition, they would prepare detailed arguments for why the site was a good choice and no others would work, but the public was not consulted. If people didn't like it, they could go to court or complain to their elected officials, but there were few successful challenges.

Now there are. Indeed, in many industries, they're the rule rather than the exception. It's nearly impossible to find a place to put an incinerator or landfill today, for instance, and although no one has ordered a nuclear plant in the United States for two decades, many in the industry believe that siting it would be the most difficult part of the process if someone did. There are various reasons for the change. The public has become less interested in the economic benefits that may accrue from a new plant and more concerned about threats to the environment, health, and quality of life. People in general have become less trusting of government and big business and quicker to challenge them. And new laws and court rulings have made it easier to challenge decisions that might once have been unassailable.

As a result, officials in a number of places have been trying out a new approach: negotiation. The particulars vary from case to case, but the underlying philosophy is the same. By getting citizens involved in the decision early on and letting them play a major role in the details of the siting agreement—which can include various forms of compensation for accepting a facility—officials hope to minimize resistance and find sites for even the least desirable occupants.

Looked at from another angle, negotiation is an attempt to relegitimize siting decisions. The decide-announce-defend approach no longer seems acceptable to much of the public, particularly when it is used for sites with hazardous technologies, and this loss of legitimacy has led to the opposition that now greets so many siting decisions. With negotiation instead, the public should be more willing to accept difficult and controversial siting decisions. That's the theory, anyway. The practice has proved a little different.

Massachusetts, for example, passed a law in 1980 setting up a system of negotiation for siting hazardous-waste facilities. Before that, state law had effectively given local areas veto power over plans for such facilities, and developers had difficulty finding any part of the state that would accept them. The 1980 act was an attempt to fix that while still allowing local areas a great deal of influence over siting decisions. The process laid out in the law is complex, but the core of it lies in negotiations between a developer and one or more communities that have been identified as feasible sites for a facility to treat or store hazardous waste. During the negotiations, the community is represented by a committee consisting of the mayor and other officials plus several residents. The developer may offer various compensations to convince a community to accept the facility. If the negotiations do not lead to an agreement within a certain time, the parties submit to binding arbitration.

In the first four years after the negotiation act had been passed, five developers sought to place hazardous-waste facilities in Massachusetts. None of them even reached the point of negotiations, and critics of the statute argued that it would have been smarter simply to give the state the power to preempt local authority in siting decisions. But in states like Massachusetts, which has a strong home-rule tradition, local communities might not stand for such preemptive decisions. This was the reasoning behind the 1980 act, which attempted to find a balance between local rule and state authority. It might take a few years of experience and perhaps a few revisions in the law, but in the long term negotiation seemed more likely to succeed.

A similar spirit moved the U.S. Congress to create the office of the Nuclear Waste Negotiator in 1987. But the difficulties in siting a nuclear-waste repository dwarf those for a hazardous-waste facility. It is not simply that the public is much more concerned with nuclear wastes than with the non-nuclear hazardous wastes generated by the chemical and metal-producing industries. An equally large problem has been created over the years by the AEC, the Department of Energy, and the Congress in their handling of nuclear wastes.

In the earliest days of nuclear power, the AEC spent little time worrying about what to do with all the radioactive material that would be generated in commercial reactors, and what little work it performed was poorly thought out and not particularly useful. Then in 1970, when the AEC did announce a comprehensive waste-management program, it turned into a major fiasco. In a hurry to solve the problem of accumulating nuclear wastes, the AEC announced that it would convert an old salt mine near Lyons, Kansas, into a high-level waste repository. But the commission had leaped before looking. When geologists examined the mine more closely, they found a number of things, such as nearby gas and oil boreholes, that offered groundwater easy access into the mine. Thoroughly embarrassed by the revelations, the AEC backed off from their proposal. Ever since, this episode has been pointed to as evidence that the federal government cannot be trusted to dispose of nuclear wastes. There have been other embarrassments since then, such as a string of problems and near-disasters involving nuclear waste at the Hanford site, but it was the singling out of Yucca Mountain as the nation's long-term nuclear-waste repository that killed any remaining trust and confidence in the system.

The 1982 Nuclear Waste Policy Act attempted to set out a fair and equitable process to deal with reactor wastes. The Department of Energy planned to build two repositories, one in the western United States and the other in the East. The site-selection process would be open and would be determined by technical criteria. Even if all had gone as planned, the selection of the sites would certainly not have been smooth, but, as it turned out, the energy department and Congress would sabotage the process. In 1986, after selecting three finalists for the western site, the department announced that it was suspending the search for an eastern site. The official explanation was that only one long-term site would be needed, but most assumed that it was a purely political move aimed at helping the reelection efforts of several senators whose states were on the list of eastern sites under consideration. It was particularly galling to those in the West, since most of the country's nuclear plants are in the East. Then, in late 1987, Congress amended the 1982 act and changed the rules. It ratified the energy department's earlier action by officially calling off the site search in the East. And it narrowed the search in the West to one site: Yucca Mountain in Nevada, near the Nevada Test Site for nuclear weapons. Again, politics seemed to be the motivating factor. Nevada's congressional delegation was small and relatively uninfluential, while the other two western sites that had been under consideration were in Washington, where House Majority Leader Tom Foley lived, and Texas, home to a number of powerful politicians, including Speaker of the House Jim Wright and Vice President George Bush. In Nevada, the act became known as the "Screw Nevada Bill."

Not surprisingly, the people of Nevada have fought the selection of Yucca Mountain with every weapon they could find. Many of them have been, on the surface, technical arguments. Opponents have offered a variety of reasons why Yucca Mountain is not a suitable place to store nuclear wastes for thousands of years, most of them involving ways that groundwater could be expected to find its way into the proposed location, which is at this time extremely dry. The arguments get quite complicated, and the amount of time the waste must be stored creates some unavoidable uncertainty in predicting what might happen, but most scientists find Yucca Mountain a good site for keeping nuclear waste safe for tens of thousands of years. On the political front, in 1989 the Nevada legislature passed a bill forbidding anyone—private individual or government agency—to store high-level nuclear waste inside the state. The governor told state agencies not to approve any of the environmental permits that the Department of Energy would need to conduct tests at the site in order to determine its suitability. And the state and the energy department have filed and counterfiled various lawsuits. As for the citizens of the state, four out of five believe the state government "should do all it can to stop the repository." The result of this unprecedented loss of political legitimacy is that the officially projected opening of Yucca Mountain has been pushed back by more than a decade—to 2010—and most people expect it to be much later than that, if ever.

Ironically, the "Screw Nevada Bill" also created the office of the Nuclear Waste Negotiator. In case Yucca Mountain proved unworkable, the Nuclear Waste Negotiator was to seek out other potential sites for a permanent nuclear-waste repository. He was also to look for places to put a monitored retrievable storage (MRS) facility—a storage site for high-level nuclear waste where workers would watch the waste closely and remove it if necessary. The MRS facility was intended to be temporary. Once a permanent repository was up and running, the waste would be sent there.

It took a couple of years to find someone who would take the job of Nuclear Waste Negotiator, but by early 1991 David Leroy was trying to schedule talks with the governors of the fifty states. And having little luck. The negotiation process may yet succeed, however, because Congress did deal Leroy one high card. In addition to the states, the Nuclear Waste Negotiator can talk with the nearly 300 Indian tribes with reservations in the United States. Indian reservations are separate entities, independent of the states in which they reside, and the tribes that run them have proved much more willing to talk with Leroy than the officials of the various state governments.

Although the office of the Nuclear Waste Negotiator expired after 1994, Leroy and his successor, Richard Stallings, managed to interest at least two Indian tribes in a managed retrievable storage facility: the Goshute tribe in Utah and the Mescalero Apaches of New Mexico. As this is written, the Apaches, descendants of the legendary chief Geronimo, have agreed to build a temporary storage facility to hold fuel rods from a consortium of utilities. The plan will likely face a variety of challenges, including an effort by the state of New Mexico to keep the nuclear waste from being transported through it, but if it works out, the site could hold at least 20,000 tons of spent fuel for a planned forty years.

Some have slammed the agreement, calling the targeting of Indian tribes for potential sites as "environmental racism." Because the Indians are poor, the argument goes, they are being bribed into accepting undesirable facilities at which their richer, non-Indian fellow citizens turn up their noses. Indeed, many of the 3,000 tribe members are poor, and indeed, they will be paid relatively well—an estimated $25 million a year, by some accounts—but the charge of environmental racism is hard to make stick. For one thing, not all well-to-do non-Indians turn up their noses at the repository. The town of Oak Ridge, Tennessee—which has one of the most highly educated populations in the country—had volunteered in the mid-1980s to host a monitored retrievable storage facility. Home to Oak Ridge National Laboratory, its people had few of the visceral fears of nuclear waste that marked the public in other areas. But the state of Tennessee was much less happy with the idea, and it refused to allow it.

The real reasons that the MRS facility will be on an Indian reservation, assuming the agreement is consummated, are political and institutional. There are probably a number of communities around the country besides Oak Ridge that would be willing to have a temporary nuclear waste site built nearby. The obvious place to start looking would be in places that, like Oak Ridge, have a history of dealing with nuclear power or nuclear wastes. But any such plans would meet the same fate as Oak Ridge's. The rest of the state would not want a repository inside its borders. People would worry about nuclear waste being shipped through their towns on road or rail. They would fret about the effects of the repository on tourism, housing prices, and their ability to attract new business to the state. Some would fixate on the threat to their water supply if the nuclear waste escaped from its repository. In the end, the state would not stand for it. But states have little control over what goes on inside Indian reservations, and, just as important, people tend to accept the Indians' right to make their own rules. Thus a nuclear-waste site on an Indian reservation is much harder for a state to derail and is much more likely to be accepted by the people of the state as legitimate.

Even if the Mescalero Apaches or some other tribes agree to watch over the country's nuclear wastes for a few decades, the tricky problem of locating a permanent waste-storage repository will remain. Many believe that Yucca Mountain, no matter its technical suitability, will not be the site for that repository, since there is just too much resistance in the state. One group of experts spent several years studying Yucca Mountain and the nuclear-waste issue and offered a series of recommendations to get the permanent repository back on track. These recommendations reveal a great deal about the factors making up this intractable technological problem.

Only one recommendation was technical: re-evaluate the commitment to underground geologic disposal. Putting the waste in the seabed might be more practical, they thought, and there should be more emphasis on engineered safeguards instead of depending on an unchanging geology. One recommendation was a practical consideration: use interim storage facilities. This would buy time to solve the more difficult problem. Two were aimed at restoring confidence and trust in the agencies running the show: guarantee stringent safety standards, and restore credibility to the waste-disposal program. The other four—half of the recommendations—were focused on establishing a workable negotiation scheme: evaluate more than one site, employ a voluntary site-selection process, negotiate agreements and compensation packages, and acknowledge and accept the legitimacy of public concerns. Although they didn't put it this way, their recipe might be summed up as: buy some time, establish confidence in the system, and then learn to negotiate with the citizens who are going to have to put up with the nuclear wastes for the rest of their lives.

For now, it's mostly just theory. Nobody knows if negotiation will help make possible some of the truly difficult technological decisions, such as siting. But it seems clear that the other options don't work, at least not for something as gut-wrenching as the permanent storage of nuclear wastes. Negotiation may well be the only choice.

eight

MANAGING THE FAUSTIAN BARGAIN

A quarter of a century ago, Alvin Weinberg offered one of the most insightful—and unsettling—observations anyone has made about modern technology. Speaking of the decision to use nuclear power, the long-time director of Oak Ridge National Laboratory warned that society had made a "Faustian bargain." On the one hand, he said, the atom offers us a nearly limitless supply of energy which is cheaper than that from oil or coal and which is nearly nonpolluting. But on the other hand, the risk from nuclear power plants and nuclear-waste disposal sites demands "both a vigilance and a longevity of our social institutions that we are quite unaccustomed to." We cannot afford, he said, to treat nuclear power as casually as we do some of our other technological servants—coal-fired power plants, for instance—but must instead commit ourselves to maintaining a close and steady control over it.

Although Weinberg's predictions about the cost of nuclear power may now seem naive, the larger issue he raised is even more relevant today than twenty-five years ago: Where should society draw the line in making these Faustian technological bargains? With each decade, technology becomes more powerful and more unforgiving of mistakes. Since Weinberg's speech, we have witnessed major accidents at Three Mile Island, Chernobyl, and Bhopal, as well as the explosion of the *Challenger* and the wreck of the *Exxon Valdez*. And looking into the future, it's easy to see new technological capabilities coming along that hold the potential for far greater disasters. In ten or twenty years, many of our computers and computer-controlled devices may be linked through a widespread network that dwarfs the current telecommunications system. A major breakdown like those that occasionally hit long-distance telephone systems could cost billions of dollars and perhaps kill some people, depending on what types of devices use the network. And if genetic engineering becomes a reality on a large scale, a mistake there could make the thalidomide debacle of the late 1950s and early 1960s look tame. Do some of these technologies demand too much in return for their favors?

The response depends, of course, on a number of things, including how much one values the contributions of nuclear power or genetic engineering or whatever technology is under discussion, but the main factor is our ability to handle risk. If we are sure that we can keep our technological genies under control, then we may decide to keep rubbing more and bigger lamps. But if not, then at some point we must resist the temptation to call out yet another jinn. We must say no to the Faustian bargain.

How, then, can we make today's complex technologies safe? And how safe can we make them? Traditionally, engineers have thought of reliability and safety mainly in technical terms: If a machine was designed well, built well, and maintained well, then as long as its operators followed correct procedures, the machine should perform well. And if there were safety concerns—the possibility of a ruptured coolant pipe in a nuclear reactor, say—then the best and most certain way to deal with them was through engineering: tacking on extra safety systems, perhaps, or redesigning the system so that it would not be so vulnerable to a given threat. As for the humans running the equipment, they were unfortunately unavoidable, and the best way to deal with them was to write out detailed instructions and manuals that explained exactly what to do in every situation.

We now know that this is a limited and naive view. The safety of complex systems depends not just on their physical characteristics but also quite intimately on the people and organizations operating them. Complexity creates uncertainty, and uncertainty demands human judgment. Furthermore, complex systems can be quite sensitive to even very small changes, so that a minor mistake or malfunction can snowball into a major accident. Such uncertainty and sensitivity make it impossible to write out procedures for every possible situation—and attempts to do so can backfire. Operators who are trained to always go by the book may freeze up when an unexpected situation arises. Or worse, they may misinterpret the situation as something they're familiar with and take exactly the wrong action. Just this sort of foul-up happened at Three Mile Island, with the operators shutting off most of the emergency cooling water going into the core for what seemed like good reasons at the time.

In short, organizational reliability is just as crucial to the safety of a technology as is the reliability of the equipment. If we are to keep our Faustian bargains in check, we must be as clever with our organizations as we are with our machines.

In the past couple of decades, a great deal of research has gone into discovering what makes some organizations work better than others, and social scientists believe they have some answers. But the groups that operate such complex, hazardous technologies as nuclear power, petrochemical plants, space flight, commercial aviation, and marine transportation pose a special challenge. They must do their jobs without mistakes—or at least without the sorts of mistakes that have catastrophic consequences. At the same time, because of the complexity of the technologies, the organizations require a great deal of learning-by-doing before they understand the systems they oversee. In most industries, such learning would be done through trial and error, but that's not an option with such hazardous industries as nuclear power. Instead they must try a high-stakes strategy of "trial without error."

Can any institution be expected to pull this off? Certainly quite a few have failed. Many if not most of the great technological disasters of the past two decades, from Three Mile Island to the *Challenger* explosion, can be traced, ultimately, to institutional breakdowns. Yes, the proximate causes may have been equipment failure or operator error, but the conditions that allowed the accidents to happen were created by the organizations in charge.

As we saw in chapter 6, Yale sociologist Charles Perrow argues that accidents are unavoidable in certain hazardous technologies, such as nuclear power, nuclear weapons, chemical plants, genetic engineering, and space flight. These technologies, he says, place mutually contradictory demands upon the organizations running them: their complexity requires that operations be decentralized to ensure a flexible response to a crisis, but the ability of one part of the system to influence another requires centralized control and direction in order to keep failures from spreading.

Perrow's arguments have been quite influential, but they are not the last word. In particular, one group of researchers claims that some organizations perform much better than Perrow—and much of standard organizational theory—believes is possible: they operate hazardous and highly complex technologies with an almost complete avoidance of major accidents. In the words of Berkeley political scientist Todd La Porte and colleague Paula Consolini, these organizations "work in practice but not in theory." La Porte and several colleagues have spent a number of years studying these so-called high-reliability organizations, or HROs, and believe they now understand some of the reasons that they work so well.

The debate between the two camps is far from over, and its resolution may hold the key to what sorts of Faustian bargains we should be willing to accept in the future. If Perrow is correct, then some technologies are best left unused. If La Porte is, then we may be able to keep up our end of the bargain—perhaps not avoid accidents altogether, but learn to make them as unlikely as possible. But it won't be easy. Managing complex, hazardous technologies demands a sustained, conscious effort far greater than most organizations are accustomed to giving.

LEARNING FROM MISTAKES

The deadliest industrial accident in history struck in the early morning hours of December 3, 1984, in Bhopal, India. A lethal cloud of methyl isocyanate, or MIC, leaked from a pesticide-manufacturing plant and out into the surrounding slums and shantytowns. Although the exact numbers are not known, probably some 200,000 people were exposed to the chemical, and of those, about 4,000 were killed and another 30,000 seriously injured.

There was no single, simple reason why those 4,000 people died. The catastrophe was the result of a concatenation of poor decisions, minor mistakes, and incredibly sloppy operating procedures that, in hindsight, seem destined to lead to some sort of accident. To understand how and why it happened, begin with some basic facts: MIC is a poisonous liquid with a low boiling point (just over 100 degrees Fahrenheit) that reacts strongly with water. It is used in making carbaryl, the active substance in a particular pesticide called Sevin made by Union Carbide. The plant at Bhopal manufactured Sevin and other pesticides. It was run by Union Carbide of India, Limited (UCIL), a partially owned subsidiary of Union Carbide.

Given MIC's toxicity and volatility, the safest thing to do is never to have too much of it on hand at any one time. Then if an accident occurs, the damage will be limited. And according to an affidavit filed after the accident by Edward Munoz, a former managing director of UCIL, the Indian subsidiary had wished to do just that. It requested that only small amounts of MIC be stored at the plant. But Union Carbide, sensitive to how much money could be saved by manufacturing MIC in bulk, insisted on producing large volumes there. At the time of the accident, the storage tank that released the MIC contained about forty-two tons of the chemical. A second tank, not involved in the accident, contained another twenty tons.

When it first opened in 1969, the Bhopal plant did not manufacture chemicals itself. Instead it produced various pesticides by combining ingredients that had been manufactured elsewhere. Because this was not a particularly hazardous activity, the facility was located only about two miles from the city's commercial and transportation hub, on the north side of town. A few years later, however, sharp competition drove UCIL to begin manufacturing the pesticide ingredients at the plant in order to save on transportation costs. No one complained at the time because those chemicals were not particularly hazardous, but in 1979, when UCIL decided to add MIC to the Bhopal plant's menu, the city balked. It did no good. By using its influence with national and state authorities, the company was able to get around the city's objections.

In the early 1980s, the pesticide market in India bottomed out. The Bhopal plant had never been particularly profitable, and after 1980 it lost money yearly. Union Carbine put up the entire plant for sale in July 1984. The financial squeeze had two effects on how the plant was run. First, because the plant was losing money and was on the block, neither Union Carbide nor UCIL paid much attention to it. The corporation did not send its better managers to work at the plant, nor did it offer much management help or advice. Second, the company pushed to cut costs at the plant every way it could.

The cost-cutting destroyed whatever margin of safety the plant had once had. The number of workers operating the MIC unit was halved between 1980 and 1984. Training programs were cut to the point that many of the operators did not understand safe versus unsafe practices for running the plant. Equipment was not repaired when it broke down, and some safety equipment was turned off. In retrospect, the only surprise is that a major accident didn't happen sooner. Indeed, one Indian journalist living in Bhopal, Raj Kumar Keswani, had written articles warning of this possibility beginning in 1982. In one of them, published in June 1984, he specifically predicted that a gas leak would lead to a catastrophe.

The countdown to that catastrophe began at half past nine on the evening of December 2. Workers were flushing out some pipes with pressurized water. These pipes were connected to other pipes leading to the MIC storage tanks, with only a single valve separating them. As a backup against the valve leaking, the normal operating procedure when flushing the pipes was to insert a metal disk called a slip bind into the valve. But although the flushing of the pipes was done by the operators of the MIC unit, the insertion of the slip disk was the responsibility of the maintenance department, and the second-shift maintenance supervisor had been let go several days earlier to save money. The job of inserting the slip disk had not been assigned to any of the remaining maintenance workers, and so the slip disk was never inserted. Without it, the valve failed when water began backing up in the clogged pipes. Water passed through the faulty valve and began to flow toward the MIC storage tanks.

There was one more line of defense that should have stopped the water, a second valve that was normally kept closed. No one knows why, but it too failed. Perhaps it had been inadvertently left open after some earlier work on the tank, or perhaps it had not been sealed correctly. Whatever the reasons, the results were disastrous. More than half a ton of water splashed into the MIC. The resulting reaction between the MIC and the water sent the temperature and pressure in the tank soaring and created a roiling mixture of gas, liquid, and foam.

The third shift took over at 10:45 P.M., and at 11 the operators noticed that the pressure in one MIC tank was high, although still within normal operating range. Hints that something was wrong began to trickle in. Workers reported eye irritation, and then someone found water and MIC coming out of a pipe attached to a relief valve. The control-room operator watched as his gauges reported a steadily rising pressure in the tank, but apparently did nothing until about 12:30, at which point the pressure had gone off the scale. He then rushed to check out the situation firsthand. When he got to the MIC storage tank, he heard it rumbling and felt it radiating heat. The gauges on the tank verified what the one in the control room had said: the temperature and pressure readings had gone through the roof.

Perhaps if the plant's safety systems had been operational, some of the catastrophe could still have been averted at this point. But they weren't. The operator ran back to the control room and tried to turn on a scrubber, which is designed to neutralize poisonous gases coming out of the plant's vent pipes. For some reason, perhaps because the scrubber had been on standby for several weeks, the caustic soda that reacted with the gases was not circulating through the scrubber, so the system was useless. Later, plant supervisors tried to use a refrigeration system to cool the MIC and slow down the reaction, but the system had been shut down six months earlier and its coolant drained for use elsewhere in the plant. The operators also turned on a collection of water sprayers, but the water did not go high enough to affect the MIC billowing out of a pipe 100 feet above the ground.

The supervisors sounded an alarm to warn the surrounding community that gas was escaping from the plant. It probably would have done little good anyway, since people outside the plant had never been taught emergency responses, but for some reason the alarm was shut off after only a few minutes. An internal plant siren did sound, warning plant employees to flee upwind to safety. The people living around the plant were not so lucky. It was 2:15 A.M. before the general alarm was turned back on and 2:30 before an evacuation was ordered. By then a toxic cloud was rolling over the city.

The thousands of people who died and tens of thousands who were injured were the victims not of any one mistake or malfunction but of a complete institutional breakdown. It is impossible to point a finger at one or two people. It was the system itself that failed. This sort of systemic breakdown is most common in developing countries that have attempted to industrialize quickly and have built large plants or other facilities without creating a complete infrastructure to support them. Expertise is in short supply, training is often insufficient, and there are few places to go for suggestions or help. Indeed, in the same year as the Bhopal disaster, two other developing countries were hit by major industrial accidents—in Brazil, a gasoline pipeline ruptured and exploded, killing 500 people, and in

Mexico, an explosion in a natural gas storage plant killed 450. But neither are industrialized nations immune to such institutional failures. They too can fail quite spectacularly in managing risky technologies.

On January 28, 1986, tens of millions of people watched—over and over again, in horrified fascination—as network television broadcast the most dramatic technological failure of the past several decades. Seventy-three seconds after liftoff, the space shuttle *Challenger* exploded and fell into the Atlantic Ocean, killing all seven crew members. Perhaps the best known of the victims was the high school teacher Christa McAuliffe. She had won her seat on the shuttle in a national competition designed to get students and teachers interested in space science, and NASA had made her famous as part of its efforts to strengthen popular support for the shuttle program. But, as the accident and the subsequent investigations of it revealed, the space agency should have spent more time worrying about the performance of its machines and the safety of its personnel and less time on such grandstanding as the teacher-in-space contest.

From a purely technical point of view, the cause of the *Challenger* explosion was a malfunction of the now-famous O-rings. Because the solid-fuel rocket boosters used to assist the shuttle liftoff were too long to be shipped in one piece, they were manufactured in four parts and assembled on site. To prevent hot gases from leaking out of the gaps between these sections, each circular joint was sealed with zinc chromate putty and two O-rings. The O-rings resembled large, thin rubber bands, 37.5 feet around and about a quarter of an inch thick. To provide a tight seal between the sections of the booster, the O-rings had to be flexible, but on that January morning the temperature at the usually mild Florida site was 28 degrees, and the synthetic rubber of the O-rings hardened up. The result was fatal. Shortly after ignition, a stream of burning gas escaped past the O-rings and hit an external fuel tank like a blowtorch, triggering the explosion that sent the shuttle back toward Earth.

Although the faulty O-rings were the proximate cause of the accident, the more serious flaws lay with the people—and, particularly, the institutions—in charge of the shuttle program. The hearings that followed the accident revealed that the danger of such an O-ring malfunction had been well known to the engineers who worked on the shuttle. Furthermore, on the night before the launch, the Morton Thiokol engineers who had designed the solid-fuel rocket boosters recommended that the flight be postponed to a warmer date because of concerns over the O-rings' performance. The shuttle was launched anyway, and that decision revealed an organization that was not up to the task of dealing with a risky technology.

In the years leading up to the *Challenger* launch, NASA had been under great pressure to establish the space shuttle as a reliable and inexpensive way of getting into space. It had originally sold Congress on the shuttle by representing it as a versatile workhorse that could put satellites in orbit, allow scientists to perform research in space, and ferry astronauts to and from a space station. But by trying to do so many things, it would inevitably do none of them well. In launching satellites, for instance, the shuttle was at a disadvantage to unmanned rockets, and the only way it could hope to compete economically was to bring the cost per launch down by having many flights per year. Thus, the four-shuttle fleet had been scheduled to take fifteen flights in 1986 and nineteen in 1987.

This pressure, combined with a complacency created by the successful Apollo program, altered the way NASA approached the safety of its space shots. In the past it had demanded that contractors who supplied components prove that they met technical specifications and were safe. But things had changed by the time of the *Challenger* launch. Now it seemed that NASA would go forward unless the contractors could prove that the shuttle wasn't safe.

The night before the launch, Allan McDonald, the Thiokol engineer who managed the solid-fuel rocket booster program, set up a teleconference among Thiokol engineers involved in the design, several Thiokol managers, and managers and engineers from NASA. The purpose: to offer a warning about the O-rings. Until then, the coldest temperature at which the shuttle had been launched had been 53 degrees. Even at the relatively warm temperatures of previous shuttle flights, the O-rings had let some hot gases escape out the sides of the boosters. Now, with the temperature expected to be below freezing at launch time, the Thiokol engineers warned that the O-rings could fail with disastrous consequences. They recommended that the launch be put off until things warmed up.

The NASA managers were not pleased. If they followed the Thiokol engineers' reasoning, it would mean never launching the shuttle if the temperature dipped below 53 degrees, the coldest temperature at which it had been launched before. "My God, Thiokol, when do you want me to launch—next April?" one NASA manager, Lawrence Mulloy, has been quoted as saying.

Seeing NASA's unhappiness with the recommendation, the Thiokol managers broke from the teleconference to hold a separate meeting with their engineers. The engineers continued to argue, quite vigorously, that the launch be postponed. The managers, however, felt they had to balance these purely engineering considerations with business goals. Launching as often as possible was not only important to NASA, it was important to Morton Thiokol. The more often the shuttle was launched, the more the company earned. At one point in the discussion, one of the Thiokol vice presidents urged the company's engineering vice president to "take off your engineering hat and put on your management hat." Eventually, the four vice presidents decided to withdraw the company's objections to the launch. When they returned to the teleconference, they told the NASA managers it was okay to proceed with the launch.

How could it happen? How could Thiokol managers ignore the concerns of their own engineers? How could NASA managers be so cavalier about the safety of their astronauts? The culprit seems to be a culture of complacency that had arisen in the space program. After two dozen successful flights, the managers assumed that shuttle was safe. Even evidence of partial failure in the O-rings on previous flights was interpreted as an indication that the shuttle would be okay even if the O-rings didn't work perfectly. In this atmosphere, the NASA managers' first reaction to the warning of the Thiokol engineers was to interpret it as overly cautious and to be annoyed that their launch might be postponed by such groundless concerns. The Thiokol managers, too, found it difficult to believe that a serious threat to the shuttle existed. Only the engineers believed that the worst might happen.

Compounding the complacency problem was the rigid hierarchy that had arisen at NASA, which made communication between departments formal and not particularly effective. At another agency, or even at NASA in an earlier era, the engineers might have gone over the managers' heads with their concerns. Old-timers at the agency, interviewed after the accident, spoke nostalgically of the days when James Abrahamson, a lieutenant general in the Air Force, had run the shuttle program. "When he sensed a problem," one such old-timer said, "he awaited no 'criticality' rating; he barged into the office of even the lowliest technician to ask how to fix it." But in the huge, bureaucratic, space-shuttle-era NASA, most communication was done through memos and reports. Everything was meticulously documented, but critical details tended to get lost in the paperwork blizzard. The result was that the upper-level managers were kept informed about possible problems with the O-rings—along with dozens of other problems given an equally high criticality rating—but they never truly understood the seriousness of the issue. The way the system was set up, they could assume that the O-rings had been taken care of. No one from the top barged into someone's office to see what had been done.

NASA's problems with the space shuttle program are far removed from Union Carbide's troubles at the Bhopal plant, but the stories do have several features in common. Both institutions were much less concerned about safety than they should have been, thanks to complacency or a failure to appreciate the risks. Both dealt with complex technologies whose behavior was impossible to predict exactly, although in each case the accident could have been avoided with a little extra caution. And both organizations were facing financial pressures that pushed them to attempt more with less. For risky technologies, this is a recipe for disaster.

HIGH-RELIABILITY ORGANIZATIONS

Success is much harder to analyze than failure. When things go wrong in a chemical plant or space program, it's usually possible to figure out the causes and resolve to avoid those things in the future. But when things go right, it's difficult to know why. Which factors were important to the success, and which weren't? Was the success due to skill, or just luck? It's likely, for instance, that other chemical plants were as poorly run and as dangerous as Bhopal, but they were never struck by just the wrong string of events to create a major disaster. Nonetheless, if we are to learn to deal with hazardous technologies, our best bet is to look for organizations that manage risk successfully and see how they do it.

This is the goal of the high-reliability organization project at the University of California, Berkeley. For more than a decade, Todd La Porte, Karlene Roberts, and Gene Rochlin have been studying groups that seem to do the impossible: operate highly complex and hazardous technological systems essentially without mistakes. The U.S. air traffic control system, for instance, handles tens of thousands of flights a day around the country. Air traffic controllers are not only responsible for choreographing the takeoffs and landings of dozens or hundreds of flights per hour at airports but also for directing the flight paths of the planes so that each keeps a safe distance from the others. The success is unequivocal: for more than a decade none of the aircraft monitored on the controllers' radar screens has collided with another. Yet the intricate dance of planes approaching and leaving airports, crisscrossing one another's paths at several hundred miles an hour, creates plenty of opportunity for error.

This record of safety is not due to extremely good luck, the three Berkeley researchers conclude, but to an institution that has learned how to deal effectively with a complex, hazardous technology. The researchers call such institutions high-reliability organizations, or HROs.

Perhaps the most impressive and best studied of HROs to date are the nuclear aircraft carriers of the U.S. Navy. While it's impossible for anyone who hasn't worked on such a ship to truly understand the complexity, stress, and hazards of its operations, this description by a carrier officer offers a taste:

> So you want to understand an aircraft carrier? Well, just imagine that it's a busy day, and you shrink San Francisco Airport to only one short runway and one ramp and gate. Make planes take off and land at the same time, at half the present time interval, rock the runway from side to side, and require that everyone who leaves in the morning returns that same day. Then turn off the radar to avoid detection, impose strict

> controls on radios, fuel the aircraft in place with their engines running, put an enemy in the air, and scatter live bombs and rockets around. Now wet the whole thing down with salt water and oil, and man it with 20-year-olds, half of whom have never seen an airplane close up. Oh, and by the way, try not to kill anyone.

A *Nimitz*-class carrier flies ninety aircraft of seven different types. They have only several hundred feet in which to take off and land instead of the mile or more available at commercial airports, and so they need help. At takeoff, the planes are catapulted by steam-powered slingshots that accelerate them from standstill to 140 knots (160 miles per hour) in just over two seconds. This launching is an intricate operation. As each plane is moved into place on the steam catapult, crewmen check it one last time to make sure that the control surfaces are functioning and that no fuel leaks or other problems are visible. The catapult officer sets the steam pressure for each launch depending on the weight of the plane and wind conditions. The spacing of the launches—about one every fifty seconds—leaves no time for errors.

But it is the recovery of the planes that is truly impressive. To land, the planes approach the flight deck at 120 to 130 knots with a tail hook hanging down to catch one of four arresting wires stretched across the deck. Just as the plane touches down, the pilot gives it full throttle so that if the hook does not catch, the plane will be going fast enough to take off and come around again. If the hook does catch a wire, it slams the plane to a halt within about two seconds and 300 feet. This operation demands exquisite teamwork. As a plane approaches, the pilot radios his fuel level. With this information, the men in charge of the arresting gear calculate the weight of the plane and figure the proper setting for the braking machines hooked up to the arresting wires. If the pressure is set too low, the plane may not stop soon enough and so topple off the end of the deck into the sea. If the wire is too taut, it could pull the tail hook off or else snap and lash out across the deck, injuring or killing anyone in its path. The pressure for each of the four wires is set individually by a single seaman. Meanwhile, landing signal officers are watching the approach of the plane, advising the pilot and then—if everything appears right—okaying the landing. As soon as the plane is down and stopped, "yellow shirts" rush to it to check the hook and to get the plane out of the way of the next one. As the arresting wires are pulled back, other crewmen check them for frays. Then it all begins again. This cycle has lasted about sixty seconds.

The launching and recovery are only part of a much larger process comprising maintenance, fueling and arming, and maneuvering and parking the planes on a crowded deck. As La Porte and his student Paula Consolini write, "A smooth twenty-plane cycle takes hours to set up and involves an intricate technological ballet of men and machines carried out by a closely integrated team monitoring incoming aircraft and handing each off to the next 'station.'"

What makes this performance truly astonishing is that it is all done not with people who have been working together for years but with a crew that turns over regularly. As one observer noted, "[T]he captain will be aboard for only three years, his 20 senior officers for about two and a half; most of the more than 5,000 enlisted men will leave the Navy or be transferred after their three-year carrier stints. Furthermore, the enlisted men are predominantly teenagers, so that the average age aboard a carrier comes to a callow 20."

What sort of organization can operate so reliably under such handicaps? La Porte, Roberts, and Rochlin spent a great deal of time on several carriers both in port and at sea, during training and on active duty, and they believe they understand at least part of the answer.

On the surface, an aircraft carrier appears to be organized along traditional hierarchical lines, with authority running from the captain down through the ranks in a clearly defined pattern. And indeed, much of the day-to-day operation of the ship does proceed this way, with discipline rather strictly enforced. Furthermore, there are thick manuals of standard operating procedures, and much of the Navy training is devoted to making them second nature. These procedures codify lessons learned from years of experience into formal procedures. But, as the Berkeley researchers discovered, the carrier's inner life is much more complicated.

When things heat up, as during the launching and recovery of planes, the organizational structure shifts into another gear. Now the crew members interact much more as colleagues and less as superiors and subordinates, and negotiations and accommodations take the place of giving and following orders. The people who have the most knowledge and experience in handling a particular chore take the initiative, with others following their lead. In the words of La Porte and Consolini, the senior noncommissioned officers, who generally know the most about the operations of the carrier, "advise commanders, gently direct lieutenants, and cow ensigns."

At the same time, cooperation and communication among various units become much more important than orders passed down the chain of command and information passed back up. With a plane taking off or landing once a minute, events can happen too quickly to wait for instructions or authorizations from above. Instead, the crew members act as a team, each doing his job and watching what others are doing, and all of them communicating constantly through telephones, radios, hand signals, and written details. This constant flow of communication helps catch mistakes before they've caused any damage. Seasoned personnel continuously monitor the action, listening for anything that doesn't jibe with what they know should be happening and standing ready to correct a mistake before it causes trouble.

A third level of organizational structure is reserved for emergencies, such as a fire on the flight deck. In such cases, the ship's crew has carefully thought out and rehearsed procedures to follow, and each member has a preassigned role. The crew practices its responses over and over again so that if an emergency occurs, it can react immediately and effectively without direction.

This multilayered organizational structure asks much more from the crew members than a traditional hierarchy, but those extra demands seem to hold the key to a carrier's effectiveness. In contrast to the stereotypical bureaucracy, where following orders is the safest path and underlings are not encouraged to think for themselves, the welfare of the ship and crew is everyone's responsibility. As the Berkeley researchers note, "Even the lowest rating on the deck has not only the authority, but the obligation to suspend flight operations immediately, under the proper circumstances and without first clearing it with superiors. Although his judgment may later be reviewed or even criticized, he will not be penalized for being wrong and will often be publicly congratulated if he is right." The result is an organization in which everyone has a stake and which everyone feels a part of.

This involvement of everyone, combined with the steady turnover among the officers and crew, also helps the Navy avoid the problem of things becoming routine and boring. Because of the regular coming and going of personnel, people on the ship are constantly learning new skills and teaching what they've learned to others. And although some of the learning is simply rote memorization of standard operating procedures, it goes much further, the Berkeley researchers found. There is a constant search for better ways of doing things. Young officers come on board with new ideas they're eager to try, and find themselves debating with the senior noncommissioned officers who have been with the ship for years and know what works. The collision of fresh, sometimes naive approaches with a conservative institutional memory produces a creative tension that keeps safety and reliability from degenerating into a mechanical following of the rules.

In some fashion—La Porte, Roberts, and Rochlin aren't quite sure how—the Navy has managed to balance the lessons of the past with an openness to change, creating an organization that has the stability and predictability of a tightly run hierarchy but that can be flexible when it needs to be. The result is an ability to operate near the edge, pushing both man and machine to their limits but remaining remarkably safe.

Of course, the aircraft carrier is a unique situation, and there is no reason to think that what works there would be effective in a commercial setting with civilian employees. So it is all the more remarkable that when the Berkeley project examined a completely different sort of high-reliability organization, the researchers tracked its success to a very similar set of principles.

The Diablo Canyon nuclear power plant, operated by Pacific Gas & Electric, lies just west of San Luis Obispo, California, on the Pacific Ocean coast. Although its construction was dogged by controversy, took seventeen years to complete, and cost $5.8 billion, the plant has by all accounts been one of the country's best run and safest since its opening in 1985. This brought it to the attention of the Berkeley group and in particular to Paul Schulman, a political scientist at Mills College in Oakland and one of several researchers who have collaborated with La Porte, Roberts, and Rochlin on the HRO project.

Like the aircraft carriers, Diablo Canyon appears at first to be a rigidly run hierarchy. It has a formal chain of command leading to a plant manager who is also a vice president of Pacific Gas & Electric. And it has a thick stack—a tower, really—of regulations telling employees how to do their jobs. As Schulman describes it:

> As of May 1990, there were 4,303 separate written procedures at Diablo Canyon covering administration, operations, maintenance, radiation protection, and chemical and radiological analysis as well as the surveillance and other testing activities conducted by [the On-Site Safety Review Group, Quality Control, Quality Assurance, and the On-Site Planning and Engineering Group]. Each procedure in turn has a multiplicity of specified steps. The average procedure has undergone well over three revisions (one has 27). There are formal procedures for the drafting of procedures as well as separate procedures for altering procedures.

This is how the regulators want it. Since Three Mile Island, the Nuclear Regulatory Commission has tried to assure safety by insisting that nuclear plants follow an increasingly detailed set of rules. Plants are rated according to how many times they violate the regulations, and a pattern of violations will lead to closer supervision by the NRC and fines that, in serious cases, can run into hundreds of thousands of dollars.

But, Schulman found, Diablo Canyon has another side—a more active, probing, learning side. Despite the hierarchy and the regulations, the organization is constantly changing, constantly questioning accepted practice and looking for ways to do things better. It is not the same sort of change found on aircraft carriers, where the steady turnover of personnel creates an endless cycle of learning the same things over and over again, plus a gradual improvement of technique. Diablo Canyon maintains a relatively stable group of employees who know their jobs well. Nonetheless, in its own way, the nuclear plant is as dynamic as the carrier.

The reason, Schulman says, is that the plant has cultivated an institutional culture rooted in the conviction that nuclear plants will always surprise you. There are actually two sets of decision-making procedures at the plant. The first, and more visible, procedure is what he calls "automatic decision making." This consists of well-established rules for what to do in a particular situation. Some are carried out by computer, others by human workers, and the operators are trained to identify problems and then to figure out the proper procedures for dealing with them. In general, Schulman says, this set of rules is designed to guard against errors of omission—people not doing something that they should.

But the Diablo Canyon workers also work very hard to avoid errors of commission—actions that have unexpected consequences—and this demands more from employees than following standard procedures. In a system as complex as a nuclear plant it is impossible to specify everything in advance, so employees must do more than follow procedures blindly. They must constantly think about what they're doing in order to avoid causing the system to do something unexpected and possibly dangerous.

As an example of how this second decision-making procedure works, Schulman tells of a time when one of Diablo Canyon's two reactors was shut down for routine maintenance. One planned test demanded turning off an air-pressure system that powered some of the unit's instruments. Shutting off the air pressure seemed harmless since the reactor was shut down anyway, but some of the operators were uneasy. Perhaps, they suggested, the air-pressure system was connected in some way with the air-pressure system of the other, still-operating unit, and turning off the first one would threaten control of the other unit. The operators and engineers sat down and discussed what might possibly go wrong. The engineering drawings of the units showed that the two air-pressure systems were independent, so it seemed that the shutdown was safe, but not everyone was convinced. Finally, the supervising engineer decided he would personally inspect the entire air-pressure system to make sure there were no interconnections. Only after he finished was the maintenance department given the go-ahead to turn off the air pressure on the off-line unit.

This is not an isolated example, Schulman says, but a typical one. As one manager told him, "These systems can always surprise you, and we want people to know that." The result is an organization that recognizes the complexity it is dealing with and understands its unpredictability. Although the plant is constantly adding to its standard procedures as people learn more about the right approaches and spot new ways that things might go wrong, no one believes the organization will ever be able to write everything down in a book. Instead, the Diablo Canyon workers refer to their procedures as a "living document," one that every employee is encouraged to make contributions to. And, indeed, the plant management chooses employees partly on the basis of how well they will fit into such a flexible, learning-oriented culture. The least desirable employee, Schulman reports, is one who is too confident or stubborn. "A Radiation Protection general foreman asserted, for example: 'There is a real danger in having very headstrong people intent on their own way.' A Quality Control inspector volunteered that 'people who have a belief they are infallible can have a very negative impact here.'"

This sort of continuous learning and improvement would not be possible if the Diablo Canyon organization were strictly a hierarchical one. Hierarchies may work for systems that are "decomposable"—that is, that can be broken into autonomous units that can function on their own—but a nuclear plant is, by its nature, very tightly coupled. A modification to the steam generators can have implications for the reactor, or a change in maintenance procedures may affect how the system responds to the operators. Because of this interdependence, the various departments in the plant must communicate with and cooperate with one another directly, not through bureaucratic channels. And, Schulman says, they do.

The dozens of different departments and groups at the plant talk to each other constantly at both high and low levels. Department managers meet once a week. A Plant Staff Review Committee, with representatives from each of the plant's units, convenes at least once a week to consider proposed changes in procedures and equipment. The so-called Safety System Outage Modification Inspection groups, also with representatives from the plant's major units, study how proposed design changes might affect safety systems and reactor shutdown. Each time some regulation is breached or some automatic safety system shuts down a reactor temporarily, an interdepartmental group is formed to understand why and prevent a repetition. And so on.

Members of different departments inevitably bring different goals and perspectives to such meetings, and that's just the point. Any out-of-the-ordinary action requires approval from a variety of people throughout the plant, any one of whom can veto it. And even normal procedures, such as maintenance or repair, must be discussed and agreed upon by representatives from several departments. With this cross-plant communication, the organization hopes to avoid any actions with unanticipated consequences.

The parallels between this and the organizational structure on the carrier are striking. As with the aircraft carrier, the underlying order of the plant is hierarchical, and this hierarchical structure operates under what might be called "normal" conditions. For the carrier, this means times when it is not launching or recovering aircraft; for the nuclear plant, it is situations where the proper action can be found in the standard operating procedures. But a second layer of structure emerges in times of stress, overlaid on the first, and this one de-emphasizes rank or position, emphasizes expertise, and places a great deal of weight on communication and cooperation among units. The purpose of this second structure is the same in both organizations: to deal with the demands created by the complexity of the system. The demands may seem quite different—split-second actions and decisions on a carrier deck versus carefully considered worries about how a nuclear plant might respond to this or that measure—but in each case they arise from the tight coupling among the components of a hazardous system.

In general, the members of the Berkeley project have found a number of features that tend to occur in high-reliability organizations. They have studied not just aircraft carriers and nuclear power plants, but also air traffic control systems and the operation of large electric power grids, and they detect a pattern.

The layered organizational structure, for instance, seems to be basic to the effectiveness of these institutions. Depending on the demands of the situation, people will organize themselves into different patterns. This is quite surprising to organizational theorists, who have generally believed that organizations take on only one structure. Some groups are bureaucratic and hierarchical, others professional and collegial, still others are emergency response, but management theory has no place for an organization that switches among them according to the situation. The realization that such organizations exist opens a whole new set of questions for researchers: How are such multilayered organizations set up in the first place? And how do the members know when it's time to switch from one mode of behavior to another? But the discovery of these organizations may also have practical implications. Although La Porte cautions that his group's work is "descriptive, not prescriptive," still the research may offer some insights into avoiding accidents in hazardous technologies.

In particular, high-reliability organizations seem to provide a counterexample to Perrow's argument that some technologies, by their very nature, pose inherent contradictions for the organizations running them. Concerning such technologies as nuclear power and chemical plants, Perrow writes: "Because of the complexity, they are best decentralized; because of the tight coupling, they are best centralized. While some mix might be possible, and is sometimes tried (handle small duties on your own, but execute orders from on high for serious matters), this appears to be difficult for systems that are reasonably complex and tightly coupled, and perhaps impossible for those that are highly complex and tightly coupled." But it is not impossible at all, if the studies of Diablo Canyon and the aircraft carriers are to be believed. As paradoxical as it sounds, an organization can be both centralized and decentralized, hierarchical and collegial, rule bound and learning centered.

Besides the layered organizational structures, high-reliability organizations share another feature that helps avoid accidents. They emphasize constant communication—talk, talk, talk, far in excess of what would be thought useful in normal organizations. The purpose is simple: to avoid mistakes. On a flight deck, everyone announces what is going on as it happens. This increases the likelihood that someone will notice and react if things start to go wrong. In an air traffic control center, although one operator has responsibility for controlling and communicating with the aircraft in a particular sector, he is helped by an assistant and, in times of peak load, one or two other controllers. The controllers constantly watch out for one another, looking for signs of trouble, trading advice, and offering suggestions for the best way to route traffic.

Poor communication and misunderstanding, often in the context of a strict chain of command, have played a prominent role in many technological disasters. The *Challenger* accident was one such, with the levels of the shuttle organization communicating mostly through formal channels so that the concerns of the engineers were never brought home to the top management. In 1982, a Boeing 737 taking off from Washington National Airport crashed into a bridge on the Potomac River and killed seventy-eight people. The copilot had warned the captain of possible trouble several times—icy conditions were causing false readings on an engine-thrust gauge—but the copilot had not spoken forcefully enough, and the pilot ignored him. In 1977, a 747 flown by the Dutch airline KLM collided with a Pan Am 747 on a runway at Tenerife airport in the Canary Islands in 1977, killing 583 people. A major factor in that crash was that the KLM pilot misunderstood where the Pan Am craft was. A postcrash investigation found that the young copilot had been concerned by the senior pilot's actions, but assumed the pilot knew what he was doing and so clammed up. And the Bhopal accident would never have happened had there been communication between the plant operators, who began flushing out pipes with water, and the maintenance staff, which was responsible for inserting the slip bind into the valve in order to keep water from coming through. The operators never checked to see if maintenance had done its job.

Besides communication, high-reliability organizations also emphasize active learning, not simply the memorization of procedures. Employees should not only know why the procedures are written as they are but should be able to challenge them and look for ways to improve them. The purpose behind this learning is not so much to improve safety—although this often happens—as it is to keep the organization from regressing. Once people begin doing everything by the book, things quickly go downhill. Workers lose interest and get bored; they forget or never learn why they do things certain ways; and they begin to feel more like cogs in a machine than integral parts of a living institution. Whether it is the steady turnover of personnel on an aircraft carrier, the challenge to nuclear- plant workers to constantly find ways to improve procedures, or some other method, organizations need ways to keep their members fresh and focused on the job at hand.

Any organization that emphasizes constant learning will have to put up with a certain amount of ambiguity, Schulman notes. There is never a point at which everything can be completely specified. There will always be times when people are unsure of the best approach or disagree even on what the important questions are. This may be healthy, Schulman says, but it can also be unsettling to people who think a well-functioning organization should always know what to do. He tells of a meeting with Diablo Canyon managers at which he described some of his findings about the plant's operation. "What's wrong with us that we have so much ambiguity?" one manager wondered. The manager had completely missed the point of Schulman's research. A little ambiguity was nothing to worry about. Instead, the plant's managers should be concerned if they ever thought they had all the answers.

Schulman offers one more observation about high-reliability organizations: they do not punish employees for making mistakes when trying to do the right thing. Punishment may work—or at least not be too damaging—in a bureaucratic, procedure-driven organization where everyone goes by the book, but it discourages workers from learning any more than they absolutely have to, and it kills communication. Suppose, for example, that a worker slips and damages an instrument. If he expects to get punished for even an inadvertent mistake like this, he may simply not report it. Or he may make the problem worse, either by trying to hide the mistake or by attempting to fix it.

The reliability of such organizations as Schulman and the others study is difficult to maintain. Because it flows not from rules and regulations but from a certain culture, it can die off quickly if the external environment changes or if outsiders take over who do not understand the culture or the importance of keeping it. It was just this sort of cultural change that lay at the root of the saddest episodes in technological history: the decline of NASA after the Apollo program.

Going to the moon is usually ranked with the Manhattan Project as one of modern history's two greatest technological projects. Both were spectacular successes, at least in terms of achieving their stated goals, and both went much faster than many had thought possible. Indeed, the success of the moon program has often been held out as proof that nothing is out of reach if we try hard enough. It's not so common now, but for years after Apollo 11 people liked to ask, "If we can send a man to the moon, why can't we . . . " followed by some thorny medical, technical, or social problem such as "cure the common cold?" or "alleviate world hunger?" Of course, curing the common cold and alleviating world hunger are much more difficult problems than putting a man on the moon, but that was difficult enough. The success of the moon program stemmed from what was arguably the most creative, competent, and accomplished technological organization ever put together.

The NASA of the 1960s was much like the high-reliability organizations of twenty years later that caught the eye of the Berkeley group. Where tried-and-true procedures existed, NASA was ruthless in making sure that everything was done according to specifications. Yet the agency recognized that more was unknown than was known, and its employees were given the flexibility and freedom to find solutions. The NASA culture embraced risk in the same way that workers at Diablo Canyon or the crew of an aircraft carrier do. The employees expected mistakes, and they knew that some mistakes would prove very costly—they could kill people, or perhaps even kill the space program. This awareness of mistakes made the employees work much harder to avoid them. Furthermore, the scientists and technicians at NASA were constantly learning to an extent much greater than possible on an aircraft carrier or at a modern nuclear plant. They were at the frontier of human knowledge, building machines and doing things unlike anything the world had seen before.

Despite some failures, such as the fire in Apollo 1 that killed three astronauts, the program was exceptionally reliable. Each time another set of astronauts went up, they were trying many things that had never been done before, and almost all of the trials were successful. Giant step by giant step, the organization went from a fifteen-minute jaunt into the atmosphere in 1961 to walking on the moon eight years later.

By 1986, however, NASA had fallen apart. In February the *Challenger* crashed. In April a Titan rocket carrying an Air Force spy satellite exploded at Vandenberg Air Force Base. In May a Delta rocket with a weather satellite had to be blown up after a launch at Cape Canaveral. By June the Rogers Commission, which was formed in the wake of the *Challenger* accident, was suggesting that only major changes in how the agency was run could get it back on track. After a hiatus, the shuttle resumed its flights and there have been no more crashes, but recently NASA has suffered a series of failures that have once again called the agency's competence into question. First came the flaw in the lens of the Hubble Space Telescope, then communications problems with the Magellan satellite orbiting Venus, the failure of the main communications antenna on the Galileo spacecraft, and the loss of the Mars Observer to a still unknown equipment malfunction. Where did the old, reliable NASA go?

Actually, it had started fading even before the final Apollo flights. After the euphoria of landing a man on the moon subsided, going back again and again became almost routine. Without the constant challenge of doing something new, interest started to wane, and quality-control problems began to appear on NASA's later moon shots.

Once the Apollo program ended, NASA was left without a mission. What could it do? One option was to declare victory and close up shop, but no one seriously considered that alternative. A bureaucracy finds ways to justify its existence, and NASA was no exception. It proposed a new challenge: sending men to Mars. On the surface it seemed like the ideal follow-up to the moon mission, an assignment that would tap the very strengths that had made the Apollo program a success. It was not, for reasons that would only become clear with time.

Mars is much farther from the Earth than the moon is and consequently much more difficult to reach with a vehicle launched from the Earth's surface. Instead, it was decided first to build a space station in Earth orbit and then to set the Mars mission off from there. But building a space station demands an incredible number of trips up from the Earth's surface to ferry materials and the astronauts to assemble them. To keep this job from becoming prohibitively expensive, NASA decided to create a system in which most of the parts were used again and again—a shuttle instead of single-shot rockets. Thus was the space shuttle program born.

This mission was of a very different sort from the moon program. Throughout the 1960s, NASA had been primarily a research-and-development organization, constantly taking on new challenges. But the space shuttle, although it would need a great deal of research and development at first, was always intended to become a routine technology, one that would do the same thing over and over again, year after year. The astronauts inside the shuttle might be performing different tasks with each flight, but the shuttle was supposed to be as reliable and as boring as a 747, if not the family station wagon. This would demand a new culture at NASA, one that maintained reliability in the face of constancy instead of change.

At the same time that its mission was changing, NASA itself was facing a changing environment. Following President Kennedy's 1961 challenge to get to the moon by the end of the decade, NASA was in a nearly ideal political situation. It had gotten all the money it needed and faced little scrutiny from Congress or the White House about how it was doing its job. But by the 1970s, budgets were shrinking and, having accomplished the one goal that everyone backed, NASA now had to constantly justify its programs. Within a few years the agency had gone from a national priority to just one more bureaucracy scrambling for funding. With its officials increasingly concerned about the survival of the agency, NASA changed from an organization willing to take risks to one that was cautious and intolerant of failure.

The organizational structure and culture reflected these changes. Rules and procedures proliferated, and the increasing accountability forced on the agency from the outside gradually killed much of the freedom and flexibility that NASA engineers had once enjoyed. NASA became less like a college campus and more like a standard government bureaucracy. At the same time, the strong central control of the 1960s degenerated into a collection of semiautonomous field centers which were sometimes jealous of each other and which often failed to communicate well with one another. And NASA came to rely increasingly on outside contractors, so that its own employees accounted for less and less of the expertise needed for the project.

The result was that the NASA of 1986 was a very different—and much less effective—organization than the NASA of 1966. And even with the shock of the *Challenger* and the resulting rededication to quality and safety, the space agency never quite regained its edge. One engineer who worked for NASA in the early days put it like this: "Look, I don't know what makes this damn thing go that we've got here—that makes us so good and so great, and our capability to inspire people. But I'll tell you this, if we ever lose it I won't know how to tell you to regain it."

CREATING RELIABILITY

That old NASA engineer could have been speaking of nearly any highly effective organization, and particularly those that deal with complex and risky technologies. Todd La Porte, Paul Schulman, and other members of the Berkeley research project heard much the same things from people they interviewed at high-reliability organizations. The management at Diablo Canyon, for instance, had certain clear principles they used in running the plant that they knew seemed to work. But they had no real idea why they worked, and they didn't know whether they would continue to work in different circumstances. Would the same approach be as effective after the plant had been running another ten or twenty years? No one could say.

Assuming that high-reliability organizations do indeed perform as well as the Berkeley researchers believe and for the reasons that they have described, one big question remains: How does one create a high-reliability organization? Or, if the organization already exists, how does one transform an ordinary institution into an extraordinary one? La Porte and his group have not yet studied this question specifically, but the nuclear industry provides a preliminary answer. For fifteen years, the industry has been attempting exactly such a transformation.

When Unit 2 at Three Mile Island suffered its partial meltdown in 1979, the U.S. nuclear industry didn't need a fortune teller to see the future. Its profitability, if not its very existence, was threatened. If nothing else, the Nuclear Regulatory Commission would create whole new volumes of rules and procedures that would have to be followed to the letter at every nuclear plant in the country. And if another accident like TMI struck or, worse, if a nuclear accident actually killed some people and spread radioactivity over the countryside, the public outcry could conceivably shut down every plant in the country. Furthermore, TMI forced the nuclear industry to face some unpleasant truths about itself. Nuclear accidents were not as unlikely as industry experts had been assuring the public and themselves. At least some utilities were not operating their plants to the high standards needed to preclude accidents. And if another meltdown—or even a minor mishap—occurred at any one of the dozens of operating nuclear plants, every other plant would pay the price. They were, as political scientist Joseph Rees writes in his book of the same name, "hostages of each other."

Thus energized, the U.S. nuclear industry created its own watchdog agency, the Institute of Nuclear Power Operators, or INPO. Its mission was to see that every nuclear plant in the United States ran according to standards of excellence that were even stricter than what the NRC would demand. Plenty of industries have established standard-setting agencies, of course, but INPO was unusual in two important ways. First, its benchmarks were not the easy-to-meet, lowest-common-denominator rules that are the usual result when a number of businesses collaborate to produce standards. Instead, they were a distillation of the best, most stringent practices in the industry. And second, INPO would take an active role not just in promoting these standards but also in grading each nuclear plant's performance vis-à-vis the standards and in teaching utilities how to do better. Although its founders did not think of it this way, INPO's goal was essentially to transform every nuclear plant in the United States into a high-reliability organization.

In the wake of TMI, experts assembled a catalog of factors that had contributed to the accident: the faulty valve that started the whole thing; mistakes by the operators; instrumentation that was poorly laid out and that didn't give the operators enough information to diagnose the accident; inadequate training of the operators; and poor communication that kept the TMI operators from hearing about similar problems with the same valve at other plants. But the more perceptive observers looked past the immediate failures to detect a deeper problem: the organizational culture of the nuclear plant and of the utility that owned it. Most of the causes of the accident could be traced back, directly or indirectly, to a mindset and an organizational structure inherited from a time when the utility operated only fossil-fuel plants.

As we saw earlier, in the 1960s, 1970s, and even into the 1980s, the dominant culture in the utility industry was the culture of the coal-fired plant. The requirements of generating electricity from coal had bred a community of men who approached their jobs in a particular way. Later, a nuclear industry official would sum up that attitude like this:

> In the fossil fuel business the general philosophy is run it till it breaks. Then you shut it down, fix it, and run it again. You see, the capital investment is enormous. Every minute you don't use that capital for production purposes you are running costs, because the capital runs continuous costs, but you're not getting any return. . . . "Let's run the plant every minute we can and we'll fix it when it breaks." That was their basic attitude.

He wasn't speaking specifically of GPU, the utility that owned and operated Three Mile Island, but he well could have been. That fossil-fuel culture was found throughout the electric utility industry, and it colored the industry's attitudes toward safety, even at nuclear plants. Although industry people paid lip service to the need to prevent nuclear accidents, few actually realized how much work and commitment that entailed. Most assumed that they could rely on the engineered safety features of a plant to avert accidents, and many thought that all the rules and regulations concerning safety had gone too far. The plants were already safe enough, and it was a waste of time and money to keep adding more devices and more procedures just to give a little extra margin of safety. In contrast to high-reliability organizations, where procedures are strictly followed and employees are constantly looking for ways to improve them, the nuclear plant workers of the 1970s generally saw no reason to go further than the regulations required. Indeed, they were often willing to skimp on the rules, particularly if it seemed that they were staying within the spirit of the regulation.

When INPO began operations in 1979, just nine months after TMI, its management quickly realized that the entire nuclear industry would have to change. But into what? Part of the answer, which nearly everyone understood from the first, was that the utilities would have to adjust how they thought of and interacted with each other. They would have to transform themselves from a collection of individual utilities, each going its own way, into a *community*. If they had done so earlier, the industry might well have avoided the meltdown at Three Mile Island, since, as we saw earlier, a similar but less serious accident had occurred eighteen months before at another plant. If the utilities had been sharing information, the operators at TMI likely would have responded properly instead of making the accident worse. So now, under the aegis of INPO, the members of the nuclear community would trade advice, pass on warnings, and look out for one another—even to the point of sanctioning a member that was not keeping up its end of the bargain.

The individual plants were a different problem. What should the goals for them be? Industry leaders were beginning to understand that they would have to concentrate on safety and get rid of the old fossil-fuel habit of running a plant until it broke. But just what should they replace the fossil-fuel culture with?

As Rees describes it in *Hostages of Each Other,* the culture that INPO began to infuse in nuclear power plants was that of the nuclear navy. This was not by design, however—at least, not at first. When industry executives began looking for INPO's first chief executive, they had no special preference for a Navy man. Indeed, retired senior officers from the nuclear navy had the reputation of not doing well in commercial nuclear power. But eventually the INPO search committee settled on Admiral Eugene Wilkinson, one of Rickover's favorite officers. Rickover had chosen Wilkinson as the original captain of the *Nautilus,* the world's first nuclear submarine, and later picked him to skipper the *Long Beach,* the first nuclear surface warship. When Wilkinson joined INPO, he brought Zack Pate, another Rickover protege, to be his chief assistant. And when the two of them began to build a staff at INPO, they found it difficult to attract people from the nuclear industry—not only was there a shortage of workers at the time, but no one knew how long INPO would be around—so they relied heavily on ex-Navy personnel. As a result, INPO began life with a strongly defined culture: that of Rickover's nuclear navy.

In hindsight, that was a stroke of luck for the nuclear industry. Instead of having to create a new culture from scratch, INPO could use the nuclear navy culture as a model and work from it. And although operating a reactor on a submarine or a ship has a number of obvious differences from running a reactor in a commercial power plant, the Navy culture that Rickover had created proved surprisingly easy to adapt to civilian life. Indeed, in 1983, as part of an evaluation of the accident at Three Mile Island, Rickover himself had described how one might adapt the Navy's culture to the commercial nuclear industry. That essay, which captures quite well what Wilkinson was trying to do at INPO, lays out principles for operating nuclear reactors that Rickover had developed in the early days of the nuclear navy. And those principles, albeit in different words, describe exactly the same thing that Todd La Porte and his colleagues would identify some years later in their studies of high-reliability organizations.

First and foremost, Rickover wrote, any hazardous, evolving technology such as nuclear power "must be built upon rising standards of excellence." It is not enough to set a goal and meet it. Once a level of competence has been reached, management must set higher goals. Hand in hand with this goes an ability to learn from experience. Mistakes will be made, he wrote, but that is to be expected. The objective should be to acknowledge the mistakes, figure out why they occurred, and make sure they don't happen again. To help management in making decisions, the organization should have enough technical know-how to investigate problems and come up with solutions in-house. And to improve safety and performance, management must treat every part of the system—equipment maintenance, training, quality control, technical support, and so on—as pieces of an integrated whole. These and several other principles, Rickover wrote, could serve as the basis for a new industrial morality that would treat the hazards of nuclear power with the respect they deserved.

Underlying these principles were certain important attitudes and beliefs, Rickover wrote. "They acknowledge the complex technology. They recognize that safe nuclear operations require painstaking care. They declare that a management must be responsible—all the time." These were the attitudes that INPO set out to instill in the nuclear industry. It would not be easy. The new culture would demand a 180-degree turn from the old ways of thinking. But after the accident at Three Mile Island and the establishment of INPO, the need for such a transformation gradually became accepted among the nuclear utilities. Plant by plant, the old attitudes were exorcised and new ones adopted that owed much to Rickover and the nuclear navy. And in many cases, the results have been dramatic.

Consider, for instance, Florida Power & Light's Turkey Point plant, south of Miami. Turkey Point's two nuclear reactors were built in the 1970s next door to an existing fossil-fuel plant. "The early going at Turkey Point was okay, according to the standards of that time," says Jerry Goldberg, president of FP&L's nuclear division. But after Three Mile Island, the NRC began raising its standards, and Turkey Point didn't keep up. The problem, Goldberg says, was the same as at so many other utilities: the plant had been run according to a fossil-fuel culture, where the goal had been to keep it online as much as possible. If small problems appeared, the normal reaction was to ignore them or to do as little as possible to fix them. The attitude was, "Oh, it's the regulator valve. Give it a poke and it will keep going." And for a while it seemed to work. The plant had a relatively high availability during its early years. But that high availability extracted a price. Over time the plant accumulated a number of little things that weren't right, and these led to an increasing number of "trips," or automatic safety shutdowns. In one year Turkey Point had seventeen trips. The NRC now considers a plant to be "troubled" if it has two.

To fix things, the Florida Power & Light management brought in Goldberg, a retired nuclear navy officer. He hired half a dozen new people during his first six months, he says, but otherwise he kept most of the plant's staff and focused instead on changing their attitudes. He made it clear that they would have to transform their fossil-fuel culture into a nuclear culture, with its increased discipline and responsibilities. Those few employees who wouldn't or couldn't change were fired. The rest gradually learned to believe in the new system by watching it in action, Goldberg says. "I told them what to do, and when they saw it work they felt better." Since the fall of 1991, when Turkey Point came back on-line after a long outage to make various repairs and modifications, it has been one of the best in the country in terms of reliability and safety record.

Overall, the average performance of nuclear plants in the United States has been steadily but slowly improving since 1979 by almost any measure: availability, number of automatic safety shutdowns, workers' exposure to radiation, thermal efficiency, and others. The industry has become more adept at managing the Faustian bargain of nuclear power. Performance remains spotty, however, with some U.S. plants as good as any in the world and others among the worst.

An entrenched culture, as INPO has discovered, is not easy to alter. People get used to doing things a certain way and don't want to change. Their way is the "right way." And even when faced with evidence that it's not the best way, people insist that it's good enough and that there's no real gain in switching gears. This tendency has made the nuclear industry's job much harder. Fortunately, the experience at Turkey Point and other plants indicates that it's not necessary to get rid of the entire work force to remake the culture. It can be enough to replace the top management and perhaps a few intransigent employees. (There may even be examples of a turnaround in corporate culture that have left the top management intact, but they're exceedingly rare.)

Of course, a company or an industry must have both the incentive to change and an idea of what it wants to change into. TMI and the threat of future accidents provided the incentive to the nuclear industry, while the nuclear navy offered the model. Other industries will find their own ways, but the task may become easier as we understand more about high-reliability organizations and what makes them work.

MAINTAINING RELIABILITY

Should society continue to make such Faustian bargains as nuclear power? The answer hinges on our ability to control them, which in turn depends not so much on engineers' ability to design well—although acceptable designs are a minimum criterion—as on organizations' ability to manage well. And this is uncharted territory. Centuries of experience have given us a good idea of what we can and can't expect from engineers, but we are just learning what can be expected of organizations.

The high-reliability organizations offer a "proof of principle": they prove that, in some cases at least, it is possible to create an organization that is extremely good at managing a hazardous technology. They do, however, leave some important questions unanswered. Is it possible, for instance, to guarantee that every one of the several dozen nuclear sites in the United States is operated by a high-reliability organization? Won't a few slip through the cracks, even if the majority are excellent? And what about the hundreds of reactors around the world? If it is so difficult to create a high-reliability culture in organizations in the United States, will it be possible in Third World countries or the countries of the former Communist Bloc?

In a generation or two, the world will likely need thousands of high-reliability organizations running not just nuclear power plants, space flight, and air traffic control, but also chemical plants, electrical grids, computer and telecommunications networks, financial networks, genetic engineering, nuclear-waste storage, and many other complex, hazardous technologies. Our ability to manage a technology, rather than our ability to conceive and build it, may become the limiting factor in many cases. For that reason, the INPO experiment is an important test of how far we can go with technology. If it proves feasible to instill a high-reliability culture throughout an industry, with cooperation and self-policing pushing all members to become HROs, then large-scale use of hazardous technologies may be acceptable. Already, the World Association of Nuclear Operators is aiming to apply INPO's principles around the globe. Formed in 1989 in response to the Chernobyl accident, the group has 139 members that operate more than 400 of the world's 500 nuclear plants. It is, of course, far too early to know how successful it will be.

Even if it proves possible to consistently create reliable organizations, the question remains whether they will be able to keep their culture intact decade after decade. The NASA experience indicates that a change of mission and a change in external conditions can conspire to blunt an organization, but it may be even harder to maintain an organization's edge in the face of unchanging conditions. The Navy may be able to pull it off on its aircraft carriers because of its regular turnover of personnel, which enforces constant learning, but can this success be duplicated in a nuclear or chemical plant that is built to operate for forty to sixty years? Or, even more daunting, what about a nuclear-waste repository whose staff must be vigilant for hundreds of years? After a decade or two of vigorous learning and improvement, will there come a point at which the efforts produce so little extra enhancement that they don't seem worth it? And once the challenge of constant improvement is gone, will the culture revert to a rule-bound, unthinking bureaucracy certain that it already knows all the answers? We can only speculate for now.

One other threat is more immediate: call it the price of success. If an organization succeeds in managing a technology so that there are no accidents or threats to the public safety, the natural response from the outside—whether it is the company's upper management, government regulators, or the public—is to begin to take that performance for granted. And as the possibility of an accident seems less and less real, it becomes harder and harder to justify the cost of eternal vigilance.

High reliability is expensive. Besides having the best equipment possible, and plenty of extra equipment as backup, the organization must spend a great deal of time and money testing and maintaining equipment and training employees. And the communication that lies at the heart of a high-reliability organization can itself be very expensive. At Diablo Canyon, Schulman says, the constant meetings to talk about what might go wrong and to decide on procedures take up a tremendous amount of time. From the outside, particularly if the plant has been operating smoothly for many years, these meetings might seem unproductive and not worth the cost.

Yet if any lesson emerges from the studies of organizations and risk, it is that skimping on safety has been at the root of many of our most horrible accidents. There is the Bhopal tragedy, with its inactivated safety equipment, poor training, and inadequate staffing. NASA's problems leading up to the *Challenger* explosion can be traced to pressures to produce quickly on an inadequate budget. The 1974 crash of a Turkish Airlines DC-10, described in chapter 4, which killed all 346 people on board, was the result of a penny-wise, pound-foolish approach to safety. And so on. If we are unwilling to invest in safety and to keep on investing for as long as we insist on using hazardous technologies, then our Faustian bargains will certainly prove to be no bargains at all.

nine

TECHNICAL FIXES, TECHNOLOGICAL SOLUTIONS

The past couple of decades have been a confusing, frustrating period for engineers. With their creations making the world an ever richer, healthier, more comfortable place, it should have been a time of triumph and congratulation for them. Instead, it has been an era of discontent. Even as people have come to rely on technology more and more, they have liked it less. They distrust the machines that are supposedly their servants. Sometimes they fear them. And they worry about the sort of world they are leaving to their children. Engineers, too, have begun to wonder if something is wrong. It is not simply that the public doesn't love them. They can live with that. But some of the long-term costs of technology have been higher than anyone expected: air and water pollution, hazardous wastes, the threat to the Earth's ozone layer, the possibility of global warming. And the drumbeat of sudden technological disaster over the past twenty years is enough to give anyone pause: Three Mile Island, Bhopal, the *Challenger*, Chernobyl, the Exxon *Valdez*, the downing of a commercial airliner by a missile from the U.S.S. *Vincennes*. Is it time to rethink our approach to technology?

Some engineers believe that it is. In one specialty after another, a few prophets have emerged who argue for doing things in a fundamentally new way. And surprisingly, although these visionaries have focused on problems and concerns unique to their own particular areas of engineering, a single underlying theme appears in their messages again and again: Engineers should pay more attention to the larger world in which their devices will function, and they should consciously take that world into account in their designs.

Although this may sound like a simple, even a self-evident, bit of advice, it is actually quite a revolutionary one for engineering. Traditionally, engineers have aimed at perfecting their machines *as machines.* This can be seen in the traditional measures of machines: how fast they are, how much they can produce, the quality of their output, how easy they are to use, how much they cost, how long they last. By definition, an improvement in a given machine means an improvement in one or more of these measures, which are assumed to be the important ones for the consumer. But, as some engineers have begun to realize, there exist other measures that may have equal or greater importance.

Consider, for instance, the Bhopal chemical plant. The accident there was the child of many fathers. As we saw in the last chapter, the plant's history and the influence of the national government created a situation in which large amounts of the deadly methyl isocyanate, or MIC, were produced next door to an area where tens of thousands of people lived. The plant's unprofitability led to staffing cuts, a lack of proper training, and neglect from top management. For reasons that no one quite understands, many of the plant's safety devices were not operational at the time of the accident. And carelessness by plant employees coupled with poorly maintained equipment triggered the accident itself. How might an engineer respond to this situation?

In the case of engineers from Union Carbide, whose subsidiary operated the Bhopal plant, the response was predictable. Company officials insisted that the design of the plant was fine. They blamed their subsidiary, Union Carbide of India, for operating lapses and further suggested that the accident was caused by a disgruntled employee. Someone, they said, deliberately unscrewed a pressure gauge and hosed water into a tank containing MIC, creating the toxic cloud of gas that escaped the plant and devastated the city. And except for the role of the hypothetical disgruntled employee—whose existence was never proven—this is a perfectly defensible position. Somewhat narrow, perhaps, but defensible. *If* the people and organizations running the plant had done their jobs correctly, the accident would have never happened, so the design of the plant was fine. This attitude exemplifies the machine-centered philosophy of engineering: design a plant so that it does its job efficiently, then expect people and organizations to adapt to it.

There is another approach, however, an approach that accepts that people make mistakes and that organizations get sloppy and takes those factors into account in the engineering process. The standard industrial method for manufacturing MIC demands that large amounts of it be produced and then stored until it is transformed into some other chemical, such as the pesticide produced at Bhopal. Thus at the time of the accident, the Bhopal plant held sixty-two tons of MIC, forty-two tons of it in the one tank involved in the accident. The engineers who designed the plant had compensated for this hazard by designing a variety of safety systems which, if things had not gone so badly wrong, would have prevented the catastrophe. But an engineer who was not tied to the machine-centered school of engineering might come at the problem from a completely different direction. Is there, he might ask, some way to avoid keeping so much MIC at the plant? If so, there would be no need to rely on safety systems—and on the people and institutions that operate them.

In effect, this is exactly what chemical engineers at Du Pont did within a year of the Bhopal accident. In 1985, Du Pont developed a different process for manufacturing MIC, one that dodges the need to keep so much MIC on hand. By creating an intermediate product that can be transformed into MIC quickly and easily, the Du Pont method produces MIC only on demand. And once it is produced, it is immediately processed into the final product, so that there is generally less than a couple of pounds of MIC in the system at any one time. The system makes a repeat of Bhopal not just unlikely but impossible.

This sort of fresh thinking holds the potential for major improvements in technological design. The traditional engineering approach, rooted in the imperative of the machine, has difficulty addressing many of the problems of modern, complex technologies, since these problems tend to have large nontechnical components. The traditional engineer doesn't believe that Bhopal was his problem. The plant's design was fine; the politicians and the regulators and the plant managers just needed to get their acts together. By contrast, the new breed of engineer sees all sorts of technological problems as opportunities for engineering solutions. No matter whether a problem arises from technical or nontechnical factors, it's a potential candidate for a technical fix.

There are limits to this approach, of course. As the historian Thomas Hughes points out, technological problems generally demand technological solutions, of which the technical fix is just one component. Sometimes what appears to be a likely engineering solution is not practical for reasons that have nothing to do with its technical merits. But then again, sometimes an issue that seems to be all about politics and personalities can be resolved only with a technical breakthrough. The trick lies in seeing the technical in the context of the technological and working from there.

ENGINEERING FOR HUMANS

The accident at Three Mile Island often gets blamed on the plant operators. If they had responded correctly to the initial problem, the accident would have been stopped in its tracks, with no radiation released and no damage to the plant. Instead their actions made the accident much worse.

Given this, there appears to be an obvious solution: better training. Give the operators bigger, more detailed manuals, drill them harder, test them to make sure that they know all the proper responses, and eventually they will respond almost perfectly. This is a traditional engineer's response, one that takes the machine as a given and expects the humans operating it to adapt. And in this case it might seem to make sense—until one takes a look at the control room at Three Mile Island.

The TMI operators worked in a large room with much of the equipment arrayed on the walls. There was a central station holding some of the main controls and instrumentation, but much else that was crucial to the assessment and the operation of the plant was scattered about. In their book *The Warning: Accident at Three Mile Island,* Mike Gray and Ira Rosen offer an impression of the TMI control room at the time of the accident. The control panels, they write, were a confusing mess of "1,100 separate dials, gauges, and switch indicators," many of them hidden from immediate view, so that the operator had to get up and move around to see them. The control room also had more than 600 alarm lights. "About forty or fifty of these alarms are always lit up—chronic malfunctions either in the alarm itself or the equipment it's monitoring." When the accident hit, many more came on, creating a confusing overload of information that made it nearly impossible to figure out what was going on.

The working conditions for the TMI operators were actually little different from those at any other nuclear plant of the time. In none of them had the engineers laying out the control room given much consideration to the particular needs of people running a nuclear power plant. Instead, the engineers had arranged the controls in much the same way that controls were laid out in fossil-fuel plants. An operator accustomed to working in a coal-fired plant could walk into the TMI control room and feel right at home, with only a few unfamiliar controls—those dealing with the reactor, for example—to get used to. Perhaps this was comforting to operators with experience in coal plants, but it made little sense otherwise. The logic of a nuclear plant is very different from that of a fossil-fuel plant, and operators must read and respond to the instruments in another way altogether. This is particularly true in an emergency, when quick responses matter.

Because of the poorly thought-out controls, there was no easy way for the TMI operators to get the "big picture" of what was going on in the plant during the accident. They had to piece it together detail by confusing detail. One clue to what was happening was an indicator showing that a feedwater valve was closed. This indicator was hidden behind a maintenance tag hanging on the control panel. An operator had to look behind the tag to check on the status of the valve—but first he had to remember that the indicator was back there and that it might be important. Another clue was to be found in a set of indicators describing the state of the reactor drain tank. They could have warned the operators that the faulty electromagnetic relief valve had stuck open—the precipitating cause for the accident. But those indicators were hidden behind a control panel. A still more important indicator had never been installed. Shortly after the start of the accident, when the relief valve had done its job and let excess pressure out of the reactor coolant system, the plant controls automatically signaled the valve to close. An indicator light came on to show that the command had been sent, but there was no sensor to detect whether the valve had actually shut. So the operators, seeing that the command had been sent, assumed that the valve was shut. It was not, and it was two hours before the operators figured that out.

The postaccident analysis of Three Mile Island did not let the operators off the hook entirely. They were careless in their operations, they did not know the basic reactor physics necessary to understand exactly what was going on inside the reactor, and they made mistakes in judgment. But everyone agreed that their job was made much harder by the design and layout of the controls. At a TMI symposium sponsored by the Electric Power Research Institute, Lawrence Kanous of Detroit Edison castigated the industry's approach to setting up control rooms. "There is little gut-level appreciation of the fact that plants are indeed man-machine systems," he said. "Insufficient attention is given to the human side of such systems since most designers are hardware-oriented."

Kanous could have been speaking about any number of different technologies. Traditionally, engineers have given relatively little weight to human factors when they design instrumentation and control systems. They focus on what is important to the physical functioning of the machine and assume that the human operators are adaptable.

Donald Norman, a research fellow at Apple Computer, makes a similar point about automation. Historically, he says, engineers have not decided which functions to automate based on a consideration of human-machine interaction but instead have automated whatever they could under the assumption that more automation is better. Modern airplanes offer a good example. They are built with more and more automated controls, Norman notes. "But did the designers do a careful analysis of the cockpit to decide which tasks were best done by people and which were in need of some machine assistance? Of course not. Instead, the parts that could be automated *were* automated, and the leftovers given to humans."

Automation is fine, Norman says, when everything is working normally, but it can fail in the face of unexpected problems and sometimes make matters worse. "When an automated system suddenly stops working, often with no advance warning, the crew is thrust into the middle of the problem and required to immediately figure out what has gone wrong and what should be done. There is not always enough time."

Indeed, this may have been behind some of the much-publicized troubles of the ATR-42 and ATR-72, turboprop planes made by the French-Italian consortium ATR. On October 31, 1994, an American Eagle ATR-72 waiting to land at Chicago's O'Hare airport suddenly spiraled into the ground, killing all sixty-eight aboard. It was the worst of a series of a dozen incidents involving the ATR-72 and its smaller sister craft, the ATR-42, that had occurred in icy weather over a ten-year period. Although no one has ever determined exactly why the plane crashed, tests at Edwards Air Force Base following the accident pointed to a fatal combination of factors in which the plane's autopilot played a key role.

As the ATR-72 was circling O'Hare, the pilots had lowered the plane's flaps by fifteen degrees, apparently to smooth out the rough flight. Although the pilots didn't realize it, they were flying through a freezing drizzle, and keeping the flaps down allowed a ridge of ice to form along the wings. The tests at Edwards showed that such a ridge could create a vacuum over the top of one aileron (a control surface at the end of the wing) that would have pulled that aileron up, automatically forcing the aileron on the other wing down. Normally, this aileron movement would push the plane into a roll, but since the plane was on autopilot as it circled, the automatic controls would have fought the roll, keeping the plane on a level flight without the pilot being aware of the struggle between the roll forces and the autopilot. Once the forces became too strong, however, the autopilot would suddenly pop off, and the plane would launch into the roll that the autopilot had been resisting. The pilot would have almost no time to respond. Although much of this is conjecture, the plane's flight recorder revealed that the autopilot did indeed switch off and the plane immediately went into a 70-degree roll, flying almost on its side. The pilots fought the roll, but without success. The plane turned completely upside down and hit the ground twenty-five seconds after the first sign that something was wrong.

The autopilot on that American Eagle flight may—or may not—have caused the plane's crash by masking a problem until it was too late. No one knows. But the FAA concluded that there are some times when automation is not worth the risk. It instituted a new rule for the ATR-42 and ATR-72 planes, insisting that the autopilot be turned off during freezing rain or drizzle.

The same sort of machine-centered automation that raises questions in aircraft also creates difficulties in manufacturing, Norman notes. In an earlier day, when people controlled the processes directly, they were in contact with the machines. They could see for themselves what was going on and react accordingly. They often developed a feel for the machines that allowed them to detect and diagnose problems almost by instinct. But with automation, the operators are removed from the machines, sometimes away in separate control rooms, and they lose touch.

> Whereas before they were physically able to keep an eye on things, often catching problems before they arose, now they are connected to the real world by second- or third-order representations: graphs, trend lines, flashing lights. The problem is that the representations people receive are most often those used by the machines themselves: numbers. But while the machines may use numbers internally, human operators should receive the information in the format most appropriate to the task they perform.

As a remedy, Norman suggests that systems be used to "informate" instead of automate. As the term implies, an "informated" system is one that provides information, preferably in great and varied detail. The operators use the system to monitor the machines under their control and to answer any questions they have about what is happening. The information is supplied in forms that make sense to and are easily digested by the operators. The emphasis is on making the operator more effective, instead of—as with automated systems—making the machine more effective.

> In an automated system, workers are relegated to the role of meter watchers, staring at automatic displays, waiting for an alarm bell that calls them to action. In an informated system, people are always active, analyzing patterns, continually able to find out the state of whatever aspect of the job is relevant at the moment. This involvement empowers the worker not only to become more interested in the job, but also to make more intelligent decisions. When the equipment fails in an informated system, workers are much more knowledgeable about the source of the problems and about possible remedies.

Creating such informated systems demands making a complete shift in engineering focus, away from the machine and to the operator, as well as studying the most effective ways to provide information to the people controlling the machines. It is an expensive process, one that is difficult to justify for such safe and mature technologies as coal-fired power plants. But nuclear plants are a different matter. Having digested the lessons of Three Mile Island, the nuclear industry is exploring designing its control rooms with the needs of the operators in mind.

In Windsor, Connecticut, for instance, ABB-Combustion Engineering has developed the System 80+, a next-generation light-water reactor designed for sale both abroad and—if the nuclear market ever returns—in the United States. Its reactor and cooling system have been simplified from the company's existing System 80 design, and a number of safety features have been added, but the major difference between the old and the new models lies in the control room. Gone is the bewildering array of indicators and gauges. In their place are computer screens designed to give operators the most important information about the plant in an easily readable form while at the same time allowing an operator to call up any extra details he wants.

In the past, the company says, an operator had as many as nineteen different gauges that offered information about the pressure of the reactor's cooling system; it was up to the operator to figure out which was the most appropriate for a given situation. The new system still takes as many pressure readings, but it doesn't dump everything on the operator. Instead, computers consolidate all the information and display a single number. If the operator wants data on a specific pressure readout, he can put a finger to a spot on the touch-sensitive screen and get the numbers immediately. Gone too are the hundreds of alarms signaling big problems and small with the same insistence. The alarms have been ordered according to their priority. The ones signaling an immediate threat to the plant are still insistent; the less important ones appear on a computer screen, and everything is prioritized so the operators know which need to be taken care of first.

The goal of the new design is to provide information in a way that is natural for the human brain, giving a big picture first but allowing an operator to zero in on the specifics of any system. Humans simply can't synthesize a big picture from hundreds of gauges and indicators, all of the same general size and shape, scattered around a large room in no particular order. By replacing that old machine-centered system with one designed with human efficiency in mind, the company hopes to make it far less likely that plant operators will make mistakes because they have been misled by their instruments.

Over the past decade, such human-factors engineering has become an increasingly popular topic in a number of engineering specialties, not just nuclear engineering. Prompted by the postaccident analyses of Three Mile Island showing how much operator performance hinged on the quality of their controls, engineers have begun to apply that lesson in chemical plants, aircraft, ships, and other technologies where human operators must respond quickly and accurately. It is becoming accepted that well-designed controls are important and that they should be set up to maximize the performance of a technology's operators.

This is a major shift in engineering philosophy, but the University of California-Berkeley political scientist Todd La Porte doesn't believe it goes far enough. He points out that in large technological systems, such as nuclear-power plants, organizational behavior may be just as important as the individual actions of operators. As we saw in chapter 8, well-run nuclear plants—or at least some of them—have an unusual organizational structure. In order to avoid mistakes during normal operation, a plant has a rigid hierarchical structure with a large number of regulations prescribing exactly how employees should do their jobs. But overlaid on this hierarchy is a second, less formal structure that encourages constant testing and reevaluation of accepted practices and that has constant communication across the lines of the hierarchy. It is this second structure that comes to the fore during times of stress, when a new situation arises and it's not clear what actions should be taken. This schizophrenic organizational structure seems to be a response to the dual demands of a nuclear plant—both flexibility and rigid control.

La Porte would like to see nuclear engineers think about such organizational factors when they are designing nuclear plants, just as they have learned to take human performance into account, but he hasn't made much headway. "Engineering and technical people don't think of what they do in social terms," he says, and so they don't realize what their designs demand from organizations—much as engineers two decades ago didn't realize what their designs demanded from plant operators. La Porte thinks it may be possible to redesign nuclear plants so that they don't place such a great strain on the organizations running them, but he's not sure how it might work—and he hasn't had much help from engineers. When he has asked nuclear plant designers, "Could you design a plant that didn't need central management?" they just don't get the point.

But one day they almost certainly will. As technologies become more complex, engineers will find it increasingly necessary to take human performance and, eventually, organizational factors into account in their designs. It will not be enough to design a machine that works in some abstract world where the people are no more real than they are in textbook engineering problems. The best designers will be part engineer and part social scientist. We are already seeing some of that with the human-factors engineers who design control rooms with people and their limitations in mind, but that is only a beginning.

CHEMICAL FIXES

Engineers in the chemical industry have been faced with a different set of pressures. Traditionally, a chemical plant has been designed with one thing in mind: to churn out its products as cheaply and efficiently as possible. Much of the energy of chemical engineers has gone into discovering and developing new lines of chemical synthesis that work faster, with fewer steps, with less expensive precursors, with less energy, and with fewer impurities.

In this single-minded pursuit of useful chemicals produced cheaply, the chemical industry has tended to treat safety and environmental issues as secondary, to be handled with add-ons. The main outlines of the chemical plant would be determined according to the dictates of efficiency and convenience, then the engineers would turn to these other matters. They would tack on safety systems to help prevent the accidental release of the toxic chemicals that appeared at various steps in the synthesis. They would add scrubbers onto the plant's stacks to clean up the gases being vented into the atmosphere, and they would create sewage treatment systems to clean up waste water being dumped into a nearby river, lake, or ocean. They would plan to ship other waste products to waste dumps and landfills, where they would be the problem of somebody else.

The result has been that, without really thinking about it, the chemical industry has reached the point where it generates large amounts of very undesirable stuff. Many of the chemicals it produces and handles in its plants are either toxic or carcinogenic: hydrogen cyanide, benzene, phosgene, and polychlorinated biphenyls (PCBs), to name just a few. In general, these tend to be intermediate products, like the MIC at the Bhopal plant, which are used in the manufacture of other, less dangerous chemicals, and which are not intended to leave the confines of the plant. Still, they put plant workers at risk and are a constant threat to escape the plant and injure people outside. Furthermore, chemical plants generate a great deal of waste that must be disposed of in waste dumps or incinerators, and much of it is equally unpleasant: organic solvents used in polymerization reactions, heavy-metal catalysts that are discarded after they lose their effectiveness, and so on.

At the same time, some of the end-products themselves have proved to be undesirable, at least in the large quantities that the modern world generates. The pesticide DDT is one example. Chlorofluorocarbons, or CFCs, are another. CFCs were widely used as refrigerants and as "blowing agents"—the gases used to create the bubbles in various foam products—until scientists realized in the 1980s that CFCs were damaging the ozone layer.

In response to such threats, the chemical industry has begun to look past its traditional preoccupation with product and process. In particular, chemists and chemical engineers are taking environmental and health issues into account explicitly in the design of a new product or process. The goal of this "green chemistry" is to avoid problems in the first place instead of cleaning them up afterward.

The most dramatic example of this is the complete replacement of CFCs with substances that do not threaten the ozone layer. Until the 1970s, CFCs seemed to be the perfect chemical success story. A family of related chemicals, they are nontoxic, nonflammable, relatively easy and cheap to manufacture, and they do a number of important jobs very well. CFC-12, better known by the brand name Freon, is an excellent refrigerant and was the almost exclusive choice for car air-conditioning units and refrigerators. CFC-11 was widely used as a blowing agent to make foams, both the rigid foams used for fast-food containers and the soft foams in furniture cushions and pillows; it is particularly well-suited for the insulating foams found in refrigerators and ice chests because it conducts heat poorly and does not leak out of the foam as easily as other gases. CFC-113 is a nearly ideal cleaning agent for computer circuit boards because its low surface tension and low viscosity allow it to seep into tiny spaces that other liquids, such as water, will not go. By the mid 1980s, the worldwide market for CFCs was $2 billion.

It would not last, however. In 1974, scientists first suggested that CFCs might be making their way into the upper atmosphere and damaging the ozone layer. Ironically, one of the CFCs' strengths as commercial products—their stability—was to blame. They are so stable that when released into the lower atmosphere, they do not break down, as many other chemicals do, but remain intact until they reach the stratosphere. There, ultraviolet radiation from the sun breaks the CFC molecules apart, releasing chlorine atoms—which destroy ozone. Early worries about the ozone layer led the United States and a few other countries to ban CFCs for use in aerosol sprays in the late 1970s. But it was not until 1985 that the discovery of the "ozone hole"—a large decrease in the normal amount of ozone found over Antarctica—convinced a majority of scientists that CFCs were a major threat. In 1987, twenty-four industrialized countries signed the Montreal Protocol, agreeing to cut CFC use in half by 1998. Six months later, after a NASA report indicated that the problem was even worse than thought, Du Pont announced that it would stop making CFCs altogether, phasing out production over several years. After this dramatic move by the world's largest CFC manufacturer, other companies followed suit. It would be the most sensational technical fix in history: to save the ozone layer, an entire family of chemicals would be killed off and replaced with nonthreatening alternatives.

The decision to do it was easy. Doing it would be harder. The chemical companies needed to find products that could do the same jobs as the CFCs, and do them as well or almost as well as the chemicals they were replacing. To replace Freon, for instance, Du Pont scoured the lists of known chemicals for something with similar properties, and eventually settled on a hydrofluorocarbon called HFC-134a. It too is nontoxic and nonflammable. Its boiling point is within a few degrees of Freon's, its molecular weight within 20 percent, and it is nearly as efficient a refrigerant. But HFC-134a must operate at higher pressures, so refrigerators and air-conditioning units built to run on Freon cannot use it. Appliance manufacturers and auto makers have had to redesign their equipment to function properly with the replacement. Furthermore, HFC-134a doesn't work with the lubricants used in the generation of refrigeration systems built for Freon, so new lubricants had to be developed. HFC-134a may have been the best choice to replace Freon, but it came at a cost.

Once Du Pont and other chemical companies had settled on CFC replacements, they were faced with the question of how to manufacture them. Making Freon is a relatively simple matter—one simply mixes carbon tetrachloride (CCl_4) and hydrofluoric acid at high temperature and pressure—and decades of experience and improvement had honed the commercial production of Freon into a highly efficient process. But in the late 1980s, no one knew the best way to make HFC-134a. Du Pont considered at least two dozen options, each of which took from two to four separate steps to reach the final product. Choosing one demanded extensive laboratory work, then building a pilot plant to discover all the things that weren't apparent in the lab, such as how long various catalysts would last, how well the system would handle the various impurities that inevitably appear, and what the effects would be of reusing the intermediate compounds in the process.

CFCs are almost gone now. Chemical companies in the industrialized nations have mostly phased them out, and although production continues in some developing countries, it is scheduled to die out in another decade. The change is particularly dramatic because an entire class of chemicals will have been killed off for the good of the environment. But other, less dramatic changes will be nearly as important.

In most cases, the problems of the chemical industry do not arise from the finished product itself. Instead, a threat to the environment or public health is more likely to turn up in one or more steps in the manufacturing process. A chemical plant is nothing more than a factory that transforms raw materials into a finished product, although both "raw" and "finished" are relative terms. Sometimes the output of a factory is truly finished and ready to be put to use—gasoline, for instance, or a plastic such as polyethylene, which is melted and molded into shape. Sometimes, however, entire plants are devoted to manufacturing intermediates that are consumed as "raw materials" by other manufacturing plants. In either case, the chemical plant transforms its starting materials with a series of chemical reactions that are, in principle at least, the same sorts of things that high-school chemistry students carry out in test tubes. The difference lies mainly in the scale and in the conditions under which the reactions are carried out.

Producing large amounts of chemicals at low cost demands that the plants be big enough to take advantage of economies of scale, and some plants stretch over areas the size of a medium-sized town. The effort to lower costs also leads chemical plant designers to look for the most efficient reactions—generally those that will produce the most end product with the least input in the shortest time. To speed up reactions, engineers will turn up the heat and pressure in the reaction vessel, or they will add catalysts—agents that help the reaction along but are themselves not chemically changed.

Byproducts appear unavoidably in many of the steps in a chemical synthesis. They may be as simple and harmless as water or carbon dioxide, or they can be complex, difficult-to-dispose-of acids or other chemicals. Sometimes the byproduct can be used elsewhere or transformed into a usable product, but often it must simply be gotten rid of. Wastes of various sorts must also be dumped: catalysts that have lost their efficiency, solvents in which reactions took place, impurities that have accumulated inside reaction vessels, and so on. Generations of chemists have not worried about such things, figuring that it was someone else's job to get rid of wastes or make sure a plant operated safely. Their job, as defined by the industry, has been simply to find the most efficient way to manufacture chemicals.

That attitude is slowly changing as chemical companies have been subjected to increasingly stringent environmental regulations and as a "green" mindset has gained a foothold in the chemical industry. Now chemical engineers are beginning to see their design goal not simply as maximizing yield but as finding a process with an acceptable yield that also minimizes pollutants and safety hazards. It is a dramatically different philosophy for the chemical industry.

There are numerous lines of attack that chemical engineers have adopted in this new struggle. One is the approach taken by Du Pont with MIC—reducing the amount of a hazardous material by finding a way to manufacture it only as needed, so that as soon as it is produced it is consumed in the next stage of the process. No storage, no transportation, and no risk of large-scale release. But such a strategy is feasible in relatively few situations, and most problems demand a different tack.

In terms of volume, organic solvents comprise one of the biggest environmental problems of the chemical industry. Many chemical reactions take place in huge vats of these solvents, chemicals such as benzene, carbon tetrachloride, and CFCs. The solvent dissolves the reactants and provides an environment in which they can combine; later, the reaction product is separated from the solvent, which is used over and over again. Unfortunately, many of these solvents are toxic, carcinogenic, or threaten the environment in some other way, and disposing of them has become a major headache. Consequently, a number of chemists have begun looking for other solvents in which to carry out reactions. One possibility is carbon dioxide in its supercritical form—a strange state between gaseous and liquid. Another possibility is water. One chemist has made polymers in water instead of the usual organic solvent. Polymerization, the joining of single molecular units into long chains, is the key step in making plastics of all sorts and is thus a staple of the chemical industry. The trick to successful polymerization in water is to find the right catalyst. The standard catalysts won't work in air or water, but a new generation of catalysts could allow chemists to cut down on organic solvents.

Catalysts themselves are candidates for reworking. Many of the catalysts used today are toxic metal compounds containing such nasty ingredients as mercury, lead, and cadmium. Other catalysts are strong acids. One approach to replacing such catalysts involves dyes, which absorb visible light and give up electrons, triggering a reaction in other molecules. Another is to work with zeolites, or "molecular cages." Zeolites have been known and used for decades, but only recently have researchers discovered how to make them to order, opening up the possibility of catalyzing almost any reaction with zeolites and producing much less waste than with organic solvents.

Still another angle is to put microorganisms to work. It is possible to genetically modify a bacterium such as *E. coli* to produce commercially important chemicals. Once that is done, colonies of the bacteria are grown in large vats and fed the appropriate chemicals, and they naturally turn out the desired product, which is then harvested from the bacteria. One researcher, John Frost of Purdue, has reported producing a number of important chemicals in bacteria, including hydroquinone, benzoquinone, and adipic acid, which is used in making nylon. These three substances are normally produced from benzene, a carcinogen that is one of the most worrisome of the chemical industry's polluters. Other researchers have looked to plants as the more likely bioreactors. Modifying such plants as corn or potatoes, instead of bacteria, has the advantage that farmers have been working for centuries to maximize crop yield; growing bacteria commercially is still a new and mostly undeveloped skill.

No matter which of these approaches is taken, it's likely that yields will go down and costs will go up. Although some researchers claim that certain environmentally benign processes will actually be less expensive than the methods they replace, that won't happen often. The old methods were chosen for their efficiency with little or no regard for the environmental costs. It's unlikely that environmentally friendly replacements will be found with higher yields. Yields aren't everything, however. It's possible—some would say probable—that the new methods will indeed be cheaper if all the costs are taken into account, including clean-up costs and the costs of pollution to society in terms of health and aesthetics, not just for our generation but for generations to come.

So far, although there has been a great deal of talk about the new approach to chemistry, actual adoption by the chemical industry has been slow. It has been difficult, for instance, to convince chemists to change their priorities and take environmental and safety factors into account in their research. For years environmental chemistry had a reputation among chemical researchers as a "soft science." Its practitioners were seen as interested mostly in tracking pollution, detoxifying hazardous chemicals, and recycling materials—none of which got much respect in a culture where "real chemists" delved into the nitty-gritty of reaction dynamics and catalysis. And the few chemists who did want to look for environmentally friendly processes had a tough time getting funding approved by peer-review panels.

But the situation is changing. The realization is spreading that environmental chemistry is more than pollution cleanup, and that designing chemical reactions to minimize waste is just as "hard" as designing reactions to maximize yield. Funding is also improving. In 1992 the National Science Foundation and the Environmental Protection Agency started a program to pay for research into environment-friendly chemistry, and chemical companies themselves are carrying out similar research as the realization sinks in that, in the long run, it will cost less to make their products with a minimum of waste and risk than to keep the old methods and clean up after them. Finally, although it's harder to measure, it seems that the chemical industry is changing its attitude on pollution control from foot-dragging resistance to an active, if not eager, commitment. The Chemical Manufacturers Association, the main industry trade group, has trumpeted this new approach with a program called Responsible Care. Although the program is at least partly a public relations stunt, it does appear to reflect a change in philosophy toward designing plants with environmental goals in mind.

Even with the changes in attitude among chemists and chemical company managers, it will be decades before the chemical industry reworks itself completely. Not only must chemists find new methods for synthesizing hundreds of chemicals, but those methods must also then be scaled up from laboratory demonstrations to industrial processes. Many, if not most, of the new methods will be less efficient than the ones they are replacing, so chemical engineers will have to be extra diligent in maximizing their yield, and chemical companies will have to weigh the environmental benefits against the economic disadvantages. Furthermore, the industry has hundreds of billions of dollars invested in plants and equipment, and it can afford to retrofit only a small part of that. Most of the changes will come as new plants are designed and built according to the new philosophy. Still, all the evidence today implies that it will happen. And when it does, it will be the most extensive technical fix in history: the remaking of an entire industry to reflect the changing needs and desires of the world around it.

INHERENT SAFETY

When Du Pont chemists searched for an alternate way to synthesize methyl isocyanate, the villain of Bhopal, they were applying an old but often ignored principle of engineering: build safety in, don't tack it on. The Bhopal plant had relied on workers and add-on safety systems to prevent accidents or, if an accident occurred, to keep it from threatening anyone outside the grounds. By contrast, a plant that exploited the Du Pont synthesis method would be "inherently safe": even in the worst possible accident, there would not be enough methyl isocyanate in the plant to be dangerous to the public.

Such inherent safety is the ultimate technical fix. It eliminates the need for complex and expensive safety devices and procedures while providing a level of protection that engineered safety features cannot. Although its exact definition depends on whom you ask, the essence of inherent safety is to avoid risk rather than to engineer systems to protect against it. If an architect wishes to reduce the risks of living in a house and notices that many injuries are caused by falling down stairs, he has various options. He can install railings and extra padding beneath the carpet on the steps. He can design less dangerous staircases by adding landings at the halfway points or making the angle of ascent less steep. Or he can build a one-story house.

Many technologies cannot be made inherently safe, of course. Dams, commercial jetliners, natural gas pipelines, ocean tankers carrying oil or chemicals—these all seem to entail unavoidable risks. As long as large volumes of water are collected artificially in one place or shiploads of chemicals are carried across the ocean, accidents will pose a hazard to the environment or human health. And even when it's possible to make a technology inherently safe, it's usually expensive. Inherent safety must be designed in from the start—once a two-story house has been built, it's too late to decide to do away with stairs. So converting an existing technology into one that is inherently safe generally means starting from scratch and repeating the whole development process. Not surprisingly, there are few examples of large-scale technologies re-engineered to be inherently safe.

Still, the logic of inherent safety makes it attractive for a number of today's complex, risky technologies. If Charles Perrow's arguments, described in chapter 6, are correct, then accidents are inevitable in such complex systems as nuclear power plants, chemical plants, and genetic engineering technology. And even if it is possible to build and operate such systems safely, their complexity—and the potentially infinite number of things that can go wrong—make it difficult to convince the public that they are safe. Either way, removing the risk altogether is a tantalizing goal.

Nowhere is the allure of inherent safety stronger than for nuclear power, and that appeal has created a sharp debate within a U.S. nuclear industry trying to make nuclear power once again acceptable in this country. The industry mainstream—the nuclear utilities, the reactor makers, and the other major suppliers—has decided that while inherent safety is a laudable goal, it is not now a feasible one. It would be too expensive, and it might raise questions about the more than 100 nuclear plants operating in the country now, which are not inherently safe. Of the three next-generation nuclear reactors now under development in the United States—one by General Electric, one by Westinghouse, and one by ABB-Combustion

Engineering—none relies on inherent safety. The most daring safety innovation is to replace some "active" safety features, such as pumps, with "passive" ones, which rely on gravity or natural convection to operate. For instance, large tanks of water can be placed above the reactor so that in case of an accident, the water can be released to flow naturally around the reactor core, without depending on pumps. Once the core has been flooded, natural convection will maintain a steady circulation of water, bringing new, cool water down and allowing the water warmed by the core to flow away and release its heat to the atmosphere.

But some in the industry don't think this will be enough. As we have seen, in the 1960s and 1970s, nuclear experts assured everyone that reactors were safe and that a meltdown was an "incredible" event. Then came Three Mile Island and Chernobyl. The result, argues MIT nuclear engineer Larry Lidsky, is that "it will no longer be sufficient for the public and the utilities to be assured that the reactor designers have computed, very carefully indeed, the probability of accident and have found it to be 'vanishingly small.' The public has, with good reason, lost all faith in such pronouncements." Unless something dramatic is done to restore the public's faith in nuclear power, it simply won't be possible to build new nuclear plants in the United States or in a number of European countries. The solution, say Lidsky and others, is to develop an inherently safe reactor—one that, by its nature, cannot release radioactivity even in a major accident. Surprisingly, such reactors are possible, and over the past twenty years, a number of nuclear mavericks have offered designs for such machines.

The pioneer in the field was Kåre Hannerz, a nuclear engineer with the Swedish company ASEA (which later merged with the European conglomerate Brown-Boveri to form the giant multinational corporation ABB). He came up with the idea in the early 1970s, he says, when the antinuclear movement began to make inroads in Sweden. "It occurred to me that if nuclear power were to survive, you would have to come up with a technology that was less vulnerable to human mistake." And so he began work on a type of reactor that would be virtually immune to any sort of mishap, from operator error and equipment malfunction to deliberate sabotage or a terrorist attack. Strictly speaking, the design he came up with is not an inherently safe reactor, merely a conventional light-water reactor with a very clever passive safety system, but it is difficult to imagine a scenario in which his reactor could overheat and melt down.

Hannerz's reactor, which he called PIUS for "Process Inherent Ultimate Safety," was designed to avoid both of the two major hazards that threaten a nuclear reactor: a runaway chain reaction or a meltdown of the reactor core. All reactors operate at the edge of a runaway reaction. The fissioning uranium atoms must produce just enough neutrons to keep the reaction going, but no more. Too few neutrons, and the chain reaction peters out;

too many, and it quickly runs out of control. This is what happened at Chernobyl. In just five seconds, the reactor's power jumped to 500 times its design value, melting the fuel, setting the graphite moderator on fire, and setting off an explosion in the hydrogen gas created by a reaction between the water coolant and the cladding—the metal jacket covering the fuel. In most reactors the chain reaction is maintained at the correct level by control rods made of a material that absorbs neutrons. When pushed into the core, they slow the reaction down by swallowing some of the neutrons that would otherwise trigger fissions of uranium atoms. At Chernobyl, the operators had pulled the rods too far out as part of a test of the reactor's emergency core-cooling system.

Most modern reactors have an additional, inherent protection against runaway chain reactions. They are designed so that any time the chain reaction starts to speed up, conditions automatically change in such a way as to slow the reaction back down. The change is not dependent on equipment or operators. Suppose, for example, that the control rods are accidentally pulled out too far in a light-water reactor. As the chain reaction speeds up, it heats up the uranium fuel in the core, which in turn heats up the cooling water flowing around the core. This causes the water to expand slightly, decreasing the number of water molecules in a given volume. But since the water is moderating the chain reaction—slowing down the neutrons and allowing them to fission uranium atoms more efficiently—the decrease in the density of the water slows down the chain reaction. This negative feedback gives the reactor an inherent stability which makes it relatively immune to the threat of a chain reaction getting out of control. By contrast, the reactor at Chernobyl was inherently unstable. It used water only as a coolant, not as a moderator, and the main effect the water had on the reaction was to absorb neutrons—that is, the water performed a function similar to that of the control rods. When the fuel heated up, it caused the water to expand and absorb fewer neutrons, causing the reaction to go faster, heating the fuel more, and so on in a cycle that quickly spiraled out of control. Only the control rods could have kept the chain reaction in check.

When Hannerz created PIUS, he based it on a conventional pressurized-water reactor with its inherent defense against a runaway chain reaction. His plan was to combine that with an inherent protection against a core meltdown—something that conventional reactors don't have. If an accident shuts down the flow of cooling water around a reactor's fuel, safety systems quickly slam control rods into the core and shut down the chain reaction, but that doesn't get rid of the heat in the core. Even after the chain reaction halts, radioactive byproducts in the fuel continue to decay and release enough heat to melt the core unless the heat is carried away. Thus light-water reactors are equipped with emergency core-cooling systems that will circulate water through the core if the primary cooling system breaks down.

It looks good on paper, but, as the accident at Three Mile Island showed, safety measures that depend on pumps and valves operating correctly and on plant personnel responding appropriately don't always go as planned.

Even before Three Mile Island, Hannerz says, he had decided that "we could not depend on equipment and institutions." What was needed was a way to guarantee that the reactor would shut down safely no matter what type of accident struck the plant. "It suddenly occurred to me that you could do it by establishing a nonequilibrium system where, if you don't add mechanical energy to the system, it shuts itself down." It was an engineering insight that has changed the way many nuclear designers think about their jobs, even though the particular system that Hannerz devised has proved somewhat difficult to make into a practical machine.

PIUS consists of a normal pressurized-water reactor—core, control rods, primary cooling system, and so on—inside a large concrete "swimming pool" that is filled with cool, borated water. In case of an accident, the borated water will flood the chamber around the reactor core. The boron atoms in the water absorb neutrons, automatically shutting down the chain reaction. Meanwhile, the water carries heat away from the core by natural circulation, preventing meltdown. All this happens naturally and automatically, without valves or pumps or human intervention. And in fact, it's impossible to prevent the safety system from kicking in if the reactor strays too far from its normal operating conditions.

When PIUS is operating normally, cooling water is pumped through the reactor to a heat exchanger and back to the reactor just as in a standard pressurized-water reactor. What makes PIUS radically different is that its cooling system is connected to the pool of borated water by two pipes, and both of those pipes are completely open. The only thing that keeps the borated water from flowing into the cooling system and shutting down the reaction is the large density difference between the hot cooling water and the cool borated water. Just as oil floats on water because it is lighter, hot water floats on cool water, and PIUS is designed so that the cool borated water must flow up through "density locks" to get into the chamber around the reactor. As long as the reactor is operating normally, a hydraulic balance keeps the cold water from flooding the reactor. But if the reactor gets too hot or too cold, or if the pumps circulating the cooling water slow down or speed up too much, that balance is destroyed and the borated water comes rushing in and halts the chain reaction.

Hannerz's reactor operates on a knife edge. In a sense, its "normal" state is to be shut down, with borated water flooding the reaction chamber. It takes active intervention and control to maintain the conditions necessary for a chain reaction. A conventional light-water reactor is just the opposite. Once a chain reaction begins in its core, that becomes the normal state, and it takes active intervention and control to keep the chain reaction under control or to shut the reactor down safely. In a PIUS reactor, the worst consequence of an equipment malfunction or operator error is an unexpected shutdown. In a conventional light-water reactor, the consequence may be a runaway chain reaction or a core meltdown.

PIUS's strength is also its main weakness. Because it can only operate under a very narrow set of conditions, some critics contend that it would be inherently unreliable, shutting down every time things got slightly out of balance. We may never know. Hannerz retired in 1992, with the PIUS program needing several hundred million dollars to build a large-scale prototype to test the density locks and other equipment. The money will not come from ABB, which has committed itself to improving its line of conventional light-water reactors. The Italian government had indicated an interest in PIUS, but, as this book was written, nothing had been decided.

Even if Hannerz's PIUS machine is never built, he has inspired a number of other nuclear engineers to design reactors that will shut down safely in the event of an accident with no help from either operators or equipment. At least two such inherently safe reactors have been designed—one that is cooled by liquid metal, the other by gas—but neither has been built. At first glance they seem radically different machines, but the inherent safety in both springs from similar principles.

To avoid a runaway chain reaction, both machines take the same tack as light-water reactors. Each has a "negative temperature coefficient": when the temperature in the core increases, the chain reaction slows down. The Advanced Liquid Metal Reactor, designed by General Electric, has a core of metal fuel rods immersed in a large pool of liquid sodium. The sodium coolant is continually circulated through the core by electromagnetic pumps that have no moving parts. If the power increases, the metal fuel heats up and expands, causing the uranium atoms in the fuel to move slightly farther apart, slowing down the reaction. Even if the control rods are completely removed, the chain reaction in the core will increase only to a power level somewhat higher than normal and then stay there. Similarly, the Modular High-Temperature Gas-Cooled Reactor, or MHTGR, will remain stable with no control rods in the core. In each reactor, the control rods are used only to fine-tune the reaction rate, not as safety devices.

To prevent a core meltdown in case of a loss-of-cooling accident, both reactor designs keep the core relatively small. Many reactors today have large cores and can generate 1,200 to 1,300 megawatts of electricity. The Advanced Liquid Metal Reactor and the MHTGR would each generate less than 200 megawatts of electricity. Although this may make it difficult to get the economies of scale that come with large reactors, the smaller core provides inherent safety: there is no need for backup safety systems to remove residual heat from the core in case of a loss of cooling, since there simply is not enough residual heat generated to threaten a meltdown. The Advanced

Liquid Metal Reactor is designed to be buried in the ground, and, if its electromagnetic pumps fail, the heat from the pool of sodium is transferred to the surrounding earth by a combination of radiant heating and conduction. MHTGR also relies on transferring heat from the core to the earth around the reactor, but with a twist. Its fuel consists of tiny spheres of uranium oxide, each less than a millimeter across, sealed inside several layers of protection designed to keep the fuel from leaking out at temperatures up to 1800 degrees Celsius. Even if all the coolant is lost, the temperature of the core cannot get high enough to damage the fuel and let radioactivity escape.

The inherent safety of the two designs is not merely theoretical. In April 1986, a test on the Experimental Breeder Reactor-II in Idaho mimicked what would happen to the Advanced Liquid Metal Reactor in a major accident. The cooling system was shut down completely, so that the heat from the reactor core was no longer being transferred to the steam-generating system. The result: nothing happened. The chain reaction shut down, and within minutes the temperature of the liquid sodium around the core had settled down to a high but comfortable level. The safety of the MHTGR has been demonstrated by tests on a similar but smaller reactor in Germany. Once again, when the flow of coolant was stopped, the reaction halted and the heat of the fuel was dissipated by passive cooling.

In short, it is now possible to build nuclear reactors that will be safe in the face of anything short of a direct hit by a missile. So far, however, the nuclear industry has shown little interest, preferring improved versions of current light-water models. Why? Cost is one big reason. The inherently safe designs need extensive—and expensive—development before they are ready for the market. And because the inherently safe reactors are much smaller than normal, critics say that they'll be more costly since they can't match the economies of scale of their larger competitors. But supporters of the inherently safe reactors contend that by linking them together, with several running one turbine, it will be possible to regain some of the economies of scale. This, combined with getting rid of the need for costly safety systems and backups, could bring the costs of the plants down to where they are competitive with conventional plants. Indeed, many believe that the MHTGR could eventually generate electricity much more cheaply than current light-water reactors, particularly if the hot gas produced in the reactor were to be fed directly into a high-efficiency gas turbine. But all of this is just guesswork based on one set of assumptions or another, and, as the industry learned with the light-water reactor, such estimates can blow up in your face.

Furthermore, the mainstream nuclear industry doesn't believe that inherent safety offers any real advantage over the traditional, add-on safety approach. Reactor manufacturers such as GE and Westinghouse say that they have learned enough from their decades of experience with light-water technology to build new reactors that are just as safe as the inherently safe machines but which will depend on well-tested designs and materials. Why step into the unknown when there's nothing to gain?

The backers of inherently safe designs respond with a couple of reasons. One might be termed the "Bhopal argument." If nuclear power is once again to become widely accepted, then reactors will likely be built in countries whose technological sophistication is rather low and whose infrastructures may not provide everything needed to keep a nuclear plant in tip-top shape. Regulatory supervision might not be as close as it is in industrialized countries, and a cash crunch might cause the company operating the plant to forgo routine maintenance or perhaps even shut down some "nonessential" safety systems. As long as the safety of the plant depends on people, institutions, and equipment, there can be no guarantees. The only solution is to sell nuclear plants that are immune to such problems as poor maintenance, carelessness, and even intentional damage.

Perhaps industrialized nations can operate nuclear plants safely—although the history of the U.S. nuclear industry does include such whoppers as operators falling asleep in control rooms and a reactor being installed backwards—but will the public believe it? That's the second issue that supporters of inherently safe reactors raise. There is no way to prove that standard light-water reactors can be operated safely, even if they are redesigned to be less complex and to include various passive safety devices. The nearest thing to proof that the nuclear industry can offer is probabilistic risk assessment, and even the experts don't agree on how believable that is. So people like Larry Lidsky at MIT, who has spent the past decade working on the MHTGR, argue that the only way the U.S. public is likely to accept more nuclear power plants is if their safety can be demonstrated simply and conclusively—by, for instance, disconnecting all the controls, cutting off the flow of coolant, and having the reactor come to a safe shutdown by itself. The nuclear industry, Lidsky charges, is wedded to one technology and doesn't understand that the vote is in on that technology, and it has lost. "If there is to be a second nuclear era," he says, "we must find a way to fit nuclear power into the world as it is, not the way we want it to be."

But is the public really more likely to accept passively safe reactors than their evolutionary cousins? Many doubt it. Much of the opposition to nuclear power stems from a general unease with the technology, caused in part by its kinship with nuclear weapons, and passively safe reactors will do little to allay that. And even if reactors are built to be passively safe, the problem of nuclear waste—perhaps the largest negative to nuclear power in the public's mind—will remain. So it's not surprising that the nuclear industry sees little to gain from changing to a radically new, unproven technology.

The debate over inherent safety in the nuclear and chemical industries offers a lesson for future technologies: Given the difficulty in switching to inherently safe designs once a technology is mature, it may be a good idea to plan for inherent safety from the beginning, at least for those technologies that have the potential for great harm. Every technology starts small. A century ago, only a few automobiles—electric, steam powered, and internal combustion—puttered along the roads. Seventy-five years ago, commercial aviation was first getting off the ground. Fifty years ago, only a few small nuclear reactors existed anywhere in the world. It is at the beginning, before its momentum builds, that a technology is most susceptible to direction. As we've seen, it is impossible in those early formative years to know which technological choices are the best, but it may not be impossible to figure out which choices allow for inherent safety and which demand complex, add-on safety systems.

Of course, engineers in the early days of a new technology generally have more pressing things to think about than inherent safety—such as finding a design that does the job at a reasonable cost and that can be manufactured as soon as possible, so that the competition doesn't corner the market. But perhaps, if the lessons of history sink in, we may learn to pick out ahead of time those few nascent technologies that could pose major hazards in the future. Today, for example, it seems that genetic engineering may be one such threat. And having identified the embryonic technologies that could become dangerous, perhaps we could make our technological choices with safety in mind. Theoretically, it would have been possible fifty years ago to demand that nuclear engineers consider only reactor designs that were inherently safe. Or chemical companies could have required their chemists to find a way to synthesize methyl isocyanate that did not involve storing tons of it at a time. Practically speaking, of course, it would have never happened—at least not fifty years ago, and probably not even today.

But if technology continues to change as it has, growing in power and complexity, and if society continues to demand less and less risk from technology, there may well come a time when safety considerations predominate in the development of a technology. At that time, inherent safety may become standard engineering practice. But until then, it may continue to take a Bhopal or a Three Mile Island to make engineers think seriously about inherent safety.

THE GRAND EXPERIMENT

The technical fixes discussed above—human-factors engineering, green chemistry, inherent safety—all are attempts to deal with issues that have not traditionally been concerns of the engineer, and they come in response to the growing influence that non-technical factors have exerted on technology. When Larry Lidsky became interested in the MHTGR, for instance, it was not due to any technical superiority that the new gas-cooled reactor would have over existing light-water reactors. Instead, he saw the MHTGR as a way to revive the nuclear option in the face of public fears about the safety of nuclear power. Lidsky argues that public acceptance is essential to the commercial viability of nuclear power and that the public will not accept a nuclear technology that cannot be proven safe. Probabilistic risk assessment may be good enough for engineers, but it won't cut it with the public.

But technological problems demand technological solutions, and as impressive a machine as the MHTGR may be, it is still just a technical fix. If Lidsky is correct, it might solve one nontechnical problem—the public's distrust of nuclear power—but it leaves many others unaddressed, and for that reason the nuclear industry has no interest in the MHTGR. There are simply too many other issues that must be resolved before the reactor could become a technological solution. Consider just the regulatory process: Getting approval to build an MHTGR would be a long, hard road, and convincing the NRC that all the usual safety systems aren't needed on an MHTGR would be even tougher. Yet unless these systems are left off, the reactor won't be economically competitive with larger light-water reactors.

So instead, the U.S. nuclear industry has settled on a very different strategy. Recognizing that many of its problems have their roots in political and cultural factors, the industry has developed an ambitious plan to rework the technology and, more importantly, to change the environment for nuclear power. The campaign is unparalleled in the history of technology—no other industry has attempted to modify the entire sociopolitical context in which it operates.

The effort began soon after the accident at Three Mile Island. The outlook for nuclear industry was, as we've seen, bleak even before the accident. But the meltdown threatened to make things much, much worse. It guaranteed more regulations, more public opposition, and more uneasiness among investors about the wisdom of putting money into nuclear power.

It also galvanized the nuclear industry as nothing had before. Once the shock of TMI had worn off, the more objective members of the nuclear fraternity accepted that, at best, no new nuclear plants were likely to be ordered for the next decade. They could either curse the hiatus as lost time or accept it and put it to good use. John Taylor, who ran Westinghouse's commercial reactor business, was one who chose the latter: "I became convinced that we should begin to use those lessons [from TMI] to frame a system for the future. Fundamental things were wrong and had to be corrected, and if the market [for electrical generation] came back before the second generation was ready, nuclear power wouldn't play a role." Before the partially melted core at Three Mile Island had cooled, Taylor and others in the nuclear industry were beginning to map out the rebirth of nuclear power. It was a remarkable, perhaps unprecedented situation: a major industry given a forced sabbatical in which to remake itself in light of the lessons it had learned to that point.

What would the chastised industry do? The first priority, Taylor recalls, was to patch up the existing system. Fundamental changes could come later. The investigations that followed TMI had revealed an industry that underestimated the difficulty of running reactors safely and efficiently. As described in chapter 8, the industry's response was to set up the Institute of Nuclear Power Operations to set standards and to monitor utilities to make sure they were meeting them.

With that under way, the next step was to evaluate the nuclear technology itself. What should the next generation of reactors look like? Were the problems with light-water reactors serious enough that they should be scrapped in favor of another type? A survey of nuclear utilities found that they considered the basic design "good enough," Taylor says. The reactors had two basic problems, both stemming from their pell-mell commercialization in the late 1960s and early 1970s. First, the designs were too complex; they needed to be simplified. And second, the reactor designs needed to be standardized so that both utilities and reactor manufacturers could learn from experience and not start nearly from scratch with each reactor.

This would be the extent of the technical component to the industry's technological solution. Over the next several years the utilities and reactor makers cooperated to map out a next generation of reactors, producing a set of specific requirements that any new designs must meet. They worked closely with the NRC to make sure that the new designs would be approved quickly and easily. To date, three manufacturers have provided designs for next-generation reactors. Two are "evolutionary" designs, large reactors very similar to current state-of-the-art models but with special attention paid to simplification and safety: General Electric has weighed in with its Advanced Light Water Reactor, a pair of which have now been built in Japan, and ABB-Combustion Engineering offers its System 80+. The third reactor type, Westinghouse's AP-600, is an "advanced" design—a smaller reactor (600 megawatts) with passive safety features. All three designs are clearly superior to anything built before—simpler, safer, and likely to be more economical to operate—but they are very much part of the light-water lineage. No revolution here.

Instead, the nuclear industry had decided that the most important parts of the technological solution were organizational, institutional, political, and social. With the technical requirements for a new generation of reactors written and three designs in progress, the industry set down its "Strategic Plan for Building New Nuclear Power Plants." Developed by the

Nuclear Power Oversight Committee, a high-powered group consisting of the heads of more than a dozen utilities, reactor manufacturers, and other nuclear organizations, the plan is a remarkably candid assessment of what it will take to build a new nuclear power plant in the United States. Reading the plan leaves no doubt that technology has moved far beyond the traditional concerns of the engineer.

The plan lists fourteen "building blocks," each a goal that must be met in order to revive nuclear power. First, if the next generation of nuclear power is not to be stillborn, the current generation must be kept healthy. To do that, the plan lists four goals: improving the economic performance of existing plants, generally by managing them better; solving the problem of where and how to store spent nuclear fuel; assuring that plants will continue to be able to get rid of their low-level nuclear wastes; and guaranteeing a stable and economical fuel supply for the reactors.

A second set of goals deals with regulatory issues. Before utilities will invest in new reactors, they must be assured that the regulatory nightmares of the past will not be repeated. So the strategic plan calls for the NRC's licensing and regulatory procedures to be reformed so that they are predictable and stable, and for the NRC to approve the set of reactor design requirements developed by the nuclear utilities on which the reactor manufacturers were basing their new machines.

A third group of goals focuses on what must be done to clear the way for a utility to actually build a nuclear plant. The reactor manufacturers must finish their "first-of-a-kind" engineering on the new designs, laying out the plans in great detail. Next, the NRC should "pre-certify" the reactors, so that when a utility decides to build a nuclear plant, it won't have to get the specific plans for that plant approved but can simply show that its plant will follow the pre-certified design. In this way, the standardization of design would pay off not just in building and operating the plants, but also in shepherding them through the regulatory process. Potential sites for nuclear plants must be approved. And in order to keep the plants standardized over their operating lives, the plan calls for developing industrywide standards for maintaining the new nuclear plants. In this way, two nuclear plants that were identical when they were built will still be essentially identical after decades of operation.

The last set of goals spotlights the social, political, and economic environment for nuclear power. Public acceptance of nuclear power must be improved. Utilities must re-examine the financing, ownership, and operation of nuclear plants to guarantee that nuclear power is profitable. Public utility commissions and other state regulatory agencies must be convinced to provide a predictable environment for nuclear plants. And the federal government must be supportive of nuclear power with its laws, regulations, and programs.

For more than seven years, the industry has been pursuing this agenda. Some of the building blocks are already in place. The NRC, for instance, has revamped its licensing procedures to allow for the pre-licensing of nuclear plants. Other building blocks, such as a resolution of the high-level waste issue, are more problematic. So far, there is no indication that a U.S. utility is seriously considering ordering a nuclear plant (although it would certainly be a closely kept secret if one were), but the nuclear industry believes that its time will come again. If nothing else, the threat of global warming may eventually force the United States and other countries to burn less coal, oil, and gas, and look for alternative ways to generate electricity. When that day arrives, the nuclear industry plans to be ready.

The difference from the development of the first generation of nuclear power could not be more striking. Forty to fifty years ago, nuclear power was simply the next step in a never-ending march of technology, not a time for soul-searching and questions about humanity's future. No one really doubted that nuclear power would be a cheaper, better way to generate electricity; the only question was how long it would take to develop. No one questioned the goal of cheap, plentiful power; we needed it to run all the other marvelous technologies being invented. No one thought to ask the public's opinion of nuclear power; the decisions would be made by panels of experts. No one worried about whether the reactors could be made safe enough; engineers could do anything you asked. No one questioned whether nuclear waste could be stored safely; the engineers could handle that, too. No one spent time thinking about regulation; since government and industry worked hand in hand, it wouldn't be a problem. No one bothered to ask what sort of organization would be needed to deal with such a complex, risky technology; a bureaucracy was a bureaucracy.

For better or worse, technology has changed. Our days of innocence, when machines were solely a product of larger-than-life inventors and hardworking engineers, are gone. Increasingly, technology will be a joint effort, with its design shaped not only by engineers and executives but also psychologists, political scientists, management theorists, risk specialists, regulators and courts, and the general public. It will not be a neat system. It is probably not the best system. But, given the power and complexity of modern technology, it is likely our only choice.

Notes

Introduction: Understanding Technology

page

3 **Alfred P. Sloan Foundation grants.** For the background of the program and a list of the books in the technology series, see Karen W. Arenson, "Grants by Foundations Help Technology Books Make It to the Shelves,"*New York Times*, August 21, 1995, p. D5.

. . . focus on one particular technology. There was one exception to this rule. The series planned one book to be an anthology of writing and pictures concerning technology in general.

4 **Some next-generation nuclear plants . . .** General Electric, in a joint venture with Hitachi, Toshiba, and Tokyo Electric Power, was building two advanced boiling water reactors in Japan at the time I began working on the book. See "Japan Begins Construction of Second Advanced Plant," *Public Utilities Fortnightly* 129 (May 1, 1992), pp. 39–40.

As this book was being finished, one of these reactors had already been put in service, and the second was scheduled to be done by the end of 1996. They are the world's first next-generation nuclear reactors.

page

Nuclear Choices. Richard Wolfson, *Nuclear Choices*, MIT Press, Cambridge, 1991.

Nuclear Renewal. Richard Rhodes, *Nuclear Renewal: Common Sense About Nuclear Energy*. Viking Penguin, New York, 1993.

5 **"sociotechnical system."** I borrow the distinction between technology and sociotechnical system from Tom R. Burns and Thomas Dietz, "Technology, Sociotechnical Systems, Technological Development: An Evolutionary Perspective," in Meinolf Dierkes and Ute Hoffmann, eds., *New Technology at the Outset: Social Forces in the Shaping of Technological Innovations*, Westview, Boulder, CO, 1992, pp. 206–238. According to Burns and Dietz, technology is "a set of physical artifacts and the rules employed by social actors to use those artifacts" (p. 208). A sociotechnical system "includes rules specifying the purposes of the technology, its appropriate applications, the appropriate or legitimate owners and operators, how the results of applying the technology will be distributed and so on" (p.209).

Others choose different definitions. The historian Thomas Hughes, for instance, uses "technology" where Burns and Dietz would

page

say "sociotechnical system," and refers to the "technical component of a technology" where Burns and Dietz would merely say "technology." See Thomas P. Hughes, "Technological History and Technical Problems," in Chauncey Starr and Philip C. Ritterbush, eds., *Science, Technology and the Human Prospect*, Pergamon, New York, 1980, pp. 141-156.

5 **a number of engineers have written books . . .** A good example of this approach is Bernard Cohen, *The Nuclear Energy Option*, Plenum, New York, 1990.

6 **the steam-powered automobile engine . . .** Charles McLaughlin, "The Stanley Steamer: A Study in Unsuccessful Innovation," *Explorations in Entrepreneurial History* 7 (October 1954), pp. 37-47.

7 **how pencil designers . . .** Henry Petroski, *The Pencil*, Alfred A. Knopf, New York, 1989, p. 207.

technological development as a good thing. Merritt Roe Smith, "Technological Determinism in American Culture," in Merritt Roe Smith and Leo Marx, eds., *Does Technology Drive History?* MIT Press, Cambridge, MA, 1994, pp. 1-35.

the past century has seen a dramatic change . . . Ronald Inglehart, *The Silent Revolution: Changing Values and Policy Styles Among Western Publics*, Princeton University Press, Princeton, NJ, 1977, p. 3.

8 **momentum.** Hughes, "Technological History and Technical Problems," p. 141.

9 ***Challenger* accident.** Maureen Hogan Casamayou, *Bureaucracy in Crisis: Three Mile Island, the Shuttle* Challenger *and Risk Assessment*, Westview, Boulder, CO, 1993, pp. 57-85.

As early as the 1960s, . . . Alvin M. Weinberg, *Nuclear Reactions: Science and Trans-Science*, American Institute of Physics, New York, 1992. See also Alvin M. Weinberg, *The First Nuclear Era: The Life and Times of a Technological Fixer*, American Institute of Physics, New York, 1994.

page

10 **"technological determinism."** For a critical review of technological determinism, see Merritt Roe Smith and Leo Marx, eds., *Does Technology Drive History?* MIT Press, Cambridge, MA, 1994.

investigated such things as . . . Leo Marx and Merritt Rowe Smith, introduction to Marx and Smith, *Does Technology Drive History?*, p. x.

it has now become almost fashionable . . . There have been a number of books published recently that analyze technology from a social science perspective. They include: Wiebe E. Bijker, Thomas P. Hughes, and Trevor Pinch, eds., *The Social Construction of Technological Systems: New Directions in the Sociology and History of Technology*, MIT Press, Cambridge, MA, 1987; Marcel C. LaFollette and Jeffrey K. Stine, eds., *Technology and Choice: Readings From Technology and Culture*, University of Chicago Press, Chicago, 1991; Meinolf Dierkes and Ute Hoffmann, eds., *New Technology at the Outset: Social Forces in the Shaping of Technological Innovations*, Westview, Boulder, CO, 1992; Wiebe Bijker and John Law, eds., *Shaping Technology/Building Society: Studies in Sociotechnical Change*, MIT Press, Cambridge, MA, 1992; Wiebe E. Bijker, *Of Bicycles, Bakelites, and Bulbs: Toward a Theory of Sociotechnical Change*, MIT Press, Cambridge, MA, 1995. See also various issues of *Technology and Culture*; *Technology in Society*; *Science, Technology & Human Values*; and *Technology Review*.

12 **"social construction."** The seminal

page

work here is Peter L. Berger and Thomas Luckmann, *The Social Construction of Reality: A Treatise in the Sociology of Knowledge*, Doubleday, New York, 1966.

13 **even science.** A good summary of the social constructionist approach to scientific knowledge can be found in Trevor J. Pinch and Wiebe E. Bijker, "The Social Construction of Facts and Artifacts: Or How the Sociology of Science and the Sociology of Technology Might Benefit Each Other," in Bijker, Hughes, and Pinch, *The Social Construction of Technological Systems*, pp. 17-50.

earliest and best known example. Thomas S. Kuhn, *The Structure of Scientific Revolutions*, University of Chicago Press, Chicago, 1962.

"[T]here is widespread agreement . . . " Pinch and Bijker, "The Social Construction of Facts and Artifacts," pp. 18-19.

14 **Physicists especially dispute . . .** David Mermin, a very thoughtful physicist at Cornell University, offers a careful refutation of the social construction of physics in a two-part article: "What's Wrong with the Sustaining Myth?" *Physics Today* 49 (March 1996), pp. 11, 13; and "The Golemization of Relativity," *Physics Today* 49 (April 1996), pp. 11, 13.

In spring 1996, the dispute between physicists and social constructionists took an amusing turn—amusing, at least, for the physicists—when Alan Sokal published an article entitled "Transgressing the Boundaries: Toward a Transformative Hermeneutics of Quantum Gravity" in *Social Text*, a "postmodernist" journal devoted to social constructionist analyses. Sokal, a physicist at New York University, later revealed that it was all a hoax. He had cobbled together bad science, even worse logic, and a bunch of catch-phrases from postmodernist theorizing and had somehow convinced the editors of *Social Text* that the nonsensical article was a contribution worthy of publication. The editors were understandably not amused but neither were they contrite. They did not seem to see it as a weakness on their part—or, even less, on the part of the entire postmodernist movement—that they could not tell the difference between an obvious hoax and what passes for serious work in their field. The original article appeared in *Social Text* 46/47 (Spring/Summer 1996), pp. 217-252. Sokal revealed the hoax in "A Physicist Experiments with Cultural Studies," *Lingua Franca* (May/June 1996), pp. 62-64.

a key difference. This argument is modified from one offered by Thelma Lavine, Clarence J. Robinson Professor of Philosophy at George Mason University in Fairfax, Virginia.

opponents of its use suggested . . . W.P. Norton, "Just Say Moo," *The Progressive* (November 1989), pp. 26-29. See also Gina Kolata, "When the Geneticists' Fingers Get in the Food," *New York Times* (February 20, 1994), sec. 4, p. 14.

One: History and Momentum

17 **"They come from near and far . . . "** *New York Herald*, January 4, 1880. As quoted in Ronald W. Clark, *Edison: The Man Who Made the Future*, G.P. Putnam's Sons, New York, 1977, pp. 99-100.

some 3,000 had gathered. Matthew Josephson, *Edison*, McGraw-Hill, New York, 1959, p. 224.

a magic lamp that ran on electricity. Clark, *Edison*, p. 98.

page

18 **Edison announced he would open his laboratory.** Clark, *Edison,* pp. 98–99.
arc lighting. Martin V. Melosi, *Thomas A. Edison and the Modernization of America,* Scott, Foresman, Glenview, Illinois, 1990, pp. 61–62. See also Josephson, *Edison,* pp. 177–178.
"a bright, beautiful light." *New York Herald,* December 21, 1879. As quoted in Josephson, *Edison,* p. 224.

19 **"Several persons who ventured . . ."** Josephson, p. 225.
"Nearly all who came acknowledged. . . " Josephson, p. 225.
Much of it was driven by Edison. Josephson, *Edison,* p. 274.
the Pearl Street station. Josephson, *Edison,* p. 264.

20 **"Mary Gates's boy's company."** Paul Freiberger and Michael Swaine, *Fire in the Valley: The Making of the Personal Computer,* Osborne/McGraw-Hill, Berkeley, CA, 1984, pp. 272–273.

21 **the historian of technology.** The best examples of this scholarship can be found in a series of books on technological history put out by the MIT Press. Books in the collection, called "Inside Technology," attempt to describe the development of individual technologies in the context of a larger social and political framework. The series includes: Pamela E. Mack, *Viewing the Earth: The Social Construction of the Landsat Satellite System,* MIT Press, Cambridge, MA, 1990; H.M. Collins, *Artificial Experts: Social Knowledge and Intelligent Machines,* MIT Press, Cambridge, MA, 1990; Donald MacKenzie, *Inventing Accuracy: A Historical Sociology of Nuclear Missile Guidance,* MIT Press, Cambridge, MA, 1990; Stuart S. Blume, *Insight and Industry: On the Dynamics of Technological Change in Medicine,* MIT Press, Cambridge, MA, 1992; Wiebe E. Bijker, *Of Bicycles, Bakelites, and Bulbs: Toward a Theory of Sociotechnical Change,* MIT Press, Cambridge, MA, 1995.
the early history of electrical power. I first ran across the work of Thomas Hughes in "Technological History and Technical Problems," a chapter in Chauncey Starr and Philip C. Ritterbush, eds., *Science, Technology and the Human Prospect,* Pergamon, New York, 1980, pp. 141–156. This is a good, short summary of Hughes's thinking on technology and, in particular, on the development of the electrical power system in the United States. The more complete reference is Thomas P. Hughes, *Networks of Power: Electrification in Western Society, 1880–1930,* Johns Hopkins University Press, Baltimore, 1983. This important book won the Dexter Prize, awarded by the Society for the History of Technology.

22 **Edison spoke to a reporter.** As quoted in Melosi, *Thomas A. Edison,* p. 63.
"a minor invention every ten days . . . " Melosi, *Thomas A. Edison,* pp. 40–47.
most of the world's supply. Clark, *Edison,* p. 100.

23 **Edison built a dynamo.** Josephson, *Edison,* pp. 207–209; Clark, *Edison,* p. 111.
"feeder and main" and "three-wire" refinements. Josephson, *Edison,* pp. 230–232.
insulation, meter, fuse. Clark, *Edison,* p. 110.
the passenger ship S.S. *Columbia.* Clark, *Edison,* p. 131.
Edison followed up that success . . .

page

As quoted in Josephson, *Edison*, p. 261.

24 **dinner and a demonstration.** Josephson, *Edison*, pp. 245–246.

high-pressure boilers. John G. Burke, "Bursting Boilers and the Federal Power," in John G. Burke and Marshall Eakins, eds., *Technology and Change*, Knopf, New York, 1979.

Mrs. Vanderbilt "became hysterical." As quoted in Josephson, *Edison*, p. 261.

the gas companies warned . . . Clark, *Edison*, p. 143.

25 **Edison was forced to put his generating plants . . .** See, for instance, the discussion in Josephson, *Edison*, p. 347.

alternating current. In direct current, the flow of electrons moves in only one direction. With alternating current, the electric current periodically reverses its direction, typically cycling fifty or sixty times each minute.

Nikola Tesla patented an AC motor. Melosi, *Thomas A. Edison*, p. 88.

as the historian Thomas Hughes points out . . . Hughes, "Technological History and Technical Problems."

26 **a series of fatal accidents.** Josephson, *Edison*, p. 256.

Edison didn't believe . . . Josephson, *Edison*, p. 346.

W.H. Preece argued . . . As quoted in Clark, *Edison*, p. 92.

After seeing his bulb, . . . Josephson, *Edison*, p. 243.

claim prior invention. Josephson, *Edison*, p. 355.

27 **"We regard it as fortunate . . . "** As quoted in Josephson, *Edison*, p. 343.

a number of companies competed. Josephson, *Edison*, p. 343.

"The feline and canine pets . . . " Josephson, *Edison*, p. 347.

page

"A Warning." Josephson, *Edison*, pp. 347–348; Melosi, *Thomas A. Edison*, p. 87.

28 **he traveled to Richmond.** Hughes, *Networks of Power*, p. 108; and Hughes, "Technological History and Technical Problems," p. 145.

a complex and bizarre plot. Details about the Brown plot come from Josephson, p. 348; Clark, *Edison*, p. 160; Melosi, *Thomas A. Edison*, pp. 87–88.

High-voltage AC motors. Josephson, *Edison*, pp. 349–350.

the first large-scale hydroelectric plant. Melosi, *Thomas A. Edison*, pp. 89–90.

29 **merger forming the Edison General Electric Company.** Josephson, *Edison*, pp. 352–353; Melosi, *Thomas A. Edison*. pp. 90–91.

merger forming General Electric. Josephson, *Edison*, pp. 362–66.

transition from AC to DC. Hughes, *Networks of Power*, pp. 120–125; see also Hughes, "Technological History and Technical Problems," p. 147.

30 **London was served . . .** Chapter 9 in Hughes, *Networks of Power*, pp. 227–261.

Hughes emphasizes this point. Hughes, "Technological History and Technical Problems," p. 142.

modern technology has a tremendous amount of "momentum." Hughes, "Technological History and Technical Problems," pp. 141–142. See also Hughes, *Networks of Power*, p. 140.

31 **It was late 1938.** Many of the details here about the discovery of fission and about the Manhattan Project to build the atomic bomb are drawn from Richard Rhodes, *The Making of the Atomic Bomb*, Simon & Schuster, New York, 1986.

page

33 **"bombs with a destructiveness vastly greater . . . "** Rhodes, *Making of the Atomic Bomb.*

34 **not all uranium is created equal.** Rhodes, *Making of the Atomic Bomb,* pp. 284–288.

only about one of every 140 atoms is U-235. A third isotope, U-233, exists but is extremely rare.

35 **That first reactor.** Rhodes, *Making of the Atomic Bomb,* pp. 434–436.

37 **etymology of "scram."** Daniel Ford, *The Cult of the Atom,* Simon and Schuster, New York, 1982, p. 28.

details on the first chain reaction. H.L. Anderson, "The First Chain Reaction," in *The Nuclear Chain Reaction—Forty Years Later: Proceedings of a University of Chicago Commemorative Symposium,* The University of Chicago Press, Chicago, 1984. See also "The First Pile," *Bulletin of the Atomic Scientists,* December 1962, pp. 19–24.

the Hanford Engineering Works. Rhodes, *Making of the Atomic Bomb,* pp. 497–499.

39 **Teflon.** Rhodes, *Making of the Atomic Bomb,* pp. 492–496.

"Eventually the Y-12 complex counted . . . " Rhodes, *Making of the Atomic Bomb,* p. 490.

40 **three Nobel Prize winners.** Enrico Fermi, Ernest O. Lawrence and Harold Urey.

seven others who would later receive . . . Luis Alvarez, Hans Bethe, Felix Bloch, Richard Feynmann, Glenn Seaborg, Emilio Segrè and Eugene Wigner.

"I was very enthusiastic. . . . " Interview with Chauncey Starr, August 17, 1994.

The manufacturing complex created by the Manhattan Project. Bertrand Goldschmidt, *Atomic Adventure,* Pergamon, Oxford, 1964, p. 35.

page

"What do we do now?" Starr interview, August 17, 1994.

41 **According to Starr's crystal ball, . . .** Starr interview, August 17, 1994.

"Crazy ideas and not-so-crazy ideas . . . " Alvin M. Weinberg, *The First Nuclear Era: The Life and Times of a Technological Fixer,* AIP Press, New York, 1994, pp. 38–39.

42 **three massive enrichment complexes.** Richard G. Hewlett and Jack M. Holl, *Atoms for Peace and War, 1953–1961,* University of California Press, Berkeley, 1989, pp. 18–19.

43 **the new atomic knowledge should be made public.** Rhodes, *Making of the Atomic Bomb,* pp. 644–646, 760–764.

44 **a psychological legacy.** The best analysis of attitudes toward nuclear power is in Spencer R. Weart, *Nuclear Fear: A History of Images,* Harvard University Press, Cambridge, MA, 1988.

German scientists. Samuel A. Goudsmit, *Alsos,* Henry Schuman, New York, 1947. See also Irving Klotz, "Germans at Farm Hall Knew Little of A-Bombs," *Physics Today* 46:10 (October 1993), pp. 11ff.

a different impression. Women in particular were apt to associate nuclear power with "aggressive masculine imagery: weapons, mysteriously powerful machines, domination of nature, contamination verging on rape." Weart, *Nuclear Fear,* p. 367.

45 **The Navy's interest in nuclear reactors.** For a history of the Navy's role in atomic energy before and during World War II, see Richard G. Hewlett and Francis Duncan, *Nuclear Navy, 1946–1962,* University of Chicago Press, Chicago, 1974, pp. 15–21.

submarines. Hewlett and Duncan, *Nuclear Navy,* p. 10.

page

46 **thermal-diffusion method.** Rhodes, *Making of the Atomic Bomb*, pp. 550–554.
Hyman Rickover. There are a number of good histories of Rickover and the nuclear navy. Perhaps the best is Norman Polmar and Thomas B. Allen, *Rickover*, Simon and Schuster, New York, 1982. See also Francis Duncan, *Rickover and the Nuclear Navy: The Discipline of Technology*, Naval Institute Press, Annapolis, MD, 1990; and Hewlett and Duncan, *Nuclear Navy*.

47 **Rickover pushed the development.** Hewlett and Duncan, *Nuclear Navy*, pp. 164–167.

48 **"During the first trial, . . . "** Hewlett and Duncan, *Nuclear Navy*, p. 217.

49 **Rickover didn't dither.** Hewlett and Duncan, *Nuclear Navy*, pp. 273–274.
Judged solely as a submarine-propulsion technology, . . . See, for instance, the history in Duncan, *Rickover and the Nuclear Navy*.

50 **a scaled-up version of the machine in the *Nautilus*.** Hewlett and Duncan, *Nuclear Navy*, p. 243.
The cost of generating electricity at Shippingport. Owen Ely, "Cost Race Between Fuel-Burning Plants and Atomic Reactors Getting Hotter." *Public Utilities Fortnightly*, August 27, 1964, pp. 47–48.
Shippingport performed well. Duncan, *Rickover and the Nuclear Navy*, p. 205.

51 **It demonstrated the feasibility of nuclear power.** Shippingport was the world's first nuclear reactor devoted only to civilian power production. In England, the Calder Hall graphite reactor had begun producing power a year before Shippingport started operations, but that was a dual-purpose plant, producing both power and plutonium.

page

the AEC was spending several times as much. Hewlett and Holl, *Atoms for Peace and War*, p. 576.

52 **it was natural they would stick with light-water machines.** General Electric modified and simplified the naval reactor somewhat for the commercial market, offering a boiling-water, instead of a pressurized-water, reactor. Westinghouse stayed with the pressurized light-water reactor.
Rickover made sure to document . . . Hewlett and Duncan, *Nuclear Navy*, pp. 255–256; Duncan, *Rickover and the Nuclear Navy*, p. 205.

Two: The Power of Ideas

53 **Bell's invention of the telephone fell flat.** See, for instance, Ira Flatow, *They All Laughed*, HarperCollins, New York, 1992, pp. 71–88.

54 **high-temperature superconductors.** Robert Pool, "Superconductors' Material Problems," *Science* 240 (April 1, 1988), pp. 25–27.
Xerox. Douglas K. Smith and Robert C. Alexander, *Fumbling the Future: How Xerox Invented, Then Ignored, the First Personal Computer*, William Morrow, New York, 1988, pp. 27–28.

55 **the first typewriters were ignored.** Flatow, *They All Laughed*, pp. 197–204.
early work on the laser. Nathan Rosenberg, *Exploring the Black Box: Technology, Economics and History*, Cambridge University Press, Cambridge, 1994, p. 3.
When Marconi began to work on radio . . . Rosenberg, *Exploring the Black Box*, p. 4.
"we are in the middle of . . . " As quoted in "What Are We Doing On-Line?" *Harper's Magazine*, August 1995, pp. 35–46.

page

56 **much of what we think and believe is socially constructed.** Peter L. Berger and Thomas Luckmann, *The Social Construction of Reality: A Treatise in the Sociology of Knowledge*, Doubleday, New York, 1966. See also Wiebe E. Bijker, Thomas P. Hughes and Trevor Pinch, eds., *The Social Construction of Technological Systems: New Directions in the Sociology and History of Technology*, MIT Press, Cambridge, MA, 1987.

57 **the invention of the jet engine.** Edward W. Constant II, *The Origins of the Turbojet Revolution*, Johns Hopkins University Press, Baltimore, MD, 1980, pp. 178–207.

58 **In Germany, . . .** Constant, *Turbojet Revolution*, p. 232.

60 **the companies that employ them.** I thank Michael Golay, head of the nuclear engineering department at MIT, for pointing out the role of companies in keeping engineers conservative.

61 **"a *fait accompli* by the end of 1939."** Constant, *Turbojet Revolution*, p. 208.
periodic upheavals . . . Thomas S. Kuhn, *The Structure of Scientific Revolutions* (2nd ed.), University of Chicago Press, Chicago, 1962.

62 **Copernicus assumed that the orbits . . .** David Lindley, *The End of Physics*, Basic Books, New York, 1993, pp. 8–10.
scientists before and after a paradigm shift. Kuhn, *Structure of Scientific Revolutions*, p. 111.

63 **Some of the Manhattan Project scientists.** Alvin M. Weinberg, *The First Nuclear Era: The Life and Times of a Technological Fixer*, American Institute of Physics, New York, 1994, pp. 38–39.

64 **Rutherford and Soddy.** The details about Rutherford and Soddy's work are drawn from Richard Rhodes, *The Making of the Atomic Bomb*, Simon & Schuster, New York, 1986, pp. 42–43.
"It may therefore be stated . . . " Ernest Rutherford, *The Collected Papers*, vol. I, Allen and Unwin, London, 1962, p. 606. As quoted in Rhodes, *The Making of the Atomic Bomb*, p. 43.

65 **Soddy informed the readers.** Spencer R. Weart, *Nuclear Fear: A History of Images*, Harvard University Press, Cambridge, MA, 1988, p. 6.
"A race which could transmute matter . . . " Frederick Soddy, *The Interpretation of Radium*, Murray, London, 3rd ed., 1912. As quoted in Weart, *Nuclear Fear*, p. 6.
a novel of a future. H.G. Wells, *The World Set Free*. E.P. Dutton & Company, New York, 1914.
"It would mean a change . . . " Wells, *World Set Free*, pp. 36–38.

66 **"But none of these possibilities . . ."** As reprinted in "50 and 100 Years Ago," *Scientific American*, November 1971, p. 10.
As Rutherford wrote in 1936, . . . E.N. da C. Andrade, *Rutherford and the Nature of the Atom*, Doubleday, Garden City, New York, 1964, p. 210. As quoted in Alvin M. Weinberg, *Nuclear Reactions: Science and Trans-Science*, American Institute of Physics, New York, 1992, p. 221.

67 **"All the things which H.G. Wells predicted . . . "** Spencer R. Weart and Gertrud Weiss Szilard, eds., *Leo Szilard: His Version of the Facts*, MIT Press, Cambridge, MA, 1978, p. 53.
"It was not clear," Fermi said, . . . As quoted in Weinberg, *The First Nuclear Era*, p. 41.
Americans had embraced technology. Merritt Roe Smith, "Technological Determinism in American Culture," in Merritt Roe Smith and

page

Leo Marx, eds., *Does Technology Drive History?* MIT Press, Cambridge, MA, 1994, pp. 1-35.

68 **"All forms of transportation . . . "** David Dietz, *Atomic Energy in the Coming Era*, Dodd Mead, New York, 1945, pp. 12-23. As quoted in Daniel Ford, *The Cult of the Atom*, Simon and Schuster, New York, 1982, pp. 30-31.

69 **"This new force . . . "** As quoted in Richard G. Hewlett and Oscar E. Anderson Jr., *The New World, 1939/1946: Volume I, A History of the United States Atomic Energy Commission*, Pennsylvania University Press, University Park, 1962, pp. 436-437.

The Atomic Energy Act of 1946. Public Law 585, 79th Congress, 60th Statute, pp. 755-75. *United Stated Code* 42, pp. 1801-1819.

70 **an "unrestrained and fertile developmental environment."** Steven L. Del Sesto, *Science, Politics and Controversy: Civilian Nuclear Power in the United States, 1946-1974*, Westview, Boulder, CO, 1979, pp. 28-29.

Lewis Strauss. Richard G. Hewlett and Jack M. Holl, *Atoms for Peace and War, 1953-1961*, University of California Press, Berkeley, 1989, p. 20.

71 **"Transmutation of the elements . . . "** Recounted in Ford, *Cult of the Atom*, p. 50.

a permanent nuclear advocacy group in Congress. See, for instance, Del Sesto, *Science, Politics and Controversy*, pp. 24-28.

72 **The saga of the nuclear plane.** W. Henry Lambright, *Shooting Down the Nuclear Plane*, Bobbs-Merrill, Indianapolis, 1967. See also Harold P. Green and Alan Rosenthal, *Government of the Atom: The Integration of Powers*, Atherton, New York, 1963, pp. 242-247.

page

"An atomic plane," Ward said, . . . "The Fight for the Ultimate Weapon," *Newsweek*, June 4, 1956, pp. 55-60. See also Weinberg, *The First Nuclear Era*, pp. 95-108.

73 **one early report.** "Atoms Aloft," *Time*, September 17, 1951, pp. 59-60.

"[A] nuclear aircraft was an oxymoron." Weinberg, *The First Nuclear Era*, p. 95.

74 **"The atomic-powered aircraft reminds me of a shite-poke."** Green and Rosenthal, *Government of the Atom*, p. 242.

"If Russia beats us in the race . . . " "Senator Jackson on A-Bombers," *Newsweek*, June 4, 1956, pp. 56-57.

The Russians "are placing considerable emphasis . . . " As quoted in Green and Rosenthal, p. 243.

75 **"I can only estimate . . . "** As quoted in "Extraordinary Atomic Plane: The Fight for an Ultimate Weapon," *Newsweek*, June 4, 1956, pp. 55-60.

the reactor "had little chance . . . " Weinberg, *The First Nuclear Era*, p. 108.

the portable reactor. Hewlett and Holl, *Atoms for Peace*, pp. 519-520.

the nuclear cargo ship *Savannah*. Hewlett and Holl, *Atoms for Peace*, pp. 506-508.

the nuclear ramjet. Hewlett and Holl, *Atoms for Peace*, pp. 518-519.

76 **a petition.** Alice Smith, *A Peril and a Hope: The Scientists' Movement in America, 1945-7*, University of Chicago Press, Chicago, 1965.

few stayed in weapons work. See Rhodes, *Making of the Atom Bomb*, p. 759.

Szilard switched to biology. Rhodes, *Making of the Atom Bomb*, pp. 26, 749-750.

They shared a conviction, he said, . . . David E. Lilienthal, *Change,*

page

Hope and the Bomb, Princeton University Press, Princeton, NJ, 1963, pp. 109–110.

77 **a novel way to dig the canal.** John W. Finney, "A Second Canal?" *The New Republic*, March 28, 1964, pp. 21–24.

"People have got to learn to live . . . " As quoted in Hewlett and Holl, *Atoms for Peace*, p. 290.

nuclear explosions might be used to push rockets. Hewlett and Holl, *Atoms for Peace*, pp. 528.

the project was named Plowshare. Hewlett and Holl, *Atoms for Peace*, p. 529.

three ambitious Plowshare schemes. "Energy for Peace: Dr. Johnson's Magic," *Newsweek*, February 8, 1960, p. 67.

78 **Project Sedan.** Details for the 1962 test, called Project Sedan, can be found in "Digging With H-Bombs," *Business Week*, May 18, 1963, pp. 154, 156. See also "Atomic Earth Mover," *Newsweek*, July 16, 1952, p. 74; and "When Nuclear Bomb is Harnessed for Peace," *U.S. News & World Report*, December 10, 1962, p. 16.

Atom blasting could do the job. "When Nuclear Bomb is Harnessed for Peace," p. 16.

a 253-mile waterway. "An Atomic Blast to Help Build a U.S. Canal?" *U.S. News & World Report*, May 20, 1963, p. 14.

a two-mile channel through the Bristol Mountains. "Nuclear Ditch-Digging," *Business Week*, December 21, 1963, pp. 84–85.

a new Panama Canal. "Another 'Panama Canal': A-Blasts May Do the Job," *U.S. News & World Report*, June 10, 1963, pp. 74–75.

79 **"Our studies," said John S. Kelly, . . .** "Another 'Panama Canal,'" p. 74.

page

A U.S. Army study. Finney, p. 23.

a string of several hundred bombs would be set off. Initially, plans to blast out a new canal involved setting off a couple of dozen nuclear explosions to create a series of craters and then joining the craters with conventional excavating techniques. Later, Plowshare scientists became more ambitious and decided they could do the whole job with simultaneous blasts.

"Permanent population may have to keep away . . . " "Nuclear Energy: Ploughshare Canals," *Time*, January 31, 1964, p. 36.

a 20-kiloton hydrogen bomb. "The H-Bomb Goes Commercial," *Business Week*, December 16, 1967, pp. 70–72.

80 **several other companies were working with the AEC.** "Oil Industry Buys Ticket on Gasbuggy," *Business Week*, October 26, 1968, pp. 77–78.

The dream of peaceful atom blasting . . . See chapter 1 of Glenn Seaborg and Benjamin S. Loeb, *The Atomic Energy Commission Under Nixon*, St. Martin's, New York, 1993; and "Plowshare: A Dying Idea," *U.S. News & World Report*, June 9, 1975, p. 53.

$160 million spent over 18 years. "Plowshare: A Dying Idea," p. 53.

AEC annual budget. See Appendix 2 in Hewlett and Holl, *Atoms for Peace*.

81 **casting around for problems that fit.** See, for example, Luther J. Carter, "Rio Blanco: Stimulating Gas and Conflict in Colorado," *Science* 180 (May 25, 1973), pp. 844–848.

the commission turned to the civilian side. Del Sesto, *Science, Politics and Controversy*, pp. 44–48.

page

81 **the world's first atomic electricity.** Del Sesto, *Science, Politics and Controversy*, pp. 44-45.
the second to produce electricity. Weinberg, *The First Nuclear Era*, p. 120.

82 **"Today no prudent businessman . . . "** "What is the Atom's Industrial Future?" *Business Week*, March 8, 1947, pp. 21-22ff. This was the first of a four-part series that the magazine ran in March 1947, calling it an "Atomic Progress Report to *Business Week* Readers."
"Right now, hardly anyone is interested . . . " "The Atomic Era—Second Phase," *Business Week*, July 8, 1950, pp. 58-65.
"Atomic energy is the most significant development . . . " "What an Atomic Bid Cost Goodyear," *Business Week*, October 4, 1952, p. 108.
four separate industrial study groups. Hewlett and Holl, *Atoms for Peace*, p. 21.
Developing the test reactors had proved to be . . . Del Sesto, *Science, Politics and Controversy*, pp. 48-49.

83 **Cole wrote a letter to Strauss.** Green and Rosenthal, *Government of the Atom*, pp. 252-253.
"It is not enough . . . " Dwight Eisenhower, "An Atomic Stockpile for Peace," Delivered before the General Assembly of the United Nations, December 8, 1953. *Vital Speeches of the Day* 20:6 (January 1, 1954), pp. 162-165.

Three: Business

85 **a six-page article.** H. Edward Roberts and William Yates, "Altair 8800," *Popular Electronics*, January 1975, pp. 33-38.
MITS, a company in Albuquerque. Paul Freiberger and Michael Swaine, *Fire in the Valley: The Making of the Personal Computer*, Osborne/McGraw-Hill, Berkeley, CA, 1984, p. 28.
To read the computer's output, . . . Freiberger and Swaine, *Fire in the Valley*, pp. 38-46.
"You buy the Altair, . . . " Steven Levy, *Hackers*, Doubleday, Garden City, NY, 1984, p. 185.

86 **a personal computer that had many of the features . . .** Douglas K. Smith and Robert C. Alexander, *Fumbling the Future: How Xerox Invented, Then Ignored, the First Personal Computer*, William Morrow, New York, 1988, pp. 102-103.

87 **By the beginning of 1975, . . .** Smith and Alexander, *Fumbling the Future*, pp. 105-113.
"Members were notorious . . . " Bro Uttal, "The Lab That Ran Away From Xerox," *Fortune*, September 5, 1983, pp. 97-102.

88 **One stumbling block was cultural.** Smith and Alexander, *Fumbling the Future*, pp. 148-149.
Xerox was extremely conventional and conservative. Smith and Alexander, *Fumbling the Future*, pp. 154-157.
Xerox had no group . . . Smith and Alexander, *Fumbling the Future*, pp. 176-177.
Xerox did bring out a product based on the Alto. Smith and Alexander, *Fumbling the Future*, pp. 227-240.

89 **the Star's developers made obvious blunders.** Smith and Alexander, *Fumbling the Future*, pp. 236-237.
the sales force had no idea how to approach it. Smith and Alexander, *Fumbling the Future*, pp. 238-240.

90 **Wozniak hadn't thought of his machine as a commercial product.**

page

Michael Moritz, *The Little Kingdom: The Private Story of Apple Computer*, William Morrow, New York, 1984, p. 124.

plans changed quickly in July 1976. Moritz, *The Little Kingdom*, pp. 142-144.

91 **The embryonic company.** Freiberger and Swaine, *Fire in the Valley*, p. 213.

Wozniak began work on the Apple II. Moritz, *The Little Kingdom*, p. 156.

Jobs soon hooked up with Mike Markkula. Freiberger and Swaine, *Fire in the Valley*, pp. 213-215.

92 **"Dozens of companies had come and gone . . . "** Freiberger and Swaine, *Fire in the Valley*, p. 214.

Jobs had courted Regis McKenna. Freiberger and Swaine, *Fire in the Valley*, p. 219.

In 1977 Apple began to grow quickly. Freiberger and Swaine, *Fire in the Valley*, pp. 220-227.

93 **The most important program was VisiCalc.** Moritz, *The Little Kingdom*, pp. 234-235.

The defining moment for Apple. Moritz, *The Little Kingdom*, pp. 318-324.

94 **IBM's 5100 and 5110 computers.** David Mercer, *The Global IBM: Leadership in Multinational Management*, Dodd, Mead, New York, 1987, pp. 104-105.

a proposal to develop an IBM personal computer. James Chpolsky and Ted Leonsis, *Blue Magic: The People, Power and Politics Behind the IBM Personal Computer*, Facts on File Publications, New York, 1988, p. 20.

The IBM Personal Computer made no attempt . . . Mercer, *The Global IBM*, pp. 106-107.

95 **"We didn't do three years of research . . . "** As quoted in Freiberger and Swaine, *Fire in the Valley*, p. 283.

page

96 **At Apple, some resisted . . .** Freiberger and Swaine, *Fire in the Valley*, pp. 237-238.

the preexisting culture and concerns of the firm. Freiberger and Swaine, *Fire in the Valley*, p. 278.

an independent business unit. F.G. Rodgers, *The IBM Way*, Harper & Row, New York, 1986, pp. 208-209.

97 **IBM's standard strategy for emerging technologies.** Chpolsky and Leonsis, *Blue Magic*, p. 48.

To get everything done in a year, . . . Mercer, *The Global IBM*, pp. 106-107.

the main processor chip was the Intel 8088, . . . Chpolsky and Leonsis, *Blue Magic*, p. 68.

the decision to give the PC an open architecture. Chpolsky and Leonsis, *Blue Magic*, p. 21.

98 **IBM is no longer the leading seller.** Jim Carlton, "Study Says Compaq Has Surpassed IBM In Personal Computer Unit Shipments," *Wall Street Journal*, December 23, 1994, p. A3.

an attempt to reassert some control. Mercer, *The Global IBM*, pp. 116-117.

99 **VEPCO's sales.** Details on VEPCO's sales efforts come from its annual reports. Throughout the mid-1960s, each report had a section focusing on the company's sales effort, but that section had disappeared by the 1970s.

101 **the same things were true for most . . .** Leonard S. Hyman, *America's Utilities: Past, Present and Future*, 4th ed. Public Utilities Reports, Arlington, VA, 1992, pp. 107-119.

These giant plants. Hyman, *America's Utilities*, p. 110.

102 **Residents of Clarke County pledged . . .** Keith Stickley, "$100,000 Pledged to Fight Power

page

Line," *Winchester Evening Star*, January 30, 1963.

102 **two hundred property owners.** "Middleburg Group Will Fight VEPCO," *London Times-Mirror*, February 7, 1963.

anti-power-line song. "Words of a Song Composed to Fight VEPCO's Proposal," *Clarke Courier*, February 7, 1963.

103 **an analysis of the situation.** "Merits of Nuclear Station Discussed," *Richmond News Leader*, June 15, 1966.

VEPCO had already made its decision. "Nuclear Station Slated in Surry," *Richmond Times-Dispatch*, June 26, 1966.

The selection had been based mainly on economics. "Coal Price Rise Prompted A-Plant," *Charleston Gazette*, June 27, 1966.

the company had asked the railroads. "Nuclear Plant Can Grow, Says VEPCO," *Daily News Record* (Harrisonburg), June 28, 1966.

104 **VEPCO figured it would save about $3 million a year.** "Plant Due to Save About $3 Million," *Richmond News Leader*, June 27, 1966.

concern that the plant's cooling water . . . Rush Loving, "More Units Possible at A-Plant," *Richmond Times-Dispatch*, June 28, 1966.

VEPCO had been sending speakers. "Plant Due to Save."

Surry was a rural county. Nita Sizer, "A-Plant Plans Please Surry Chairman," *Norfolk Virginian-Pilot*, June 28, 1966.

It had no library, . . . Nita Sizer, "1st Vepco A-Plant to Generate in '71," *Norfolk Virginian-Pilot*, October 31, 1966.

"If we had picked the whole world over, . . ." Sizer, "A-Plant Plans."

a number of factors operating below the surface. Steven L. Del Sesto, *Science, Politics and Controversy: Civilian Nuclear Power in the United States, 1946–1974*, Westview, Boulder, CO, 1979, pp. 55–56.

105 **the drive to civilian nuclear power was going too slowly.** Richard G. Hewlett and Jack M. Holl, *Atoms for Peace and War, 1953–1961*, University of California Press, Berkeley, 1989, p. 342.

the Gore-Holifield bill. Hewlett and Holl, *Atoms for Peace*, p. 344–345.

106 **"Civilian atomic power is bound to benefit . . . "** "Either Way It Looks Like a Lift for Power Reactors," *Business Week*, December 22, 1956, pp. 32–33.

trouble getting sufficient liability insurance. "Either Way It Looks Like a Lift," p. 33.

the Price-Anderson Act. Del Sesto, *Science, Politics and Controversy*, p. 58.

If the various incentives . . . "Either Way It Looks Like a Lift."

Executives in the electric power industry. Ellen Maher, "The Dynamics of Growth in the U.S. Electric Power Industry," in Kenneth Sayre, ed., *Values in the Electric Power Industry*, University of Notre Dame Press, Notre Dame, IN, 1977, pp. 149–216.

107 **A Vepco Shareowner inspects . . .** Virginia Electric and Power Company, 1962 Annual Report, p. 2.

The list is long. Virginia Electric and Power Company, 1962 Annual Report, p. 2.

Duquesne's management judged . . . Hewlett and Holl, *Atoms for Peace*, p. 197.

Similar reasoning motivated VEPCO. Virginia Electric and Power Company, *1959 Annual Report*, p. 7.

108 **A number of other companies followed suit.** These five nuclear plants were subsidized under the third

page

round of the Power Reactor Demonstration Program, under which the AEC would "finance all research and development costs, as well as waive entirely uranium lease charges for up to seven years." Del Sesto, *Science, Politics and Controversy*, pp. 59–61.

108 **a standoff between the two types of people . . .** James M. Jasper, *Nuclear Politics: Energy and the State in the United States, Sweden and France*, Princeton University Press, Princeton, NJ, 1990, p. 49.

Neither group could sway the other. Jasper identifies a third "policy style" used by people in making decisions about technology and justifying those decisions to others. In addition to the technological enthusiasts and the cost benefiters, there are the "ecological moralists," who place great value on the "natural world" and seek policies that will disturb nature as little as possible or—when it has already been disturbed—will return it to its pristine state. At the time, however, there were essentially no ecological moralists in the management of the electrical utilities. *Nuclear Politics*, pp. 31–35.

109 **a 515-megawatt nuclear plant at Oyster Creek.** "GPU Announces Big Low-Cost Atomic Power Plant for 1967," *Public Utilities Fortnightly*, January 16, 1964, 41–42.

Jersey Central took the unprecedented step. "The Jersey Central Report," *Forum Memo to Members* (newsletter of the Atomic Industrial Forum), March 1964, pp. 3–7.

Even Philip Sporn was convinced. Irvin C. Bupp and Jean-Claude Derian, *Light Water: How the Nuclear Dream Dissolved*, Basic Books, New York, 1978, 45–46.

page

Sporn now ran through the numbers . . . Owen Ely, "Debate Over 'Breakthrough' in Cost of Atomic Power at Oyster Creek Plant," *Public Utilities Fortnightly*, October 8, 1964, pp. 95–97.

GE published a "price list." "GE Price List for Atomic Power Plants," *Public Utilities Fortnightly*, November 19, 1964, pp. 53–54.

110 **numbers of orders during "the Great Bandwagon Market."** Arturo Gándara, *Electric Utility Decisionmaking and the Nuclear Option*, RAND Study for the National Science Foundation, 1977, pp. 60–61.

111 **"The atom is the power of the future."** "Atomic Energy: The Powerhouse," *Time*, January 12, 1959, pp. 74–86.

General Electric expected to lose . . . Allan T. Demaree, "G.E.'s Costly Ventures Into the Future," *Fortune*, October 1970, pp. 88–93ff.

112 **"Civilization is moved forward by restless people."** "Atomic Energy: The Powerhouse," p. 86.

395-megawatt San Onofre reactor. Some readers may have noticed that the Camp Pendleton reactor was earlier referred to as 370 megawatts. The design had changed somewhat over the two and a half years it took to get a lease for the land. For these early plants, the power rating was never sure until it had actually been built and tested. The San Onofre 1 reactor at Camp Pendleton, for instance, was ultimately rated at 436 megawatts.

a giant, 1,000-megawatt reactor. "Con Ed Plans 1,000-Mw Reactor in N.Y. City; Pendleton Site Cleared" *Nucleonics Week*, December 13, 1962, pp. 1–2.

112 **a 490-megawatt plant in Malibu Beach.** "Westinghouse Wins

page

LADWP Bidding," *Nucleonics Week*, January 31, 1963, pp. 1-3.
It was later canceled. Richard L. Meehan, *The Atom and the Fault* (paperback ed.), MIT Press, Cambridge, 1986, pp. 41-42.

113 **"We had a problem like a lump of butter sitting in the sun."** Demaree, "G.E.'s Costly Ventures," p. 93.

114 **It would not lose too much money, the company thought.** Demaree, *"G.E.'s Costly Ventures,"* p. 93.
Sporn, normally a skeptic, swallowed . . . Ely, "Debate Over Breakthrough," pp. 96-97.

115 **"I find it hard to convey to the reader . . . "** Alvin Weinberg, *The First Nuclear Era: The Life and Times of a Technological Fixer*, American Institute of Physics, New York, 1994, p. 135.
"The competition was rather desperate in those days." As quoted in Mark Hertsgaard, *Nuclear Inc.: The Men and Money Behind Nuclear Energy*, Pantheon, New York, 1983, p. 43.
Babcock & Wilcox and Combustion Engineering. Weinberg, *The First Nuclear Era*, p. 135.
nuclear plants cost twice as much to build as estimated. Bupp and Derian, *Light Water*, p. 79.
GE and Westinghouse stopped offering turnkey contracts. The last turnkey sale actually came in April 1967 with the Indian Point 2 plant, but Consolidated Edison was exercising an option on a contract it had signed in July 1966. Gándara, *Electric Utility Decision-Making*, p. 53.

116 **GE and Westinghouse lost as much as a billion dollars.** Gándara, *Electric Utitlity Decision-Making*, p. 53.
"We lost money on the turnkeys largely because . . . " Interview with Bertram Wolfe, August 15, 1994.
labor costs were much higher than expected. Demaree, "G.E.'s Costly Ventures," p. 93.
a two-unit, 2,196-megawatt nuclear plant at Browns Ferry. Tom O'Hanlon, "An Atomic Bomb in the Land of Coal," *Fortune*, September 1966, pp. 132-133.

117 **Duke Power said it would enter the nuclear era.** "A Midsummer Avalanche," *Nuclear Industry*, July 1966, pp. 3-15.
In 1968, orders for new plants plummeted. Gándara, *Electric Utitlity Decision-Making*, pp. 60-62.
But utilities were finding . . . Gándara, *Electric Utitlity Decision-Making*, p. 62.

118 **the Texas Railroad Commission prohibited utilities . . .** *An Historical Overview of the Comanche Peak Steam Electric Station*, a booklet published by TU-Electric, Dallas, TX, p. 7.
the nuclear industry was assuring its customers . . . Bupp and Derian, *Light Water*, p. 82.
a second bandwagon market even larger than the first. Gándara, *Electric Utitlity Decision-Making*, p. 8.

Four: Complexity

119 **human technology.** A nice, readable account of technology throughout human history is John Purcell, *From Hand Ax to Laser: Man's Growing Mastery of Energy*, Vanguard, New York, 1982.
a Boeing 747. Barry Lopez, "On the Wings of Commerce," *Harper's Magazine*, October 1995, pp. 39-54.
six million individual parts. About half of the six million pieces are fasteners—rivets, etc.
Such complexity . . . A good summary of how complexity makes modern technology different from earlier technologies can be found in

page

Tom R. Burns and Thomas Dietz, "Technology, Sociotechnical Systems, Technological Development: An Evolutionary Perspective," in Meinolf Dierkes and Ute Hoffmann, eds., *New Technology at the Outset: Social Forces in the Shaping of Technological Innovations*, Westview, Boulder, CO, 1992, pp. 206–238. See especially pp. 211-224.

122 **the steam engine.** A comprehensive history of the steam engine up to the 1820s is John Farey, *A Treatise on the Steam Engine*, Longman, Rees, Orme, Brown, and Green, London, 1827. It was reprinted in 1971 by David & Charles, Devon. See also R.A. Buchanan and George Watkins, *The Industrial Archaeology of the Steam Engine*, Allen Lane, London, 1976.

Thomas Newcomen. Many of the details of Newcomen's life and his invention of the steam engine are taken from L.T.C. Rolt and J.S. Allen, *The Steam Engine of Thomas Newcomen*, Moorland, Hartington, UK, 1977.

Coal was a natural substitute. It was in the early part of the eighteenth century, not too long after the first operation of Newcomen's steam engine, that Abraham Darby discovered coke-smelting. By freeing iron production of its reliance on charcoal, Darby opened the way to making large amounts of cheap iron, which in turn led to iron becoming a major component in structures and tools.

123 **Newcomen's first successful engine.** Rolt and Allen, *Steam Engine*, p. 46.

he had all the scientific information he needed. Buchanan and Watkins, *Industrial Archaeology*, p. 5.

124 **Many of them were working . . .** Buchanan and Watkins, *Industrial Archaeology*, pp. 10–12.

page

James Watt. Details about Watt are taken from H.W. Dickinson, *James Watt*, Babcock & Wilcox, Cambridge, 1935; Buchanan and Watkins, *Industrial Archaeology*, pp. 14–20; and Purcell, *From Hand Ax to Laser*, pp. 246–252.

125 **With each stroke, as . . . the piston was moving up.** Actually, in Watt's original design, he turned the piston and cylinder upside down so that the piston was moving down as the cylinder filled with steam and moving up on the power stroke. For simplicity, however, the Watt engine is described here as if it had the same orientation as the Newcomen engine.

"I have now made an engine . . . " Dickinson, *James Watt*, p. 39.

126 **The piston and its cylinder proved even trickier.** Dickinson, *James Watt*, pp. 43, 87.

"The people . . . are steam mill mad." Dickinson, *James Watt*, p. 124.

127 **the rocking beam.** Dickinson, *James Watt*, pp. 143–144.

Watt himself never built . . . Buchanan and Watkins, *Industrial Archaeology*, p. 17.

128 **engine pressures rose steadily.** Buchanan and Watkins, *Industrial Archaeology*, pp. 51-52.

Each Watt engine was built at the spot. Dickinson, *James Watt*, p. 91.

129 **"On Thursday they had attempted to set . . . "** Dickinson, *James Watt*, p. 103.

A Newcomen machine was a rugged thing. Rolt and Allen, *Steam Engine*, chapters 3 & 5.

a Watt engine needed more careful handling. See, for instance, Dickinson, *James Watt*, pp. 103–105.

Watt himself recognized . . . Dickinson, *James Watt*, p. 132.

130 **after 1800 there was a steady increase in the pressure.** See

page

Buchanan and Watkins, *Industrial Archaeology*, chapter 4, for details on increasing the pressure and efficiency of steam engines.

Engineers devised a number of stratagems. Buchanan and Watkins, *Industrial Archaeology*, p. 52.

131 **This in turn has allowed turbines . . .** Buchanan and Watkins, *Industrial Archaeology*, pp. 79–80.

Within a few decades of their development, turbines . . . For a good, short history of steam turbines, see Edward W. Constant II, *The Origins of the Turbojet Revolution*, Johns Hopkins University Press, Baltimore, MD, 1980, pp. 63–82.

50-horsepower Watt engines. Buchanan and Watkins, *Industrial Archaeology*, p. 19.

132 **water leaking onto some computer chips.** Douglas Lavin, "Chrysler Recalls Neon Cars to Fix Computer Units," *Wall Street Journal*, eastern ed., February 7, 1994, p. C6.

a problem with . . . the anti-lock brakes. Charles E. Ramirez, "ABS Seals Triggered Neon Recall," *Automotive News*, February 21, 1994, p. 32.

a problem with the car's brake brackets. Douglas Lavin, "Chrysler's Neon Had Third Defect, U.S. Agency Says," *Wall Street Journal*, eastern ed., April 8, 1994, p. A4.

General Motors' Quad 4 engine. Liz Pinto, "GM Works to Fix Bugs, Reputation of Quad 4 Engine," *Automotive News*, April 5, 1993, pp. 1ff.

1.8 million Accords and Preludes. Oscar Suris and Gregory N. Racz, "Honda Doubles Size of Recall to Biggest Ever," *Wall Street Journal*, eastern ed., June 16, 1993, p. B1.

auto manufacturers recalled a total of 11 million vehicles. Douglas Lavin, "In the Year of the Recall, Some Companies Had to Fix More Cars Than They Made," *Wall Street Journal*, eastern ed., February 24, 1994, p. B1.

GM's problems with the Quad 4 engine. Pinto, "GM Works to Fix."

the culprit was the rubber compound. Ramirez, "ABS Seals Triggered."

133 **the 1974 crash of a Turkish Airlines DC-10.** Stephen H. Unger, *Controlling Technology: Ethics and the Responsible Engineer*, Wiley, New York, 1994, pp. 16–20.

134 **a pressure test of the first DC-10 fuselage.** Unger, *Controlling Technology*, p. 18.

McDonnell Douglas resisted major changes. Unger, *Controlling Technology*, pp. 18–19.

the bugs that infest computer programs. See, for example, Bev Littlewood and Lorenzo Strigini, "The Risks of Software," *Scientific American*, November 1992, pp. 62–66ff.

when the plane crossed the equator, . . . Ivars Peterson, "Warning: This Software May Be Unsafe," *Science News*, September 13, 1986, pp. 171–173.

135 **the bug triggered a series of shutdowns.** Ivars Peterson, "Finding Fault," *Science News*, February 16, 1991, pp. 104–106.

The details of that shutdown. Details about the failure of AT&T's system come from Leonard Lee, *The Day the Phones Stopped*, Donald I. Fine, New York, 1991, pp. 71–97.

136 **Even a program with only a few hundred lines of code.** Littlewood and Strigini, "Risks of Software," p. 62.

137 **a software-verification process that took nearly three years.** Peterson, "Finding Fault," pp. 104, 106.

The best approach may be . . . Littlewood and Strigini, "Risks of Software," p. 75.

the 1974 Rasmussen report. U.S.

page

Nuclear Regulatory Commission, "Reactor Safety Study: An Assessment of Accident Risks in U.S. Commercial Nuclear Power Plants" [Rasmussen report], WASH-1400, NUREG 75/014. Nuclear Regulatory Commission, Washington, DC, 1975. Available from National Technical Information Service, Springfield, VA.

138 **As the AEC presented it to the public, . . .** Daniel Ford, *The Cult of the Atom,* Simon and Schuster, New York, 1982, pp. 157–158.

the reactor core at Unit 2 of Three Mile Island melted down. For details on the accident at Three Mile Island, see chapter 6 of this book. The official government report on the accident was John G. Kemeny *et al., Report of the President's Commission on the Accident at Three Mile Island,* Pergamon, New York, 1979.

the steady improvement of these machines . . . Nathan Rosenberg, *Inside the Black Box: Technology and Economics,* Cambridge University Press, Cambridge, 1982, pp. 125–135.

139 **"With the advent of the jet engine, . . . "** Rosenberg, *Inside the Black Box,* p. 126.

there was no body of knowledge . . . See, for instance, Nathan Rosenberg, *Exploring the Black Box: Technology, Economics and History.* Cambridge University Press, Cambridge, 1994, pp. 18–19.

140 **he described the difference . . .** Theodore Rockwell, *The Rickover Effect: How One Man Made a Difference,* Naval Institute Press, Annapolis, MD, 1992, pp. 158–159.

141 **the early AEC programs . . .** Robert Perry *et al., The Development and Commercialization of the Light Water Reactor,* RAND Study for the National Science Foundation, Santa Monica, CA, 1977, p. 82.

142 **U.S. utilities had sharply boosted . . .** Federal Power Commission, *The 1970 National Power Survey, Part I.* U.S. Government Printing Office, Washington, DC, 1971, pp. I-5-3, I-5-4.

the new, large plants were less reliable. Edison Electric Institute, *Report on Equipment Availability for the Twelve-Year Period 1960–1971,* November 1971. See also John Hogerton, "The Arrival of Nuclear Power," *Scientific American* 218:2 (February 1968), pp. 21–31.

"We have one coal-fired unit . . . " "The Pathfinder for Nuclear Power," *Business Week,* March 11, 1967, pp. 77–78.

143 **Babcock & Wilcox.** Harold B. Meyers, "The Great Nuclear Fizzle at Old B&W," *Fortune,* November 1969, pp. 123–125ff.

144 **many of the difficulties GE encountered . . .** Interview with Bertram Wolfe, August 15, 1994.

145 **Once a reactor has operated for a while, . . .** Richard Wolfson, *Nuclear Choices.,* MIT Press, Cambridge, MA, 1991, p. 187.

It's hard to imagine . . . Carroll L. Wilson, "Nuclear Energy: What Went Wrong," *Bulletin of the Atomic Scientists,* June 1979, 13–17.

treat the rest of the plant according to nuclear . . . standards. "Although some of the difficulties [in building nuclear plants] were of the 'concrete and steel' type, most involved the far more complex piping, wiring, and control mechanisms required for nuclear plants. The requirements for precision fitting and for fail-free equipment were considerably more demanding for

page

nuclear installations than for older fossil-fuel plants, and the difficulties of satisfying demands for high safety standards accounted for many of the cost and schedule overruns that marked the 1970s." From Perry, *Development and Commercialization*, p. 83.

146 **When the first reactors were built, . . .** Ford, *Cult of the Atom*, pp. 66–67.

These constant changes in safety requirements . . . Perry, *Development and Commercialization*, p. 57.

"the regulatory division could never 'catch up.'" Perry, *Development and Commercialization*, p. 53.

147 **The utilities . . . got paid according to their expenses.** The cost overruns got so high on some nuclear plants that some public utility commissions disallowed part of the costs, forcing the utilities to swallow the losses, but nobody saw this as a possibility when the plants were ordered.

eventually as large as 1,200 megawatts. The increase in size would have continued past 1,200 if the AEC had not put a ceiling on the size reactor it would approve.

the French nuclear industry. James M. Jasper, *Nuclear Politics*, Princeton University Press, Princeton, NJ, 1990, pp. 74–97.

Electricité de France, Framatome, Alsthom-Atlantique. Kent Hansen, Dietmar Winje, Eric Beckjord, Elias P. Gyftopoulos, Michael Golay, and Richard Lester, "Making Nuclear Power Work: Lessons From Around the World." *Technology Review*, February/March 1989, pp. 31–40. See also Jasper, *Nuclear Politics*, p. 91.

148 **Framatome did increase the size of its reactors.** Jasper, pp. 252–253; also "World List of Nuclear Power Plants," *Nuclear News*, September 1993, pp. 43–62.

Its plants cost $1 billion to $1.5 billion apiece. Jasper, *Nuclear Politics*, p. 251.

EdF has consistently built plants in six years or less. Simon Rippon, "Focusing on Today's European Nuclear Scene," *Nuclear News*, November 1992, pp. 81–82ff.

they are very reliable. Hansen, "Nuclear Power Work."

Duke developed an in-house expertise. Duke did work with the architect-engineering firm Bechtel on the first plant, but since then it has it relied completely on its own resources.

Duke's construction costs have been comparable to EdF's. James Cook, "Nuclear Follies," *Forbes*, February 11, 1985, pp. 82–100. See the boxed story about Duke Power on p. 93.

its capacity factors . . . have been among the best. E. Michael Blake, "U.S. Capacity Factors: Soaring to New Heights," *Nuclear News*, May 1993, pp. 40–47ff.

At the Trojan plant . . . The story of the Trojan shutdown is in itself a study in the effects of complexity. After steam-generator problems forced Portland General Electric to shut Trojan down from March 1991 to February 1992, the company planned to run the plant until 1996, by which time it expected to have new generating capacity. But in November 1992 a leak in the steam generator forced the reactor to shut down again, and after the leak was fixed the NRC refused permission to restart the reactor because the NRC staff could not agree on whether the reactor could be operated safely until 1996. Given the uncertainty about when or if the

page

reactor could go back on-line, the utility decided it would be cheaper to shut the plant down altogether. See "PGE Decides to Close Plant Now, Not in 1996," *Nuclear News*, February 1993, pp. 28–29.

148 **"If you've got a plant that's supposed to last 40 years, . . . "** Wolfe interview, August 15, 1994.

Five: Choices

149 **it was common to hear reactors described as nuclear "furnaces."** See, for instance, "1951—The Payoff Year," *Business Week*, July 28, 1951, pp. 99–108; and "Atomic Furnaces in the Service of Peace," *Business Week*, July 28, 1951, p. 136.

150 **workers in the Manhattan Project had dreamed up . . .** Alvin M. Weinberg, *The First Nuclear Era: The Life and Times of a Technological Fixer*, American Institute of Physics, New York, 1994, pp. 38–39.

several hundred types of reactors were conceivable. Alvin M. Weinberg, "Survey of Fuel Cycles and Reactor Types," *Proceedings*, First Geneva Conference, vol. 3, p. 19. As reported in Philip Mullenbach, *Civilian Nuclear Power: Economic Issues and Policy Formation*, Twentieth Century Fund, New York, 1963, p. 39.

How do we make choices between competing technologies? For a review of the state of the art in understanding choices between competing technologies, see W. Brian Arthur, "Competing Technologies: An Overview," in G. Dosi *et al.*, eds., *Technical Change and Economic Theory*, Pinter, London, 1988, pp. 590–607.

According to classical economic theory, . . . W. Brian Arthur, "Positive Feedbacks in the Economy," *Scientific American*, February 1990, pp. 92–99.

page

151 **As in Darwin's theory of evolution, . . .** There are actually a number of parallels between economics and evolutionary theory, since the selective pressures of the marketplace are similar in many ways to the selective pressures of the animal and plant worlds. See, for example, Richard W. England, ed., *Evolutionary Concepts in Contemporary Economics*, University of Michigan Press, Ann Arbor, 1994.

And just as economists have realized that the market doesn't necessarily make the best selection, so have evolutionary theorists realized that natural selection doesn't necessarily result in the most fit species. See, for instance, Robert Pool, "Putting Game Theory to the Test," *Science* 267 (March 17, 1995), pp. 1591–1593; and Robert Pool, "Economics: Game Theory's Winning Hands," *Science* 266 (October 21, 1994), p. 371.

a weaker technology may get "locked in." W. Brian Arthur, "Competing Technologies, Increasing Returns, and Lock-In by Historical Events," *The Economic Journal* 99 (March 1989), pp. 116–131. See also Arthur, "Positive Feedbacks in the Economy."

Consider, for example, the automobile. Many of the details of the competition between steam-powered and gas-powered automobiles are taken from Charles McLaughlin, "The Stanley Steamer: A Study in Unsuccessful Innovation," *Explorations in Entrepreneurial History* 7 (October 1954), pp. 37–47.

152 **The electric cars . . . held the world's land speed record.** *The Ran-*

page

dom House Encyclopedia, Random House, New York, 1983, p. 1694.
it is only now . . . See, for example, Mac DeMere, "Batteries Not Included," *Motor Trend,* November 1995, pp. 134–135.

153 **advantages of steam-powered vehicles.** McLaughlin, "Stanley Steamer," p. 40.
"Every steam carriage which passes . . . " William Fletcher, *English and American Steam Carriages and Traction Engines* (reprinted 1973). David and Charles, Newton Abbot, 1904, p. ix.

154 **"The Doble steam car . . . "** McLaughlin, "Stanley Steamer," p. 39.
"If the money and effort . . . " As quoted in Thomas S. Derr, *The Modern Steam Car and Background,* Los Angeles, 1945, p. 145.
Even now, engineers . . . wonder . . . Arthur, "Competing Technologies: An Overview," p. 596.
the Stanley Company . . . was a relic of an earlier time. McLaughlin, "Stanley Steamer," p. 44.

155 **"The principal factor responsible for the demise . . . "** McLaughlin, "Stanley Steamer," p. 45.

156 **a new breed of economist/historian . . .** See, for example, Nathan Rosenberg, *Exploring the Black Box: Technology, Economics and History,* Cambridge University Press, Cambridge, 1994. In particular, chapter 1 discusses "path-dependent aspects of technological change."
This new approach . . . Much of the following discussion on positive and negative feedbacks and the economy follows Arthur, "Positive Feedbacks in the Economy."

157 **positive feedbacks . . . create uncertainty.** Arthur, "Positive Feedbacks in the Economy," p. 94.

page

158 **when information is involved, increasing returns are likely.** Arthur, "Positive Feedbacks in the Economy," p. 93.
five separate ways . . . Arthur, "Competing Technologies: An Overview," p. 591.

159 **the typewriter.** Many of the details in the story of the typewriter keyboard come from Paul A. David, "Clio and the Economics of QWERTY," *Economic History* 75:2 (May 1985), pp. 332–337.
somewhere between 5 percent and 40 percent faster. One of the many optimistic descriptions of the Dvorak keyboard's promise can be found in William Hoffer, "The Dvorak Keyboard: Is It Your Type?" *Nation's Business,* August 1985, pp. 38–40.

A more pessimistic account can be found in S.J. Liebowitz and Stephen E. Margolis, "The Fable of the Keys," *Journal of Law and Economics* 33 (April 1990), pp. 1–25.

160 **It took more than a decade . . .** See, for instance, the chapter on the typewriter in Ira Flatow, *They All Laughed.* HarperCollins, New York, 1992.
a U.S. Navy report. David, "Clio," p. 332.
Abner Doble decided to give it one more try. McLaughlin, "Stanley Steamer," p. 45.

161 **Perhaps the makers of the gasoline-powered cars colluded . . .** One author, writing from a Marxian perspective, has even argued that the steam automobile was the victim of a conspiracy among powerful capitalist groups that preferred the gas-powered automobile because it would favor their plans to develop the petroleum industry: "The steam carriage was economically advanta-

page

geous but failed because it was repressed. It was repressed because the political-economic groups supporting the railway were more powerful than those supporting the steam carriage. They supported the railway because they had monopoly control over it. The mineral industries they wished to develop were closely tied to railway development, and the railway as a carrier of heavy freight fitted in with their plans to develop England as a free trading economy." David Beasley, *The Suppression of the Automobile: Skullduggery at the Crossroads,* Greenwood Press, Westport, CT, 1988, p. xv.

161 **Or perhaps . . . the typewriter makers conspired . . .** David, "Clio," pp. 332–333.

"technological lock-in." The ideas behind technological lock-in were first explored in Brian Arthur's seminal paper, "Competing Technologies and Economic Prediction," *Options,* April 1984, pp. 10–13.

162 **Can some central authority . . . prevent lock-in?** Robin Cowan, "Tortoises and Hares: Choice Among Technologies of Unknown Merit," *The Economic Journal* 101 (July 1991), pp. 801–814.

a dozen or so serious candidates had emerged. Phillip Mullenbach, *Civilian Nuclear Power: Economic Issues and Policy Formation.* Twentieth Century Fund, New York, 1963, p. 39.

163 **the coolant gas was air—a poor choice.** In 1957, the graphite in a reactor core at Great Britain's Windscale plant caught fire and gave the operators an unpleasant choice. They could cut off the flow of air to kill the fire, but this would shut down the cooling of the fuel, and the residual heat from the chain reaction would wreck the core. Or

page

they could continue to cool the core, at the price of having the graphite continue to burn. After several days of hoping the fire would burn itself out, the operators gave in, shut off the air, and flooded the reactor with water. See, for instance, John Jagger, *The Nuclear Lion,* Plenum, New York, 1991, p. 117.

CANDU reactors. American Nuclear Society, *Controlled Nuclear Chain Reaction: The First 50 Years,* American Nuclear Society, Lagrange Park, IL, 1992, pp. 52–62.

CANDU reactors . . . use a slightly enriched fuel. *Controlled Nuclear Chain Reaction,* p. 52.

164 **The few such reactors that have been built . . .** Harold Agnew, "Gas-Cooled Nuclear Power Reactors," *Scientific American,* June 1981, pp. 55–63.

It was difficulties with this system . . . Richard G. Hewlett and Francis Duncan, *Nuclear Navy, 1946–1962,* The University of Chicago Press, Chicago, 1974, pp. 273–274.

165 **The Canadians, for example, . . .** *Controlled Nuclear Chain Reaction,* p. 60; and interview with Chauncey Starr, August 17, 1994.

the homogeneous reactor. Alvin M. Weinberg, *The First Nuclear Era The Life and Times of a Technological Fixer,* American Institute of Physics, New York, 1994, pp. 109–131.

the breeder reactor. There are various possible breeders. Shippingport, in its later days, was transformed into a light-water breeder reactor with the addition of a thorium "blanket" that absorbed neutrons and created U-233. (Francis Duncan, *Rickover and the Nuclear Navy: The Discipline of Technology,* Naval Institute Press, Annapolis,

page

MD, 1990, pp. 219–223). The most popular design has been the fast breeder—"fast" because it does not use a moderator to slow the neutrons down—and the United States, the former Soviet Union, France, and Japan have all built fast breeders of differing designs.

165 **the resulting uranium-239 atoms spontaneously transmute.** The transmutation of uranium-239 into plutonium-239 actually takes place in two steps. The U-239 emits an electron to turn into neptunium-239, which then emits a second electron to form Pu-239.

that is what the Atomic Energy Commission set out to do. See, for instance, chapter 3 in Steven L. Del Sesto, *Science, Politics and Controversy: Civilian Nuclear Power in the United States, 1946–1974*, Westview, Boulder, CO, 1979.

166 **Other countries . . .** *Controlled Nuclear Chain Reaction*, pp. 52–97.

the "multi-armed bandit." Cowan, "Tortoises and Hares." See also Robin Cowan, "Backing the Wrong Horse: Sequential Technology Choice Under Increasing Returns," Ph.D. diss., Stanford University, Stanford, CA, 1987.

167 **The best strategy, mathematicians have proved, . . .** For details on optimal strategies for multi-armed bandit problems, see J.C. Gittins and D.M. Jones, "A Dynamic Allocation Index for the Sequential Design of Experiments," in J. Gani, K. Sarkadi, and I. Vincze, *Progress in Statistics* North-Holland, Amsterdam, 1974; and J.C. Gittins and D.M. Jones, "A Dynamic Allocation Index for the Discounted Multiarmed Bandit Problem," *Biometrika* 66 (1979), pp. 561–565.

168 **what happens if a central authority gets involved?** Cowan, "Tortoises and Hares."

"If increasing returns are very strong . . . " Cowan, "Tortoises and Hares," p. 809.

169 **a treaty creating Euratom.** Besides France, Italy and West Germany, Euratom also contained Belgium, Luxembourg and the Netherlands. These six were also the founding members of the European Economic Community, and the Euratom agreement was signed at the same time as the agreement that formed the Common Market.

encouraging European countries to invest . . . Details of the relationship between Euratom and the United States are drawn mostly from Irvin C. Bupp and Jean-Claude Derian, *Light Water: How the Nuclear Dream Dissolved*, Basic Books, New York, 1978; and Richard G. Hewlett and Jack M. Holl, *Atoms for Peace and War, 1953–1961*, University of California Press, Berkeley, 1989.

170 ***A Target for Euratom.*** Louis Armand, Franz Estel, and Francesco Giordani, *A Target for Euratom*, reprinted in U.S. Congress, Joint Committee on Atomic Energy, *Hearings*, Proposed Euratom Agreements, 85th Congress, 2d Session, 1958, pp. 38–64.

the reasons lay as much in internal politics . . . James M. Jasper, *Nuclear Politics*, Princeton University Press, Princeton, NJ, 1990, pp. 74–97.

172 **The Integral Fast Reactor.** Yoon I. Chang, "The Total Nuclear Power Solution," *The World & I*, April 1991, pp. 288–295.

the high-temperature gas-cooled reactor. See, for example, Agnew, "Gas-Cooled Nuclear Power Reactors."

a number of the smaller reactors

page

could be strung together. Lawrence M. Lidsky, "Safe Nuclear Power," *The New Republic*, December 28, 1987, pp. 20–23.

172 **The PIUS reactor.** Charles W. Forsberg and Alvin M. Weinberg, "Advanced Reactors, Passive Safety and Acceptance of Nuclear Energy," *Annual Review of Energy* 15 (1990), pp. 133–152.

173 **"Perhaps the moral to be drawn . . ."** Alvin M. Weinberg, *The First Nuclear Era*, pp. 130–131.

a strategic plan. Nuclear Power Oversight Committee, *Strategic Plan for Building New Nuclear Power Plants*, Nuclear Power Oversight Committee, 1990.

176 **the chances are about three in ten . . .** Cowan, "Tortoises and Hares," p. 808.

Although Cowan cautions . . . Interview with Robin Cowan, March 25, 1996.

Six: Risk

177 **a conference on xenograft transplantation.** Xenograft Transplantation: Science, Ethics, and Public Policy. Conference sponsored by the Institute of Medicine, Bethesda, Maryland, June 25–27, 1995.

178 **the audience began to debate . . .** The comments are paraphrased from a discussion session that took place on the second day of the Xenograft Transplantation conference.

179 **In late 1995, Deeks performed . . .** Lawrence K. Altman, "Baboon Cells Fail to Thrive, but AIDS Patient Improves," *New York Times*, February 9, 1996, p. A14.

180 **a program to guarantee an adequate supply of milk.** Keith Schneider, "Biotechnology's Cash Cow," *New York Times Magazine*, June 12, 1988, pp. 44–47+.

page

four separate drug companies announced . . . Schneider, "Biotechnology's Cash Cow," p. 46.

each had a slightly different version. Wade Roush, "Who Decides About Biotech?" *Technology Review*, July 1991, pp. 28–36.

181 **"We saw this as a continuation . . ."** As quoted in Schneider, "Biotechnology's Cash Cow," p. 47.

its first organized opposition. Laura Tangley, "Biotechnology on the Farm," *BioScience*, October 1986, pp. 590–593.

a study done by an agricultural economist. Robert J. Kalter, "The New Biotech Agriculture: Unforeseen Economic Consequences," *Issues in Science and Technology*, Fall 1985, pp. 125–133.

it struck a chord. Roush, "Who Decides?" p. 31.

182 **in Europe . . .** Jeremy Cherfas, "Europe: Bovine Growth Hormone in a Political Maze," *Science* 249 (August 24, 1990), p. 852.

In 1985, Rifkin had made headlines. Reginald Rhein, "'Ice-minus' May End Killer Frosts—and Stop the Rain," *Business Week*, November 25, 1985, p. 42.

In 1986, he had challenged . . . Leon Jaroff, "Fighting the Biotech Wars," *Time*, April 21, 1986, pp. 52–54.

"This is for the big marbles." Schneider, "Biotechnology's Cash Cow," p. 46.

As *Science* magazine reported, . . . Ann Gibbons, "FDA Publishes Bovine Growth Hormone Data," *Science* 249 (August 24, 1990), pp. 852-853.

183 **when other scientists examined Epstein's scenario, . . .** Roush, "Who Decides?" p. 32.

page

183 **Cows were indeed more likely . . .** Gibbons, "FDA Publishes Bovine Growth Hormone Data," p. 853.

the issue of antibiotics in milk was a red herring. Judith Juskevich and Greg Guyer, response to letter by David Kronfield, *Science* 251 (January 18, 1991), pp. 256–257.

both milk and meat from rBGH cows were safe. Ann Gibbons, "NIH Panel: Bovine Hormone Gets the Nod," *Science* 250 (December 14, 1990), p. 1506.

five large supermarket chains. Roush, "Who Decides?" p. 31.

184 **The protest campaign, Rifkin told the *New York Times*, . . .** Keith Schneider, "Grocers Challenge Use of New Drug for Milk Output," *New York Times*, February 4, 1994, p. A1.

The companies professed . . . Schneider, "Grocers Challenge Use."

both Maine and Vermont passed labeling laws. Keith Schneider, "Maine and Vermont Restrict Dairies' Use of a Growth Hormone," *New York Times*, April 15, 1994, p. A16.

Monsanto . . . responded by suing some of the labelers. Keith Schneider, "Lines Drawn in a War Over a Milk Hormone," *New York Times*, March 9, 1994, p. A12.

Jerry Caulder blamed the continuing controversy on . . . Roush, "Who Decides?" p. 34

185 **"I think a lot of it was swept under the rug."** Gina Kolata, "When the Geneticists' Fingers Get in the Food," *New York Times*, February 20, 1994, sec. 4, p. 14.

"I know the government says things . . . " Schneider, "Lines Drawn in a War."

the conspiracy charges got a big boost. Roush, "Who Decides?" p. 32.

page

"This creates an inherent conflict of interest." W.P. Norton, "Just Say Moo," *The Progressive*, November 1989, pp. 26–29.

Often the public didn't buy it. Kolata, "When the Geneticists' Fingers."

"I distrust scientists too, and I am a scientist." Kolata, "When the Geneticists' Fingers."

187 **In a March 1994 letter . . .** Michael Hansen, Jean Halloran, and Hank Snyder, "The Health of Cows," *New York Times*, March 7, 1994, p. A16.

"builders" and "conservers." The builders and conservers are closely related to—but not identical with—two of the categories that James Jasper uses in *Nuclear Politics* (Princeton University Press, Princeton, NJ, 1990) to analyze decision making in technological organizations: technological enthusiasts and ecological moralists. The difference is that the first categories are defined with respect to attitudes toward science, while the second ones deal with attitudes about technological development. A scientist could, for instance, be a builder but not a technological enthusiast, pooh-poohing any talk of rBGH's risks but still not believing rBGH should be put to work as a technology.

188 **"What they fail to address . . . "** Roush, "Who Decides?" p. 36.

Nuclear power is inherently risky. It is perhaps more accurate, technically speaking, to say that nuclear power is inherently hazardous. To specialists, "hazardous" implies the ability to cause damage if something goes wrong, while "risky" carries the extra implication that damages are somewhat likely. "Risk" is a probabilistic concept. If, for instance, an

page

accident at a nuclear plant could kill 1,000 people, the plant would certainly be considered hazardous. But if such an accident were expected to happen only once every million years of operation, then the expected risk from such an accident would be only one death per thousand years of operation, and the plant would not seem to be particularly risky.

Normally, this book reserves "risky" or "high-risk" to describe technologies that are risky in this technical, probabilistic sense. But since the book is intended for a general audience and since the distinction is not particularly important for nonspecialists, I do not follow this rule religiously and will sometimes describe a technology—such as nuclear power—as risky when some might argue that it is merely hazardous.

189 **the Atomic Energy Commission was exploding bombs.** Operation Plumbbob, which ran from May 1957 to February 1958, tested bombs above ground at the Nevada site. Richard G. Hewlett and Jack M. Holl, *Atoms for Peace and War, 1953–1961,* University of California Press, Berkeley, 1989, pp. 389, 579.

many tests created measurable amounts of radioactive fallout. Hewlett and Holl, *Atoms for Peace,* pp. 153–154.

190 **There was a tension between the two goals.** See, for example, Robert Gillette, "Nuclear Safety (I): The Roots of Dissent," *Science* 177 (September 1, 1972), pp. 771–774ff.

Advisory Committee on Reactor Safeguards. The best book on the ACRS, even though it covers only a portion of its history, is David Okrent, *Nuclear Reactor Safety: On the History of the Regulatory Process.* University of Wisconsin Press, Madison, 1981.

191 **professional and nonprofessional technological enthusiasts.** Jasper, *Nuclear Politics,* pp. 30–31.

"The practice of engineering." Jasper, *Nuclear Politics,* p. 30.

"exclusion radius." Okrent, *Nuclear Reactor Safety,* p. 18.

192 **"containment" building.** Spencer R. Weart, *Nuclear Fear: A History of Images,* Harvard University Press, Cambridge, MA, 1988, p. 284.

"Like Rickover, they say, 'prove it.'" "ACRS Qualms on Possible Vessel Failure Startle Industry," *Nucleonics* 24:1 (January 1966), pp. 17–18.

What, for example, were the chances . . . Okrent, *Nuclear Reactor Safety,* pp. 85–98.

193 **"China syndrome."** Okrent, *Nuclear Reactor Safety,* pp. 101–103.

194 **The ACRS and AEC agreed . . .** Okrent, *Nuclear Reactor Safety,* pp. 103–133.

the AEC was cutting back on research . . . Weart, *Nuclear Fear,* pp. 306–307.

195 **Shaw turned his attention to the breeder.** Gillette, "Nuclear Safety (I)," p. 772.

Shaw's autocratic personality . . . Weart, *Nuclear Fear,* p. 308.

It began in the spring of 1971 . . . Details on the role of the Union of Concerned Scientists in making public the controversy over nuclear safety are drawn mainly from two sources: Joel Primack and Frank von Hippel, *Advice and Dissent: Scientists in the Political Arena,* Basic Books, New York, 1974, pp. 208–235; and Joel Primack and Frank von Hippel, "Nuclear Reactor Safety: The Origins and Issues of a Vital Debate," *Bulletin of the Atomic Scientists,* October 1974, pp. 5–12.

The three intervenors. Ford was

page

educated as an economist. Mackenzie and Kendall were physicists.

196 **the AEC decided to hold national hearings.** The most complete description of these hearings, although told from one side, can be found in Daniel Ford, *The Cult of the Atom,* Simon and Schuster, New York, 1982.

A number of engineers . . . Weart, *Nuclear Fear,* p. 319.

197 **"Instead of claiming . . . "** Alvin M. Weinberg, *The First Nuclear Era,* p. 193.

198 **a numerical target for reactor safety.** Reginald Farmer, the British scientist who first suggested applying probabilistic risk assessment to nuclear power, offered one premature death for every hundred years of reactor operation as a reasonable goal. See Okrent, *Nuclear Reactor Safety,* pp. 182–183.

Rasmussen report. U.S. Nuclear Regulatory Commission, "Reactor Safety Study: An Assessment of Accident Risks in U.S. Commercial Nuclear Power Plants" [Rasmussen report], WASH-1400, NUREG 75/014, Nuclear Regulatory Commission, Washington, DC, 1975. Available from National Technical Information Service, Springfield, VA.

It hired . . . Rasmussen. Ford, *Cult of the Atom,* pp. 133–173.

199 **AEC officials were boasting . . .** Ford, *Cult of the Atom,* pp. 157–158.

The APS assembled a study group. Ford, *Cult of the Atom,* pp. 163–166.

the dramatic repudiation of the report. Ironically, the TMI accident was in some ways a validation of the probabilistic risk assessment of the Rasmussen report, for that report had called attention to exactly the sort of chain of events that led to the core meltdown at TMI. But because the report's probability estimates for a meltdown had received most of the attention, TMI was widely seen as a complete repudiation of the Rasmussen study.

200 **"There is a difference between believing . . . "** As quoted in Mike Gray and Ira Rosen, *The Warning: Accident at Three Mile Island,* W. W. Norton, New York, 1982, p. 269.

the Kemeny Commission concluded . . . John G. Kemeny *et al., Report of the President's Commission on the Accident at Three Mile Island,* Pergamon, New York, 1979.

Utilities tend to run . . . Joseph Rees, *Hostages of Each Other: The Transformation of Nuclear Safety Since Three Mile Island,* University of Chicago Press, Chicago, 1994, pp. 16–19.

201 **It began at 4 a.m.** For a blow-by-blow description of the Three Mile Island accident, see Ellis Rubinstein, "The Accident that Shouldn't Have Happened," *IEEE Spectrum* 16:11 (November 1979), pp. 33–42.

202 **The NRC had known about a similar accident.** Gray and Rosen, *The Warning,* pp. 33–69.

there had been a steady leak of reactor coolant. Rubinstein, "The Accident that Shouldn't Have Happened," p. 38.

adding safety devices will decrease safety. Aaron Wildavsky, *Searching for Safety,* Transaction Publishers, New Brunswick, NJ, 1988, pp. 125–140.

The control room had more than 600 alarm lights. Gray and Rosen, *The Warning,* p. 75.

203 **Charles Perrow has argued . . .** Charles Perrow, *Normal Accidents: Living With High-Risk Technologies,* Basic Books, New York, 1984.

he argues that the requirements for successful control . . . Perrow, *Normal Accidents,* pp. 334–335.

page

204 **Bernard Cohen lays out the scientist's case for nuclear power.** Bernard Cohen, *The Nuclear Energy Option*, Plenum, New York, 1990.

205 **1 millirem of radiation reduces . . .** Cohen, *Nuclear Energy Option*, pp. 52–54.

"It would seem to me . . . " "Plants 'Safe,' Executive Declares," *Public Utilities Fortnightly*, January 2, 1964, pp. 60–61.

One study after another . . . See, for instance, Paul Slovic, Baruch Fischhoff, and Sarah Lichtenstein, "Facts and Fears: Understanding Perceived Risk," in Richard C. Schwing and Walter A. Albers, *Societal Risk Assessment: How Safe Is Safe Enough?* Plenum, New York, 1980, pp. 181–216.

nuclear power is responsible for at most tens of deaths per year. Cohen, *Nuclear Energy Option*, pp. 134–135.

Puzzled by the discrepancy, . . . The discrepancy between the experts and the lay public is far greater for nuclear power than for any other technology, but the discrepancy exists elsewhere also. See Slovic, Fischhoff, and Lichtenstein, "Facts and Fears," p. 191.

206 **In one study, for example, . . .** Slovic, Fischhoff and Lichtenstein, "Facts and Fears," pp. 190–194.

the official figure. NRC, "Reactor Safety Study: An Assessment."

207 **the difference between the experts and the public . . .** The ideas in the next several paragraphs follow the arguments presented in Steve Rayner and Robin Cantor, "How Fair Is Safe Enough? The Cultural Approach to Societal Technology Choice," *Risk Analysis* 7:1 (1987), pp. 3–9.

208 **radioactivity came to be seen as "unnatural."** Weart, *Nuclear Fear*, pp. 184–195.

"Neo-Luddites judge . . . " Daniel Grossman, "Neo-Luddites: Don't Just Say Yes to Technology," *Utne Reader*, March/April 1990, pp. 44–49.

209 **If the answer to all three questions is yes, . . .** Rayner and Cantor, "How Fair Is Safe Enough?" p. 4.

The engineer's culture is an aggressively rational one. A good example of this rationalistic approach to risk can be found in Cohen's *Nuclear Energy Option*, pp. 137–144. By examining how much Americans spend to avoid various risks, Cohen calculates an implicit dollar value for a single human life: $90,000 for each life saved by women getting Pap smears; $120,000 per life saved by installing smoke alarms in every home; less than $100,000 per life saved by screening for cancer; and an estimated $2.5 billion per life saved by increasing the safety of nuclear plants. This is an irrational approach, Cohen argues, since a life saved is a life saved—we shouldn't be willing to spend a lot more to reduce certain risks over others.

Perhaps the most striking examples . . . Mary Douglas and Aaron Wildavsky, *Risk and Culture*, University of California Press, Berkeley, 1982.

"sects." Douglas and Wildavsky use "sect" in a very specialized way, not merely to refer to any religious group that is out of the mainstream. Some fringe religious groups, for instance, do have a hierarchy. They would not be sects in this sense. See Douglas and Wildavsky, *Risk and Culture*, pp. 102–125.

"The sectarian cosmology expects . . . " Douglas and Wildavsky, *Risk and Culture*, p. 127.

page

210 **The Audubon Society, for instance, . . .** Interview with Jan Beyea, National Audubon Society, November 10, 1993.

210 **"The Clamshell Alliance . . . "** Douglas and Wildavsky, *Risk and Culture*, p. 150.

In 1990, two political scientists, . . . Richard P. Barke and Hank C. Jenkins-Smith, "Politics and Scientific Expertise: Scientists, Risk Perception, and Nuclear Waste Policy," *Risk Analysis* 13:4 (1993), pp. 425–439.

211 **This is even more true among nonscientists.** Stanley Rothman and S. Robert Lichter, "Elite Ideology and Risk Perception in Nuclear Energy Policy," *American Political Science Review* 81:2 (June 1987), pp. 383–404.

213 **engineers and geologists.** Richard L. Meehan, *The Atom and the Fault*, MIT Press, Cambridge, MA, 1984.

Seven: Control

215 **Antinuclear groups denounced the settlement.** Geoffrey Aronson, "The Co-opting of CASE," *The Nation*, December 4, 1989, pp. 678ff.

Texas Utilities bemoaned the years of discord. TU Electric published several booklets offering its version of events, including "An Historical Overview of the Comanche Peak Steam Electric Station," TU Electric, Dallas, TX, 1989.

216 **It seemed like a good idea at the time.** "An Historical Overview," pp. 6–7.

218 **Ellis does not seem to be the sort . . .** Many of the details about Juanita Ellis versus Comanche Peak are drawn from Dana Rubin, "Power Switch," *Texas Monthly* 18:10 (October 1990), pp. 144–147ff.

"an anti-licensing expert . . . " Rubin, "Power Switch," p. 147.

"The utility seemed reluctant." Rubin, p. 188.

219 **Ellis was the only thing keeping the public involved.** Without intervenors, citizens could still observe the proceedings—they are open to the public—and review the documentation, but could not ask questions or raise issues.

"They slurred Juanita . . . " Rubin, "Power Switch," p. 188.

220 **the NRC judged that they were not serious enough . . .** "An Historical Overview," pp. 25, 29.

Peter Bloch. Bruce Millar, "NRC Judge Finds Release in Meditation," *The Washington Times*, February 4, 1988, p. B4.

221 **"The Licensing Board's procedures . . . "** As quoted in "An Historical Overview," pp. 43–44.

222 **Counsil and Ellis announced an agreement.** Rubin, "Power Switch," p. 191.

Americans, with their exceptional ideas . . . The best treatment of "American exceptionalism" can be found in two books by Seymour Martin Lipset: *American Exceptionalism: A Doubled-Edged Sword*, W.W. Norton & Company, New York, 1996; and the earlier work, *The First New Nation: The United States in Historical and Comparative Perspective*, Basic Books, New York, 1963.

It has its roots in the Atomic Energy Act of 1954. David Okrent, *Nuclear Reactor Safety: On the History of the Regulatory Process*, University of Wisconsin Press, Madison, 1981, pp. 6–9.

223 **In France, for instance, . . .** In the American literature on nuclear power, it is common to choose France as a point of comparison because the French nuclear pro-

page

gram was so successful. But though the comparison with France is valuable, it can also be misleading. The United States is much further from France in its culture and political and legal systems than it is from England, or even Germany. Furthermore, because France has few fossil fuels, the pressure to develop nuclear power was much greater there than in other countries. Thus, comparisons with France tend to exaggerate the country-to-country differences in how nuclear power developed.

223 **the public has essentially no access . . .** Dorothy Nelkin and Michael Pollak, "The Antinuclear Movement in France," *Technology Review*, November/December 1980, pp. 36–37.

Regulatory policy is set in discussions among . . . Jack Barkenbus, "Nuclear Power and Government Structure: The Divergent Paths of the United States and France," *Social Science Quarterly* 65:1 (1984), pp. 37–47.

. . . as the NRC must. The Advisory Committee on Reactor Safeguards makes a public report on the safety of each reactor as part of the hearing for an operating license.

In other countries the regulators and the regulated . . . Kent Hansen, Dietmar Winje, Eric Beckjord, Elias P. Gyftopoulos, Michael Golay, and Richard Lester, "Making Nuclear Power Work: Lessons From Around the World," *Technology Review*, February/March 1989, 31–40.

224 **"Many industry people think . . . "** Hansen *et al.*, "Making Nuclear Power Work," p. 35.

regulating "from the outside." James M. Jasper, *Nuclear Politics*, Princeton University Press, Princeton, NJ, 1990, p. 17.

page

225 **the government crafted . . .** Spencer R. Weart, *Nuclear Fear: A History of Images*, Harvard University Press, Cambridge, MA, 1988, pp. 306–307.

226 **safety had been reduced to following the regulations.** President's Commission on the Accident at Three Mile Island, *The Need for Change: The Legacy of TMI*, U.S. Government Printing Office, Washington, DC, 1979.

"Nuclear utility officials were so consumed . . . " Joseph Rees, *Hostages of Each Other: The Transformation of Nuclear Safety Since Three Mile Island*, University of Chicago Press, Chicago, 1994, pp. 19–20.

regulation focused more on ends than on means. Jack N. Barkenbus, "Nuclear Regulatory Reform: A Technology-Forcing Approach," *Issues in Science and Technology* 2 (Summer 1986), pp. 102–110.

British reactor operators had 45 pages . . . Barkenbus, "Nuclear Regulatory Reform," p. 104.

"In Britain . . . " Walter Marshall, "The Sizewell B PWR," *Nuclear Europe* 2 (March 1982), p. 17.

"The Canadian approach to nuclear safety . . . " As quoted in Barkenbus, "Nuclear Regulatory Reform," pp. 104–105.

It's difficult to know how much . . . One of the debates that goes on in political science and sociology is the relative importance of culture and institutions. Does culture determine the types of institutions a society has, or do the institutions shape culture? It's really a chicken-or-egg sort of question, however. Clearly institutions tend to reflect the culture that gives birth to them, but they also influence that culture in turn by steering social and political interactions and by creating expectations

page

for how things should be done. The U.S. nuclear licensing system, with its provisions for intervenors, was clearly a product of the American beliefs that citizens should have access to the system and that it should be accountable to them. The possibility of intervening so directly—and sometimes effectively—in licensing decisions, however, created an opposition culture and a perception that the AEC and the federal government were trying to cram nuclear power down the throats of unwilling citizens.

227 **nuclear power was developed by a few large concerns.** Hansen *et al.*, "Making Nuclear Power Work," pp. 36–37.

U.S. regulators never knew what to expect. See, for instance, James Cook, "Nuclear Follies," *Forbes*, February 11, 1985, pp. 82–100.

Mary Sinclair waged a successful 15-year fight . . . Anne Witte Garland, "Mary Sinclair," *Ms.* 13 (January 1985), pp. 64–66ff. See also Frank Graham, "Reformed Nuke," *Audubon* 93 (January 1991), p. 13.

Margaret Erbe faced down the Zimmer plant. James Lawless, "Moscow 'Radicals' Stop a Nuclear Plant," *Sierra* 72 (January/February 1987), pp. 125–130.

"Our object is to stop Seabrook." Todd Mason and Corie Brown, "Juanita Ellis: Antinuke Saint or Sellout?" *Business Week*, October 24, 1988, pp. 84ff.

228 **the licensing delays . . . wounded it noticeably.** John L. Campbell, *Collapse of an Industry: Nuclear Power and the Contradictions of U.S. Policy*, Cornell University Press, Ithaca, NY, 1988, p. 85.

when the state of Massachusetts objected . . . Matthew Wald, "Seabrook Feels the Chernobyl Syndrome," *New York Times*, July 27, 1986, sec. 4, p. 5.

page

it was the constantly changing regulations . . . Campbell, *Collapse of an Industry*, pp. 85–86.

"At the McGuire plant, . . . " As quoted in Cook, "Nuclear Follies," p. 89.

according to a careful analysis by Charles Komanoff . . . Charles Komanoff, *Power Plant Cost Escalation: Nuclear and Coal Capital Costs, Regulation and Economics*, Van Nostrand Reinhold, New York, 1981.

. . . or to find financing for building the plants. Some argue that the institutional inability to finance nuclear power in the 1970s, rather than the increasing costs of the nuclear plants, was the primary cause of the utility industry pulling back from nuclear power. See Campbell, *Collapse of an Industry*, pp. 92–109.

229 **new regulations that made no economic sense.** Michael W. Golay, "How Prometheus Came To Be Bound: Nuclear Regulation in America," *Technology Review* (June/July 1980), pp. 29–39.

If everything had not been spelled out . . . Cf. Golay, "How Prometheus Came To Be Bound," p. 36.

230 **Either a utility complied or it didn't.** In theory, many NRC "rules" were no more than suggestions. That is, the agency would spell out a particular approach that would satisfy a broad requirement, and a utility could choose a different approach if it wished. In practice, however, this was almost never done because the utility would have to prove that its different approach would fulfill the agency's requirements. It was easier and cheaper to follow the NRC's "suggestions" to the letter.

Every four years when Americans

page

elect a president, . . . Some political scientists worry about just how legitimate election by the Electoral College is. Many Americans assume that the president is chosen by popular vote and that the candidate with the most votes is declared president. Historically, this has usually been the case, and the last exception came in 1888, when Grover Cleveland won a plurality of the popular vote but Benjamin Harrison received a majority of the electoral votes and won the election. An election today in which one candidate got a majority of electoral votes but another got a majority of the popular vote could challenge the legitimacy of the system. See James P. Pfiffner, *The Modern Presidency*, St. Martin's Press, New York, 1994, p. 26.

231 **a study of nuclear industries around the world.** Hansen *et al.*, "Making Nuclear Power Work," p. 38.

232 **scientists open the door to countless applications.** See, for instance, Robert Pool, "In Search of the Plastic Potato," *Science* 245 (September 15, 1989), pp. 1187–1189.

this power made rDNA researchers stop and think. Sheila Jasanoff, *Science at the Bar: Law, Science, and Technology in America*, Harvard University Press, Cambridge, MA, 1995, pp. 141–142.

233 **The original 1976 guidelines had called for . . .** Jasanoff, *Science at the Bar*, p. 151.

In 1983 the NIH gave the go-ahead . . . "Mutant Bacteria Meet Frosty Reception Outside the Laboratory," *New Scientist*, April 12, 1984, p. 8.

The bacteria, called "Ice-Minus," . . . The Ice-Minus bacteria were a modified form of *Pseudomonas syringae*, a bacterium normally found on crops. The normal version of *P. syringae* produces a protein that helps ice nucleate at 0°C. Without that protein, ice will not nucleate until the temperature hits -5°C. Ice-Minus was created from the *P. syringae* bacterium by removing the gene that encoded for the ice-nucleating protein. By spraying crops with Ice-Minus, the scientists hoped to displace the naturally occurring bacteria and protect the crops from frost.

But before the trial could go ahead, . . . Jasanoff, *Science at the Bar*, p. 151.

234 **"The plaintiff's arguments . . . "** Maxine Singer, "Genetics and the Law: A Scientist's View," *Yale Law and Policy Review* 3:2 (Spring 1985), pp. 315–335.

after jumping other legal hurdles erected by Rifkin . . . Mark Crawford, "Lindow Microbe Text Delayed by Legal Action Until Spring," *Science* 233 (September 5, 1986), p. 1034.

the frost-prevention experiment did go ahead . . . Jean Marx, "Assessing the Risks of Microbial Release," *Science* 237 (September 18, 1987), pp. 1413–1417.

235 **the courts' role in shaping technology is a reactive one.** Jasanoff, *Science at the Bar*, pp. 11–12.

a few researchers uncovered evidence . . . See, for instance, Raymond I. Press *et al.*, "Antinuclear Antibodies in Women with Silicone Breast Implants," *Lancet* 340 (November 28, 1992), pp. 1304–1307.

the manufacturers decided to settle out of court. Gina Kolata, "A Case of Justice, or a Total Travesty?" *New York Times*, June 13, 1995, p. D1.

Dow Corning filed for bankruptcy.

page

"Dow Corning Broke," *Maclean's*, May 29, 1995, p. 50.

By August, the settlement was falling apart. Linda Himelstein, "A Breast-Implant Deal Comes Down to the Wire," *Business Week*, September 4, 1995, pp. 88–89ff.

236 **In June 1994 a study . . .** Sherine E. Gabriel, W. Michael O'Fallon, Leonard T. Kurland, and C. Mary Beard, "Risk of Connective-Tissue Diseases and Other Disorders After Breast Implantation," *New England Journal of Medicine* 330 (June 16, 1994), pp. 1697–1702.

an even larger study also exonerated . . . Jorge Sanchez-Guerrero, Graham A. Colditz, Elizabeth W. Karlson, and David J. Hunter, "Silicone Breast Implants and the Risk of Connective-Tissue Diseases and Symptoms," *New England Journal of Medicine* 332 (June 22, 1995), pp. 1666–1670.

a product liability case over a spermacide . . . Jasanoff, *Science at the Bar*, pp. 54–55.

His demeanor as a witness was excellent. Jasanoff, *Science at the Bar*, p. 54.

237 **the two sides can bring in equal numbers of expert witnesses.** Jasanoff, *Science at the Bar*, p. 55.

the adversarial system cuts and shapes the technical arguments. Joel Yellin, "High Technology and the Courts: Nuclear Power and the Need for Institutional Reform," *Harvard Law Review* 94:3 (January 1981), pp. 489–560. In particular, see pp. 552–553.

an example of how badly the adversarial system can twist . . . Yellin, "High Technology and the Court," p. 516–531.

238 **the district court found for the plaintiffs . . .** The district court and Supreme Court agreed that the plaintiffs had standing.

science courts. The idea of a science court was originally proposed in 1967 by Arthur Kantrowitz in "Proposal for an Institution for Scientific Judgment," *Science* 156 (May 12, 1967), pp. 763–764. See also Task Force of the Presidential Advisory Group on Anticipated Advances in Science and Technology, "The Science Court Experiment: An Interim Report," *Science* 193 (August 20, 1976), pp. 653–656.

The science court, as it is usually envisioned, . . . Jasanoff, *Science at the Bar*, p. 66.

they are convinced the legal system would work better if . . . See, for instance, the discussion of *Daubert v. Merrell Dow Pharmaceuticals* in Jasanoff, *Science at the Bar*, pp. 62–65. This is the key decision setting out the basis for treating scientific evidence in court.

239 **a U.S. Appeals Court ruled that . . .** J. Samuel Walker, *Containing the Atom: Nuclear Regulation in a Changing Environment, 1963–1971*, University of California Press, Berkeley, 1992, pp. 363–383.

The case had little effect on the nuclear industry. Jasper, *Nuclear Politics*, p. 55.

Past this, the courts did little to challenge . . . Yellin, "High Technology and the Courts," p. 550.

the 1974 Rasmussen report on reactor safety. Daniel Ford, *The Cult of the Atom*, Simon and Schuster, New York, 1982, pp. 133–173.

240 **The result of creating a system of science courts . . .** Jasanoff, *Science at the Bar*, p. 66.

241 **"Litigation becomes an avenue for working out . . . "** Jasanoff, *Science at the Bar*, p. 140.

page

241 **David Leroy held the loneliest job in government.** Matthew L. Wald, "Hired to Be Negotiator, but Treated Like Pariah," *New York Times*, February 13, 1991, p. B5.

242 **The Swedish nuclear industry has devised . . .** Luther J. Carter, *Nuclear Imperatives and Public Trust*, Resources for the Future, Washington, DC, 1987, pp. 302-306.

people find a nuclear waste site more frightening . . . Cf. James Flynn, James Chalmers, Doug Easterling, Roger Kasperson, Howard Kunreuther, C.K. Mertz, Alvin Mushkatel, K. David Pijawka, and Paul Slovic, *One Hundred Years of Solitude: Redirecting America's High-Level Nuclear Waste Policy*, Westview, Boulder, CO, 1995, p. 71.

One telephone study asked respondents . . . Paul Slovic, Mark Layman, and James H. Flynn, "Lessons from Yucca Mountain," *Environment* 33:3 (April 1991), pp. 7-11ff.

243 **"I've never come across any industry . . . "** As quoted in Carter, *Nuclear Imperatives*, pp. 9-10.

When the Walt Disney Co. announced . . . Spencer S. Hsu, "The Debate Over Disney Intensifies," *Washington Post*, January 7, 1994, p. D1; Linda Feldman, "Disney Theme Park Sparks New Civil War in Virginia," *Christian Science Monitor*, June 28, 1994, p. 10.

244 **Massachusetts passed a law in 1980 . . .** Bernard Holznagel, "Negotiation and Mediation: The Newest Approach to Hazardous Waste Facility Siting," *Boston College Environmental Affairs Law Review* 13 (1986), pp. 329-378.

In the first four years . . . Holznagel, "Negotiation and Mediation," pp. 377-378.

local communities might not stand for preemptive decisions. Holznagel, "Negotiation and Mediation," p. 352.

245 **In the earliest days of nuclear power, . . .** Carter, *Nuclear Imperatives*, pp. 54-61.

it turned into a major fiasco. Carter, *Nuclear Imperatives*, pp. 65-71.

it was the singling out of Yucca Mountain . . . Flynn *et al.*, *One Hundred Years of Solitude*, pp. 33-44.

In 1986, after selecting three finalists . . . Flynn *et al.*, *One Hundred Years of Solitude*, pp. 40-42.

246 **most scientists find Yucca Mountain a good site.** See, for example, Chris G. Whipple, "Can Nuclear Waste Be Stored Safely at Yucca Mountain?" *Scientific American*, June 1996, pp. 72-79.

in 1989 the Nevada legislature passed a bill . . . Slovic, Layman, and Flynn, "Lessons from Yucca Mountain," p. 7.

As for the citizens of the state, . . . Slovic, Layman, and Flynn, "Lessons from Yucca Mountain," p. 7.

247 **the Goshute tribe in Utah.** Tom Uhlenbrock, "Tribe Offering Repository Site," *St. Louis Post-Dispatch*, October 16, 1994, p. A10.

the Mescalero Apaches of New Mexico. George Johnson, "Nuclear Waste Dump Gets Tribe's Approval in Re-Vote," *New York Times*, March 11, 1995, p. A6.

Some have slammed the agreement . . . "Stop Environmental Racism," *St. Louis Post-Dispatch*, May 11, 1991, p. B2.

Indeed, many of the 3,000 tribe members are poor . . . Robert Bryce, "Nuclear Waste's Last Stand: Apache Land," *Christian Science Monitor*, September 2, 1994, p. 6.

247 **The town of Oak Ridge, Tennessee . . .** Carter, *Nuclear Imperatives*, pp. 412-413.

People would worry about . . .

page

These are all real fears of Nevadans with regard to Yucca Mountain. See Flynn *et al.*, *One Hundred Years of Solitude,* pp. 10–11.

248 **they offered a series of recommendations.** Flynn *et al.*, *One Hundred Years of Solitude,* pp. 97–101.

Eight: Managing the Faustian Bargain

249 **society had made a "Faustian bargain."** The much-quoted "Faustian bargain" comment comes from Alvin Weinberg's Rutherford Centennial lecture at the annual meeting of the American Association for the Advancement of Science, in Philadelphia, December 27, 1971. The text of the speech was later printed as "Social Institutions and Nuclear Energy," *Science* 177 (July 7, 1972), pp. 27–34.

250 **Just this sort of foul-up happened at Three Mile Island.** The operators had been trained to avoid a situation called a "solid pressurizer," and some confusing instrument readings convinced them that this was occurring. It wasn't, and the operators' decision to cut back on the amount of emergency cooling water flowing into the reactor allowed the core to become uncovered and melt down partially. See Ellis Rubinstein, "The Accident That Shouldn't Have Happened," *IEEE Spectrum* 16:11 (November 1979), pp. 33–42.

251 **they must try a high-stakes strategy of "trial without error."** Todd R. La Porte and Paula M. Consolini, "Working in Practice But Not in Theory: Theoretical Challenges of 'High-Reliability Organizations,'" *Journal of Public Administration Research and Theory,* January 1991, pp. 19–47. See also Karl E. Weick, "Organizational Culture as a Source of High Reliability," *California Management Review* 39:2 (Winter 1987), pp. 112–127.

page

accidents are unavoidable in certain hazardous technologies. Charles Perrow, *Normal Accidents: Living With High-Risk Technologies,* Basic Books, New York, 1984, p. 334.

these organizations "work in practice but not in theory." La Porte and Consolini give credit for the phrase to Walter Heller in "Working in Practice," p. 19.

The debate between the two camps is far from over. Strangely, the question of whether there is a debate between the two camps is itself a matter of debate. Perrow and others of his school portray the dialogue as a clear debate between optimistic and pessimistic views of managing hazardous technologies. See, for instance, Scott Sagan, *The Limits of Safety: Organizations, Accidents and Nuclear Weapons,* Princeton University Press, Princeton, NJ, 1993. But La Porte sees his and his colleagues' work as complementary, not contradictory, to that of the Perrow school. They, too, accept that certain technologies by their nature carry a high risk of accident and put a strain on the organizations that are managing them. But they observe that some organizations do a much better job than others of handling hazardous technologies, and they ask what the characteristics are of these high-reliability organizations. See, for instance, two articles in *Journal of Contingencies and Crisis Management* 2:4 (December 1994): Todd R. La Porte, "A Strawman Speaks Up: Comments on *The Limits of Safety,*" pp. 207–211; and Todd R. La Porte and Gene Rochlin, "A Rejoinder to Perrow," pp. 221–227.

Debate or no, there is a strong disagreement between the two

page

camps about how much difference organizational structure can make in avoiding accidents. According to Perrow, there is little that can be done. According to La Porte, some organizations do much better than expected, and it is useful to study how they do it and to learn what the costs are of maintaining this high reliability. If anything, La Porte says, he is more pessimistic than Perrow because he understands on a more fundamental level just how difficult it is to maintain a highly reliable organization. Furthermore, he sees his work as descriptive, not prescriptive—he offers no road maps to a more reliable organization. Nonetheless, if society is going to operate such hazardous technologies as nuclear power, studies such as La Porte's may eventually point the way to the most promising ways of setting up organizations so as to minimize the chance of accidents.

252 **The deadliest industrial accident in history.** Steven Unger, *Controlling Technology: Ethics and the Responsible Engineer* (2nd ed.), Wiley, New York, 1994, p. 72.

Union Carbide insisted on producing large volumes there. Unger, *Controlling Technology*, p. 69.

At the time of the accident, the storage tank . . . Paul Shrivastava, *Bhopal: Anatomy of a Crisis*, Ballinger, Cambridge, MA, 1987, p. 45.

When it first opened in 1969, . . . Shrivastava, *Bhopal*, pp. 39–42.

253 **Union Carbine put up the entire plant for sale.** Shrivastava, *Bhopal*, pp. 41, 51–52; Unger, *Controlling Technology*, p. 68.

The cost-cutting destroyed . . . Shrivastava, *Bhopal*, pp. 49–50; Unger, *Controlling Technology*, pp. 68–69.

page

one Indian journalist living in Bhopal . . . Unger, *Controlling Technology*, pp. 69–70.

The countdown to that catastrophe began . . . Details for the Bhopal accident were drawn primarily from Shrivastava, *Bhopal*, pp. 46–57, and Unger, *Controlling Technology*, pp. 69–71.

254 **two other developing countries were hit . . .** Shrivastava, *Bhopal*, pp. 13, 20.

255 **From a purely technical point of view, . . .** Unger, *Controlling Technology*, pp. 92–93.

NASA had been under great pressure. Maureen Hogan Casamayou, *Bureaucracy in Crisis: Three Mile Island, the Shuttle Challenge and Risk Assessment*, Westview, Boulder, CO, 1993, pp. 57–85. See also the excellent new book by Diane Vaughn, *The Challenger Launch Decision: Risky Technology, Culture, and Deviance at NASA*, University of Chicago Press, Chicago, 1996.

256 **Thus the four-shuttle fleet had been scheduled . . .** Unger, *Controlling Technology*, p. 97.

it seemed that NASA would go forward unless . . . Unger, *Controlling Technology*, pp. 96–97. See also Hugh Sidey, "'We Have to Be in Space,'" *Time*, June 9, 1986, p. 18.

Seeing NASA's unhappiness . . . Casamayou, *Bureaucracy in Crisis*, pp. 52–53.

"take off your engineering hat and put on your management hat." Unger, *Controlling Technology*, p. 94.

257 **Even evidence of partial failure in the O-rings . . .** Casamayou, *Bureaucracy in Crisis*, p. 42.

Old-timers at the agency . . . Ed Magnuson, "Fixing NASA" *Time*, June 9, 1986, p. 23.

Everything was meticulously docu-

page | page

mented, but . . . Magnuson, "Fixing NASA," p. 17.

258 **for more than a decade none of the aircraft . . .** La Porte and Consolini, "Working in Practice," p. 21.

high-reliability organizations. For a discussion of exactly what constitutes an HRO, see Gene I. Rochlin, "Defining 'High Reliability' Organizations in Practice: A Taxonomic Prologue," in Karlene H. Roberts, ed., *New Challenges to Understanding Organizations*, Macmillan, New York, 1993, pp. 11–32.

"So you want to understand an aircraft carrier?" As quoted in Gene Rochlin, Todd R. La Porte and Karlene H. Roberts, "The Self-Designing High Reliability Organization: Aircraft Carrier Flight Operations at Sea," *Naval War College Review* (Autumn 1987), pp. 76–90.

259 **At takeoff, the planes are catapulted . . .** La Porte and Consolini, "Working in Practice," p. 39. See also Karlene H. Roberts, "Introduction" in Karlene H. Roberts, ed., *New Challenges to Understanding Organizations*, Macmillan, New York, 1993, pp. 1–10.

Launching is an intricate operation. John Pfeiffer, "The Secret of Life at the Limit: Cogs Become Big Wheels," *Smithsonian* 20 (July 1989), pp. 38–48.

it is the recovery of the planes that is truly impressive. La Porte and Consolini, "Working in Practice," pp. 38–39.

"A smooth twenty-plane cycle takes hours . . . " La Porte and Consolini, "Working in Practice," p. 39.

260 **What makes this performance truly astonishing . . .** Pfeiffer, "The Secret of Life at the Limit," pp. 43–44.

they believe they understand at least part of the answer. For a more complete discussion, see Rochlin, La Porte, and Roberts, "The Self-Designing High-Reliability Organization."

The people who have the most knowledge . . . La Porte and Consolini, "Working in Practice," p. 32.

cooperation and communication become much more important. Rochlin, La Porte, and Roberts, "Self-Designing High-Reliability Organization," p. 85.

261 **A third level of organizational structure . . .** La Porte and Consolini, "Working in Practice," pp. 34–35.

"Even the lowest rating on the deck . . . " Rochlin, La Porte, and Roberts, "Self-Designing High-Reliability Organization," pp. 83–84.

a constant search for better ways of doing things. Rochlin, La Porte, and Roberts, "Self-Designing High-Reliability Organization," pp. 82–83.

262 **Diablo Canyon appears to be a rigidly run hierarchy.** Paul R. Schulman, "The Negotiated Order of Organizational Reliability" *Administration & Society* 25:3 (November 1993), pp. 353–372.

The reason, Schulman says, . . . Interview with Paul Schulman, August 16, 1994.

263 **Schulman tells of a time . . .** Paul R. Schulman, "The Analysis of High Reliability Organizations: A Comparative Framework," In Karlene H. Roberts, ed., *New Challenges to Understanding Organizations*, Macmillan, New York, 1993, p. 43.

"These systems can always surprise you, . . . " Schulman interview, August 16, 1994.

no one believes the organization will ever . . . Schulman, "The Negotiated Order of Organizational Reliability," pp. 363–364.

264 **The dozens of different depart-**

page

ments . . . Schulman, "The Negotiated Order of Organizational Reliability," p. 361.

265 **a whole new set of questions.** La Porte and Consolini, "Working in Practice," pp. 29-36.

"descriptive, not prescriptive." Interview with Todd La Porte, June 21, 1996.

"Because of the complexity, . . . " Perrow, *Normal Accidents*, p. 334.

They emphasize constant communication. La Porte and Consolini, "Working in Practice," pp. 32-34.

266 **The copilot had warned the captain . . .** Pfeiffer, "The Secret of Life at the Limit," p. 48.

A post crash investigation found . . . Karl E. Weick, "The Vulnerable System: An Analysis of the Tenerife Air Disaster," in Karlene H. Roberts, ed., *New Challenges to Understanding Organizations*, Macmillan, New York, 1993, pp. 173-198.

267 **Any organization that emphasizes constant learning . . .** Schulman interview, August 16, 1994.

they do not punish employees for making mistakes. Schulman interview, August 16, 1994.

268 **it had started fading before the last Apollo flights.** David Collingridge, *The Management of Scale: Big Organizations, Big Technologies, Big Mistakes*, Routledge, New York, 1992, p. 22.

269 **Thus was the space shuttle program born.** Collingridge, p. 22.

NASA changed from an organization willing to take risks . . . Howard E. McCurdy, *Inside NASA: High Technology and Organizational Change*, Johns Hopkins University Press, Baltimore, MD, 1993, p. 163.

The organizational structure and culture reflected these changes. McCurdy, *Inside NASA*, pp. 163-172.

page

270 **"Look, I don't know what makes . . . "** As quoted in McCurdy, *Inside NASA*, p. 174.

271 **"hostages of each other."** Joseph Rees, *Hostages of Each Other: The Transformation of Nuclear Safety Since Three Mile Island*, University of Chicago Press, Chicago, 1994, pp. 43-45.

272 **"In the fossil fuel business . . . "** As quoted in Rees, *Hostages of Each Other*, p. 18.

273 **INPO began life with a strongly defined culture.** Rees, *Hostages of Each Other*, pp. 61-63.

Rickover himself described how to adapt . . . Hyman G. Rickover, "An Assessment of the GPU Nuclear Corporation and Senior Management and Its Competence to Operate TMI-1." A report written for the GPU Corporation, November 19, 1983.

274 **"They acknowledge the complex technology."** Rickover, "Assessment of GPU," p. 2.

"The early going at Turkey Point was okay, . . . " Interview with Jerry Goldberg, September 9, 1993.

275 **it has been one of the best in the country.** Richard A. Michal, "Turkey Point's Turnaround Draws Industry's Notice," *Nuclear News*, January 1995, pp. 33-34.

Overall, the average performance of nuclear plants . . . See, for instance, National Research Council, *Nuclear Power: Technical and Institutional Options for the Future*, National Academy Press, Washington, DC, 1992, pp. 49-55.

the experience at . . . other plants. See, for instance, the description of problems at the Peach Bottom plant in Rees, *Hostages of Each Other*, pp. 110-118.

276 **World Association of Nuclear Oper-**

page

ators. Rees, *Hostages of Each Other*, p. 3.

277 **At Diablo Canyon, Schulman says, . . .** Schulman interview, August 16, 1994.

the result of a penny-wise, pound-foolish approach to safety. Unger, *Controlling Technology*, pp. 16–20.

Nine: Technical Fixes, Technological Solutions

280 **Company officials insisted . . .** Gary Stix, "Bhopal: A Tragedy in Waiting," *IEEE Spectrum* 26 (June 1989), pp. 47–50.

The standard industrial method for manufacturing MIC. Joseph Haggin, "Catalysis Gains Widening Role in Environmental Protection," *Chemical and Engineering News*, February 14, 1994, pp. 22–30.

281 **developed a different process.** Haggin, "Catalysis Gains," p. 24.

it is immediately processed into the final product. Stix, "Bhopal," p. 48.

there is generally less than a couple of pounds of MIC . . . Haggin, "Catalysis Gains," p. 24.

technological problems demand technological solutions. Thomas P. Hughes, "Technological History and Technical Problems," in Chauncey Starr and Philip C. Ritterbush, eds., *Science, Technology and the Human Prospect*, Pergamon, New York, 1980, pp. 141–156.

282 **the TMI control room at the time of the accident.** Mike Gray and Ira Rosen, *The Warning: Accident at Three Mile Island*, W.W. Norton, New York, 1982, pp. 75–77.

"About forty or fifty of these alarms . . ." Gray and Rosen, *The Warning*, p. 75.

the same way controls were laid out in fossil-fuel plants. Robert Sugarman, "Nuclear Power and the Public Risk," *IEEE Spectrum* 16 (November 1979), pp. 59–79.

283 **those indicators were hidden behind a control panel.** Sugarman, "Nuclear Power and the Public Risk," pp. 63–64.

A still more important indicator had never been installed. Ellis Rubinstein, "The Accident That Shouldn't Have Happened," *IEEE Spectrum* 16:11 (November 1979), pp. 33–42.

"There is little gut-level appreciation . . ." As quoted in Sugarman, "Nuclear Power and the Public Risk," pp. 63–65.

Traditionally, engineers have given relatively little weight . . . Donald A. Norman, "Toward Human-Centered Design," *Technology Review* 30 (July 1993), pp. 47–53.

Modern airplanes offer a good example. Norman, "Toward Human-Centered Design," p. 49.

"When an automated system . . ." Norman, "Toward Human-Centered Design," p. 49.

284 **the much-publicized troubles of the ATR-42 and ATR-72.** Stephen J. Hedges, Peter Cary, and Richard J. Newman, "Fear of Flying: One Plane's Story," *U.S. News & World Report*, March 6, 1995, pp. 40–46.

difficulties in manufacturing. Norman, "Toward Human-Centered Design," p. 49.

285 **"Whereas before they were physically able . . ."** Norman, "Toward Human-Centered Design," pp. 49–50.

"informate" instead of automate. Norman borrows the informate/automate distinction from Shosana Zuboff, *In the Age of the Smart Machine*, Basic Books, New York, 1988.

"In an automated system, . . ." Nor-

page

man, "Toward Human-Centered Design," p. 50.
the System 80+. "The ABB Combustion Engineering System 80+." *Nuclear News*, September 1992, pp. 68–69.

286 **Todd La Porte doesn't believe it goes far enough.** Interview with Todd La Porte, August 15, 1994.

287 **Traditionally, a chemical plant has been designed . . .** Deborah L. Illman, "'Green' Chemistry Presents Challenge to Chemists," *Chemical & Engineering News*, September 6, 1993, pp. 26–30.

288 **the chemical industry has begun to look past . . .** Deborah L. Illman, "Environmentally Benign Chemistry Aims for Processes That Don't Pollute," *Chemical & Engineering News*, September 5, 1994, pp. 22–27.
the complete replacement of CFCs. Robert Pool, "The Elusive Replacements for CFCs," *Science* 242 (November 4, 1988), pp. 666–668.

289 **To replace Freon, for instance, . . .** Robert Pool, "The Old and the New," *Science* 242 (November 4, 1988), p. 667.

290 **Du Pont considered at least two dozen options.** Pool, "The Elusive Replacements for CFCs," pp. 667–668.

291 **One possibility is carbon dioxide in its supercritical form.** Ivan Amato, "Making Molecules Without the Mess," *Science* 259 (March 12, 1993), p. 1540.
One chemist has made polymers in water. M. Mitchell Waldrop, "A Safer Way to Make Plastics," *Science* 248 (May 18, 1990), p. 816.

292 **dyes.** Illman, "'Green' Technology Presents Challenge," p. 27.
zeolites. Illman, "Environmentally Benign Chemistry," p. 27.
One researcher, John Frost of Purdue, . . . Illman, "'Green' Technology Presents Challenge," p. 28.
Other researchers have looked to plants. Robert Pool, "In Search of the Plastic Potato," *Science* 245 (September 15, 1989), pp. 1187–1189.
Although some researchers claim . . . Illman, "Environmentally Benign Chemistry," pp. 23–24.
actual adoption by the chemical industry has been slow. Ivan Amato, "The Slow Birth of Green Chemistry," *Science* 259 (March 12, 1993), pp. 1538–1541.

293 **The realization is spreading . . .** Illman, "'Green' Chemistry Presents Challenge."
it seems that the chemical industry is changing . . . Ivan Amato, "Can the Chemical Industry Change Its Spots?" *Science* 259 (March 12, 1993), pp. 1538–1539.

294 **Such inherent safety is the ultimate technical fix.** See, for instance, Trevor Kletz, *Cheaper, Safer Plants*, Institute of Chemical Engineers, Rugby, Warwickshire, England, 1984, pp. 5–7.
Charles Perrow's arguments. Charles Perrow, *Normal Accidents: Living With High-Risk Technologies*, Basic Books, New York, 1984.

295 **The most daring safety innovation . . .** Michael W. Golay and Neil E. Todreas, "Advanced Light-Water Reactors," *Scientific American*, April 1990, pp. 82–89. See also the various reports of the Nuclear Power Oversight Committee (NPOC), such as *Strategic Plan for Building New Nuclear Power Plants*, 1992. NPOC is an industry group formed to shepherd in the next generation of nuclear power, and its plans represent an industry consensus on what that will require.
"it will no longer be sufficient . . ."

page

Lawrence M. Lidsky, "Nuclear Power: Levels of Safety," *Radiation Research* 113 (1988), pp. 217–226.

"It occurred to me that if nuclear power were to survive, . . . " Interview with Kåre Hannerz, July 27, 1992.

296 **This is what happened at Chernobyl.** Richard Wolfson, *Nuclear Choices*, MIT Press, Cambridge, MA, 1991, p. 203.

297 **Even before Three Mile Island, Hannerz says, . . .** Hannerz interview, July 27, 1992.

PIUS consists of . . . Charles Forsberg and William J. Reich, *Worldwide Advanced Nuclear Power Reactors With Passive and Inherent Safety: What, Why, How, and Who*, a report for the U.S. Department of Energy by Oak Ridge National Laboratory, ORNL/TM-11907, September 1991, pp. 8–10, 46–49.

298 **both machines take the same tack as light water reactors.** John J. Taylor, "Improved and Safer Nuclear Power," *Science* 244 (April 21, 1989), pp. 318–325.

The Advanced Liquid Metal Reactor. The Advanced Liquid Metal Reactor is also known as PRISM, for Power Reactor Inherent Safety Module, and as the Integral Fast Reactor.

299 **MHTGR also relies . . .** Lidsky, "Nuclear Power: Levels of Safety."

In April 1986, a test on the Experimental Breeder Reactor-II. Yoon I. Chang, "The Total Nuclear Power Solution," *The World & I*, April 1991, pp. 288–295.

The safety of the MHTGR has been demonstrated . . . Taylor, "Improved and Safer Nuclear Power," p. 324.

the MHTGR could eventually generate electricity . . . Larry M. Lidsky and Xing L. Yan, "Modular Gas-Cooled Reactor Gas Turbine Power Plant Designs," paper presented at the 2nd JAERI Symposium on HTGR Technologies, Ibaraki, Japan, October 21–23, 1992.

300 **such whoppers as operators falling asleep . . .** At Philadelphia Electric's Peach Bottom plant, night-shift operators were found to have slept and played video games on the job. At certain points, all the operators were asleep. See Joseph Rees, *Hostages of Each Other: The Transformation of Nuclear Safety Since Three Mile Island*, University of Chicago Press, Chicago, 1994, p. 111. There is also the case of Bechtel Corporation installing the reactor at San Onofre backwards. James Cook, "Nuclear Follies," *Forbes*, February 11, 1985, p. 90.

"If there is to be a second nuclear era . . . " Larry Lidsky, untitled paper presented at the Second MIT International Conference on the Next Generation of Nuclear Power Technology, October 25, 1993.

the problem of nuclear waste will remain. The Advanced Liquid Metal Reactor, also known as the Integral Fast Reactor, could help solve the nuclear-waste problem by burning off some of the worst radioactive byproducts of fission. But, as described in chapter 5, because the reactor could be used as a breeder and produce plutonium, funding for its development was killed by Congress.

302 **it was not due to any technical superiority.** Once Lidsky had worked on the MHTGR for a while, he also became convinced that it would be economically superior to light-water reactors, but even so his strongest case remains the public-acceptance issue. See Lawrence M. Lidsky, "Safe Nuclear Power," *The*

page

New Republic, December 28, 1987, pp. 20–23.

302 **"I became convinced . . . "** Interview with John Taylor, October 10, 1991.

303 **a pair of which have now been built in Japan.** As this is written, in mid-1996, one GE advanced reactor has started up and the other is scheduled to go on-line before the book's publication date.

three designs in progress. There were actually four designs at first. GE had a passively safe reactor planned, the 600-megawatt Simplified Boiling-Water Reactor, but it halted the project in early 1996.

"Strategic Plan for Building New Nuclear Power Plants." Nuclear Power Oversight Committee, "Strategic Plan for Building New Nuclear Power Plants," November 1990.

page

304 **the NRC to approve . . .** By mid-1996, the NRC had issued design approval for both GE's Advanced Boiling-Water Reactor and ABB-Combustion Engineering's System 80+, and the commission was in the process of deciding on final certification for the two reactor designs. See "ABB-CE Receives Design Approval for System 80+." *Nuclear News*, September 1994, p. 28.

Index